ETHNOBOTANY OF BHIL TRIBE

Other Related Books

Title	Author
Ayurvedic Medicinal Plants of India, Vol. 1-2	*Bhutiya, Ramesh*
Bibliography of Indian Ethnobotany, 2002	*Jain, S.K.*
Biodiversity and Sustainable Utilization of Biological Resource	*Sahu, T.R.*
Biodiversity and Vegetation of Pachmarhi Hills	*Singh, V.P.*
Biodiversity Conservation	*Binoj Kumar, M.S.*
Biodiversity of Ranthambhore Tiger Reserve, Rajasthan	*Singh, V.*
Ethnic Aphrodisiac Plants	*Sood, S.K.*
Ethnic Indian Plants in Cure Diabetes	*Sood, S.K.*
Ethnobotany and Medicinal Plants of India & Nepal Vol. 3	*Singh, V.*
Flavouring and Fragrant Resources of India	*Sood, S.K.*
Floristic Biodiversity of Barda Hills and its Surrounding	*Nagar, P.S.*
Flowers and Vegetables of India	*Sud, R.K.*
Glossary of Medicinal Plants used in Ayurveda	*Saroya, A.P. Singh*
Handbook of Indian Medicinal Plants	*Joshi, M.C.*
Handbook of Legumes of World Economic Importance	*Duke, J.A.*
Herbal Technology Recent Trends and Progress	*Daniel, M.*
Herbs of Tribal Land, Jharkhand, India	*Jaipuriar, M.K.*
Herbs, Culinary, Medicinal, Aromatic	*Sud, R.K.*
Identification of Common Indian Medicinal Plants	*Naik, V.N.*
Indian Folk Medicines & other Plant-based Products	*Singh, V.*
Indigenous Knowledge: An Application	*Sahu, T.R.*
Leafy and Edible Plants of North-East India	*Kumar, S.*
Manual of Ethnobotany	*Jain, S.K.*
Medicinal and Poisonous Plants of India	*Caius, J.F.*
Medicinal Plants of India (Vol. 1-3)	*Jadhav, Dinesh*
Multipurpose Plants	*Srivastava, P.K.*
Non-conventional Legumes & Useful Plants of North East India	*Kumar, S.*
Phytotherapeutic Wisdom of Indian Rurals Aboriginals	*Singh, V.*
Sacred and Magico-Religious Plants of India	*Sood, S.K.*
The Poisonous Plants of Bombay	*Kirtikar, K.R.*
Traditional Folk Veterinary Medicines	*Katewa, S.S.*

www.scientificpub.com

ETHNOBOTANY OF BHIL TRIBE

A CASE STUDY AMONG THE BHILS OF RATLAM DISTRICT (MADHYA PRADESH)

V.P. SINGH

Retd Professor & Head, Department of Plant Sciences (IEMPS),
Vikram University, Ujjain (M.P.)

&

DINESH JADHAV

Guest Faculty, Department of Botany,
Govt. Arts and Science PG College, Ratlam (M.P.)

SCIENTIFIC
PUBLISHERS (INDIA)
www.scientificpub.com

Published by:
Scientific Publishers (India)
5-A, New Pali Road, P.O. Box 91,
Jodhpur – 342 001 (India)

E-mail: info@scientificpub.com
Website: www.scientificpub.com

Print, 2019

ISBN: 978-81-7233-699-8

Reprinted from Journal of Economic & Taxonomic Botany
Additional Series # 33 (2011)
ISSN: 0970-3306

Typeset by Rajesh Ojha

Printed in India

Dedicated
to
The Tribal People of Ratlam district
Who revealed their secrets
for the welfare of mankind.

PREFACE

Man has depended on plants from time immemorial. The human life and culture have directly or indirectly been influenced by their surrounding environment. The primitive people are well acquainted with the properties and uses of plants of their surroundings. They have inherited rich traditional knowledge of surrounding plants used as food, fodder, fibres, woods, fuel, medicine, beverage, tannin, dye, gum, resin, cosmetic, crafts & relegious ceremonies. Some other plants are also used as narcotic, hallucinogenic and poisonous. Such knowledge of economic and other properties of surrounding plants acquired by the tribal community through experience of ages has been passed on the word of mouth from generation to generation as a part of their cultural heritage. Ratlam district lying between 23°05' – 23°55' N latitude and 74°30' – 75°42' E longitude covers an area of 4861 sq. km. The district is situated on Malwa plateau at 493.62 m above the sea level. Bhils are the main inhabitants of different villages of the district. According to 2001 census, the population of Bhil tribe in the district was 2.26 lac, constituting about 23.27% of the total population. The ethnobotanical works in organized way were started by Botanical Survey of India in 1960 in this country. Since then, uses of plants by the tribals are being recorded for a variety of purposes. But, records of the past revealed that no ethno-botanical studies have been conducted in this area. During extensive field trips in the tribal pockets in the district, the authors recorded multipurpose ethno-botanical uses of plants. The present ethnobotanical work was carried out by taking periodical and extensive ethnobotanical surveys, assessment of field and tribal knowledge bank of Ratlam district from 2004-2007. For making the work more comprehensive and useful to the readers, the subject matter of this book encompasses information on general Introduction, Review of the previous works, Geographical position, topography and administrative aspects of study area, The people setting, Methodology; Enumeration of ethnobotanical uses of plants, Discussions and conclusions. The investigation has brought to light indigenous knowledge on the utilization of 210 plant species [medicinal-186; vegetables- 27; wild edible fruits-36; fodder/forage - 18; beverage and drinks-3; gum and resin-3; magico-religious beliefs and offerings-25; agriculture implements, handle, tools, boats-5; fibres-11; detergent-3; dye-3; tannin-12; oil yielding plants-3; house, building construction, thatching and furniture-11; taboos and totems-6; musical instruments-11; fuel-6;

biofencing-10; fish poison-10; other economic uses-7] belonging to 178 genera and 71 families. Further, the dicots are represented by 180 species belonging to 150 genera and 59 families and the remaining 30 species belong to 28 genera and 12 families of monocots. Statistically, 210 plant species of ethnobotanical interest belong to different habits, viz, 96 herbs, 55 trees, 37 shrubs, 16 climbers, 5 grasses and one angiospermic parasite. It is hoped all such information provided would be utilized judiciously by all concerned for sustainable utilization of biodiversity and for well being of the humanity.

Lastly, attempts have been made to provide complete bibliography reflected in the text (references), the index to valid botanical names (along with attached local names) in appendix-I, condensed information regarding ailments/diseases and plants used for cure in appendix-II, and other uses and their plant resources in appendix-III. It has made the work (book) more comprehensive and usable upto grass-root level. The beautiful photographs have further added new dimensions to the book.

The authors are thankful to various tribal medicinemen 'Bhopa' and villagers of the tribal inhabited localities of Ratlam district for sharing their traditional knowledge based on the surrounding plants with us. Also our thanks go to all the Volunteers of 'Sewa Bharti NGO' of Ratlam district who helped us in this research work.

Lastly, we take this opportunity to express our gratitude to Shri Pawan Kumar Ji, Managing Editor, Scientific Publishers (India), Jodhpur for accepting the Manuscript and also for guidance and helpful suggestions. Without his encouragement and support this book would not have been completed.

V.P. Singh
Dinesh Jadhav

LIST OF PHOTOGRAPHS

CONTENTS

1 *Acacia catechu*

2 *Acacia leucophloea*

3 *Acacia nilotica*

4 *Achyranthes aspera*

5 *Adansonia digitata*

6 *Argyreia nervosa*

7 *Ailanthus excelsa*

8 *Acacia pennata*

9 *Aloe vera*

Plate-1

10 *Argemone mexicana*

11 *Asparagus racemosus*

13 *Boerhavia repens*

12 *Barleria prionitis*

14 *Bombax ceiba*

15 *Butea monosperma*

16 *Calotropis gigantea*

17 *Calotropis procera*

18 *Cannabis sativa*

Plate-2

19 *Capparis zeylanica*

20 *Senna occidentalis*

21 *Senna tora*

22 *Celosia argentea*

23 *Chenopodium album*

24 *Chlorophytum tuberosum*

25 *Clerodendrum multiflorum*

26 *Datura stramonium*

27 *Echinops echinatus*

Plate-3

28 *Eclipta prostrata*

29 *Ficus benghalensis*

30 *Ficus racemosa*

31 *Gloriosa superba*

32 *Gossypium herbaceum*

33 *Holarrhena pubescens*

34 *Hygrophila schulli*

35 *Ipomoea carnea* Jacq. subsp. *fistulosa*

36 *Jatropha curcas*

Plate-4

37 *Lantana camara*

38 *Lawsonia inermis*

39 *Madhuca indica*

40 *Moringa oliefera*

41 *Mucuna prurience*

42 *Ocimum tenuiflorum*

43 *Oroxylon indicum*

44 *Phyllanthus fraternus*

45 *Pithecellobium dulce*

Plate-5

46 *Plumbago zeylanica*

47 *Pongamia pinnata*

48 *Ricinus communis*

49 Solanum nigrum

50 Solanum virginiannum

51 *Tamarindus indica*

52 *Tectona grandis*

53 *Terminalia cuneata*

54 *Tinospora cordifolia*

Plate-6

55 *Ziziphus mauritiana*

56 Bhil women & children in local weekly market 'Haat'

57 Bhil couple in front of their hut ('Tapra')

58 Bhil family in front of their hut ('Tapra')

59 Village ('Phalia') of Bhil Tribe

60 Temple of Bhil Tribe

Plate-7

INTRODUCTION

What is Ethnobotany

Indigenous knowledge is as old as human civilization, but the term Ethnobotany was first coined by an American botanist, John Harshburger, in 1896 to study the plants used by the primitive and aboriginal people. Since then, it has been defined as the traditional knowledge of indigenous communities about surrounding plant diversity, and as the study of how the people of a particular culture and region make use of indigenous plants. Ethnobotany is the study of the relationship between plants and people. Ethnobotany is considered as a branch of Ethnobiology. Ethnobotanical studies deal with the complex relationships between people, plants and cultures. The focus of ethnobotany is mainly on how plants have been or are used, managed and perceived in human societies.

Ethnobotany is the study of people of a particular culture and region making use of indigenous plants. The ethnobotanists explore how plants are used for food, shelter, medicine, clothing, hunting, religious ceremonies etc. It is the science which deals with the relationship between a given society and its environment, in particular the plant world. The study of direct interaction between human and plant population through its culture, learning the use of plants, human impact on the plant communities with which they interact etc. are the main aspects which are focused in ethnobotany.

Concepts of Ethnobotany

According to Harshburger (1896) *'Ethnobotany is the study of plants used by the primitive and aboriginal's people'*. Although this definition still provides the root of the ethnobotany, but to describe the field in broader sense ethnobotanists have given their own definitions from time to time, these are:

"Ethnobotany is the study of direct interrelationship between human and plant" (Johes, 1941).

"Ethnobotany is the study of the relationship, which exists between people of primitive societies and their environment" (Schultes, 1962).

"Ethnobotany is concerned with a wide range of interest of plants in cultural and ecological context" (Ford, 1994).

"Ethnobotany is the part of ethnoecology which concerns plants" (Martin, 1995).

"Ethnobotany is considered to encompass all studies which concern the mutual relationship between plants and traditional peoples" (Cotton, 1996).

"Ethnobotany is the total natural and traditional relationships and interaction between man and his surrounding environment or plant wealth" (Jain, 1989). This interaction has been classified by Jain (1989) into 2 categories: (1) Abstract - The abstract relationship of man with plants includes faith in the good or bad powers of plants, taboos, avoidances, sacred plants, worship and folklore, (2) Concrete - The concrete relationship of man with plants includes mainly the material use such as in food, medicine, house-building, agricultural operations, other domestic uses, trade or barter, plants in fine arts and culture like paintings, carvings and house decoration and the acts of domestication, conservation, improvement or destruction of plants.

Scope of Ethnobotany

The scope of ethnobotany has expanded enormously and we have now many specialized subdivisions: Archaeoethnobotany, Ethnoecology, Ethnomedicine, Ethnogynaecology, Ethnomusicology, Ethnomycology, Ethnonarcotics, Ethnopharmacology, Ethnotaxonomy, Ethnotoxicology, Ethnocosmetics, Ethnopediatrics, Palaeoethnobotany etc. These are discussed below:

Archaeoethnobotany is the study of the plant materials from archaeological sites for studies on migration of human cultures, and origin, dispersal and domestification of crops etc.

Ethnoecology is the study of past and present interrelationships between human societies and their living and non-living environments.

Ethnomedicine is an area of medical anthropology that deals with the study of traditional medicines—not only those with relevant written sources e.g. Ayurveda and Unani, but also those whose knowledge and practices have been orally transmitted over the centuries.

Ethnogynaecology is the study which deals with various diseases among women in tribal societies, related to sterility, conception, abortion, etc., and the use of abortifacients.

Ethnomusicology is the study and cultural aspects of music and dance of tribals and aborigines.

Ethnomycology is the study of the historical uses and sociological impact of fungi.

Ethnonarcotics is the study of the use of narcotics, snuffs, hallucinogens etc. in primitive societies.

Ethnopharmacology is the scientific study correlating ethnic groups, their health, and how it relates to their physical habits and methodology in creating and using medicines.

Ethnotaxonomy deals with the naming and classification of plants and their cultivars, and animals and their races by human societies in their languages.

Ethnotoxicology deals with various toxic plants used as fish poison, arrow poison etc. by tribal people.

Ethnocosmetics deals with the study of cosmetics of ancient past which were used by ladies to make them beautiful.

Ethnopediatrics is the study of parents and infants in our own and different cultures. It explores the way of different care-taking styles regarding the health, well-being and survival of infants.

Palaeoethnobotany deals with the identification of fossilized plant materials and remains that had effect on ancient economy of people.

Advantages of the study of Ethnobotany

As we have seen, ethnobotany is a multidisciplinary science and its scope is not confined to one area but it covers a broad range of study areas, which are interconnected to each other in one sense or the other. The advantages of the study of Ethnobotany are manifold, as these provide valuable data on: (1) new economic uses of plants not recorded in literature, (2) new vernacular names of plant, (3) traditional conservation strategies of natural resources, (4) new plant crude drugs, (5) knowledge of the indigenous flora of the region, (6) traditional cultivation techniques and (7) understanding the relationship of different human societies with plant resources.

REVIEW OF THE PREVIOUS WORKS

Ethnobotany in India

Man has depended on plants from time immemorial. The history of usage of plants by man is as old as the history of development of civilization. The human life and culture have directly or indirectly been influenced by their surrounding environment. The primitive people are well acquainted with the properties and the uses of the plants of their surroundings. Plants are an integral part of human life and, therefore, used in every walk of life. The study of plants in the service of mankind has been a part of human civilization since ages. Information on the economic aspects of plants has been passed from one generation to the next generation. With the knowledge of the traditional practices, botanists of world examined the practical uses of plants, either reported or unreported. And in the light, a new branch of botany was emerged and termed as 'Ethnobotany'.

India has a population of 102 million, out of which about 50 million people belong to 550 tribal communities, representing 7.76% of the total population of the country (According to Census 1981) spread over 19% of the total area of the nation. Ethnobotanical works done in India in the last five decades have led to the publication of inventories and documentation of plants and their products used by the aboriginals for various purposes. Janaki Ammal (1954) initiated researches on Ethnobotany in India; she studied subsistence food plants of south India. Jain (1960) made intensive field studies among the tribals of central India. Thereafter, a number of sporadic studies were undertaken. After establishment of the 'Society of ethnobotanists' (SEB) during 1980, ethnobotany in India was stream - lined and a common platform was made available. A 'News Letter' was brought out and seminars, symposia and training courses and workshops were organized to bring awareness among students and field workers. In 1989, SEB started an international journal named *Ethnobotany.* During 1982-1993, an 'All India Coordinated Research Project on Ethnobiology' was initiated at Regional Research Laboratory, Jammu, under the leadership of Dr. P. Pushpangadan. It was a multi-disciplinary, multi-institutional and action-oriented dimensional perspective effort regarding the study of the tribal culture, traditions and knowledge. These activities created tremendous impact on botanists, anthropologist and medical practitioners etc in India during the last two decades. As such, similar works have been undertaken in some areas of Andaman & Nicobar Islands, Andhra Pradesh, Arunachal Pradesh, Assam, Bihar, Gujarat, Harayana, Himachal Pradesh, Jammu & Kashmir, Karanataka, Kerela, Madhya Pradesh, Maharshtra, Manipur, Meghalaya, Orissa, Punjab, Rajasthan, Sikkim, Tamil Nadu, Tripura, Uttar Pradesh, Uttranchal, Chhattishgarh, Jharkhand and West Bengal sustaining various tribal communities, *viz.* Asur, Angami Nagas, Apatianis, Abhujmarhia, Bhil, Birhor, Birjia, Bhumij, Bhag Paharia, Bangree, Bharia, Baiga, Bathudi, Bhilala, Bhoxa, Boro, Chenchu, Dangers, Garo, Gond, Gallong, Ho, Hillmiri, Halba, Jaintia, Korku, Konda, Karbis (Mikir), Khasis, Kondh, Khols, Koya, Kharia, Kunwar, Reddies, Kissan, Khairwar, Kora, Kutiakondh, Lodha, Lohara, Mahli, Munda, Mechs, Maria Gond, Mishmees, Monpas, Miris, Mal Paharia, Mankirdia, Mudia, Nicobaris, Nagesia, Nayadis, Nissis, Onge, Oraon, Paharia, Pardhi, Panika, Paraja, Ratha, Reddhi, Saora, Santal, Shompen, Sauira Paharia, Suleng, Seharia, Tharus etc.

In the last 50 years, several hundred small and large inventories of ethnobotanical plants have been published on the subject. Glimpses of Indian Ethnobotany (Jain, 1981) is the first book dealing with Indian Ethnobotany. From the available literature on Indian Ethnobotany, it is evident that more than 1500 research papers on various aspects of ethnobotany were published till 2006. Almost all this literature has been enumerated in the two bibliographies published in 1984 and 2002 (Jain *et al.,* 1984; Jain, 2002). Ethnobotanical studies in different regions and pertaining to different tribes of India have been taken up during recent years; which have served as a real and true introduction to Indian ethnobotany. Some of these notable works done in different States/Tribes are mentioned here : **A and N Islands** - Nicobarese (Dagar, 1986; Dagar & Chaghtai, 1991; Dagar & Dagar, 1992, 1996, 1999); Onge (Bhargava, 1983). **Andhra Pradesh** - Konda (Prasad *et al.,* 1999); Koya (Upadhaya & Chauhan, 2000). **Arunachal Pradesh** - Miri (Tag & Das, 2004); Adis (Das, 1997); Nishi (Bhuyan, 2003). **Assam**- Boro (Baruah & Sharma, 1984); Garo (Guha, 1939); Mikir (Borthakur, 1981; Das, & Sharma, 1958-59; Jain & Bothtakur, 1980); Miris (Mishing) (Hajra & Baishya, 1981); Tharu (Prasad *et al.,* 1990). **Bihar** - Santhals (Bodding, 1925). **Gujarat** - Kathodi (Punjani, 2006). **Harayana** - Morni (Jain, 1984). **Himachal Pradesh** - Gaddi (Brij Lal *et al.,* 1996; Lal *et al.*, 1996; Sharma & Maheshwari, 2005). **J & K** - Gujjar and Bakerwal (Gonoi & Nawchoo, 2003). **Jharkhand** - Mundas (Hembrom & Goel, 2005). **Karanataka** - Soligas (Hosagoudar & Henry, 1993; Kshirsagar & Singh, 2003); Kuruba (Udayan *et al.,* 2004). **Kerala** - Kaadar (Udayan *et al.*, 2005); Kani (John, 1981); Kuruma and Paniya (Hema *et al.,* 2006); Nayadis (Nagendra Prasad & Abraham, 1984). **MP and CG** - Abujhamarh (Sahu *et al.,* 2003;

Maheshwari & Dwivedi, 1985); Baiga (Anonymous, 1985; Bhattacharya *et al.,* 2004; Brij Lal, 1988; Roy *et al.,* 1956); Bharia (Pandey & Shrivastava, 1993); Bhil (Maheshwari *et al.,* 1986; Samvatsar & Diwanji, 2004; Srivastava, 1985); Gond (Kumar & Yadav, 2003; Roy *et al.*, 1956; Kaufmann, 1961; Kumar, 1996; Sahu, 2000); Oroan (Maheshwari *et al.*, 1990); Sahariya (Jain & Sharma, 1990). **Maharashtra** - Gond (Bodhe *et al.*, 1997); Korku (Bhogaonkar & Devarkar, 2002; Kamble & Pradhan, 1980). **Meghalaya** - Garo (Kumar *et al.*, 1980; Rao & Neogi, 1980; Rao, 1981); Khasi (Chaudhary & Neogi, 1999; Rao & Neogi, 1980; Rao, 1981); Khasi and Jaiantia (Kharkongor & Joseph, 1981); Pnar (Samati, 2004). **Nagaland-** Naga (Megoneitso & Rao, 1983; Rao & Jamir, 1982). **Orissa** - Kandha (Panda *et al.,* 2005). **Punjab** - Bhat (Lal & Lata, 1980). **Rajasthan-** Bhil (Joshi, 1982; Sebestian & Bhandari, 1984a; Singh, 1971). **Tamil Nadu** - Irular, Kota, Toda (Abraham, 1981); Irular (Ramachandran & Nair, 1981); Konikkors (Janki Ammal & Prasad, 1984); Valaiyans (Ganesan & Kesavan, 2003; Subramanian *et al.,* 2003). **Uttar Pradesh and Uttranchal** - Bhoxa (Maheshwari & Singh, 1984; Singh, 1988; Singh & Maheshwari, 1992; Singh *et al.,* 1987); Gond (Singh & Prakash, 1994); Kol (Maheshwari & Singh, 1987); Sharia (Maheshwari & Singh, 1988); Tharu (Maheshwari *et al.*, 1980; Pant & Pandey, 1995; Kumar *et al.,* 2003). **West Bengal** - Kharia (Varghese, 1996); Lodha (Pal & Jain, 1989); Rabha (Molla & Roy, 1985); Santhal (Bodding, 1925).

Ethnobotany in Madhya Pradesh

Madhya Pradesh is known to harbour a rich wealth of medicinal plants. The State lies in the central region of the Country between latitudes 21°6' – 26°30' N and longitudes 74°9' – 81°48' E, covering an area of 3,08,245 sq. km. Uttar Pradesh bounds it in the North, Chhattisgarh in the East, Maharashtra in the South and Rajasthan and Gujarat in the West. It comprises 45 districts, *viz.,* Balaghat, Barwani, Betul, Bhind, Bhopal, Chhatarpur, Chhindwara, Damoh, Datia, Dewas, Dhar, Dindori, East Nimar (Khandwa), Guna, Gwalior, Harda, Hoshangabad, Indore, Jabalpur, Jhabua, Katni, Mandala, Mandsaur, Morena, Narsinhpur, Neemuch, Omaria, Panna, Raisen, Rajgarh, Ratlam, Rewa, Sagar, Satna, Sehore, Seoni, Shahdol, Shajapur, Shivpuri, Sheopur, Sidhi, Tikamgarh, Ujjain, Vidisha, West Nimar (Khargone) (Fig. 1). With respect to forest cover, it is the first largest State in India. The forests that account for 30.9% of the total geographical area cover about 95,211 sq km area. The forests are dominated by Teak (*Tectona grandis*) and Sal (*Shorea robusta*). The major rivers are Narmada, Chambal, Kalisindh, Betwa, Son, Venganga, Mahi, Tapti, Ken and Pench. The climate of the State is generally sub-humid with an annual rainfall ranging from 75-125 cm. The minimum and maximum temperatures range between 2°C and 45°C in the months of January and May, respectively. According to 2001 census, the total population of the State is about 6,03,48,023, out of which 1,22,33,474 (20.27%) form 41 tribal communities, which are divided into more than 59 ethnic groups. The main tribes and their sub-tribes of the State are Agaria, Baiga (Binjhwar, Narotia, Bharotia, Nahar, Re-bhena, Kad-bhena), Basor, Bharia (Bhumia, Bhuihar, Pando), Bhil (Bhilala, Berela, Patlia), Gond (Pardhan, Agaria, Ojha, Nagarchi, Solhas), Halba (Halbi), Kanjar, Khairwar, Kol (Rhoeo, Rothel), Kondar, Korku (Nahala, Bodoya), Pao (Pabra), Madia (Abhujmadia, Dandami Madia, Metakoitur), Mawasi, Mobasi, Oroan, Sahariya, Sor etc. The economy of the State is primarily agriculture based. The State is famous for Soya bean cultivation and is known as the 'Soya bean State'. Nearly 75% of the population lives in villages. The life of villagers is wholly or partially dependent upon forest resources. They utilize a wide variety of plants for their basic needs, such as food, fibre, fodder, wood, medicine, gum, tannin, resin, dye and shelter.

Fig. 1. Map of Madhya Pradesh.

Ethnobotanical works done in Madhya Pradesh in last 5 decades have led to the publication of several inventories and documentations of plants and their products used by the aboriginals for various purposes (Brij Lal, 1993; Brij Lal & Maheshwari, 1993; Jain, 1963c, 1963d, 1963e, 1981b, 1997; Kadel & Jain, 2008; Khan *et al.*, 2000; Kumar & Jain, 2002; Lal, 1993; Maheshwari, 1990b, 1996; Oommacham & Masih, 1987, 1989, 1993; Masih, 1997; Mishra & Sahu, 1984; Pandey & Saini, 2007; Rai, 2004, 2007, 2008a, 2008b, 2008c, 2008d; Sahu, 1982; Sahu *et al.*,1983; Saini, 1996; Shah & Singh, 1990a, 1990b; Sikarwar *et al.*, 1994; Sikarwar, 2001; Singh *et al.*, 1999, 2003). Madhya Pradesh State is rich not only in floristic diversity but also in tribal or ethnic diversity. A large number of ethnobotanical papers on the region are scattered in various journals, magazines, books, proceedings, etc. A year-wise analysis of publications shows that there was a great boost to the ethnobotanical works in the last decade of 20th century. The attempt to collect information on these has resulted in identification of about 200 references relevant to ethnobotanical aspects. Record of the past revealed that no ethnobotanical studies have been conducted in Ratlam district of Madhya Pradesh. Ratlam district is one of the largest districts of the Ujjain Division where the tribal population forms a considerable part of total population and having very rich flora from medicinal and economical point of view. As such, in view of the considerations outlined above the present work, entitled "Ethnobotanical studies of tribal inhabited localities of Ratlam district (Madhya Pradesh)" was undertaken. The present ethnobotanical work was conducted first time to document the ethnobotanical plant species used by Bhil tribe of the district. The present ethnobotanical work is based on 210 plants used by the tribal for various ailments and for other purposes like food, fuel, fibers, house building, fodder, musical instruments and agriculture implements, etc.

GEOGRAPHICAL POSITION, TOPOGRAPHY & ADMINISTRATIVE ASPECTS

Location and Extent of study area

Ratlam district lies in the western part of the Malwa Plateau in Madhya Pradesh. Ratlam has an important place in the Railway Map of Madhya Pradesh. Its position as the Divisional Headquarters of Western Railway has considerably contributed to its growth as a Railway Town. The district forms part of the Commissioner's Division of Ujjain. The mathematical limits of the district extend from 23°05' North to 23°55' North and from 74°30' East to 75°42' East. It has the shape of a wide crescent with its concave on the east and widest part in the south. The greatest length of the district extends about 122 km from southwest to northeast. The district is bounded by Mandsaur district on the north, Jhabua and Dhar on the south, Ujjain on the east and Chittorgarh and Banswara district of Rajasthan State on the west. Shajapur district of Madhya Pradesh and Jhalawar district of Rajasthan also touch the north eastern boundary of Ratlam district. The elevation generally varies from 434 to 549 m above mean sea level. The district comprises an area of 4861 sq. km inhabited by 1214536 persons according to Census 2001. The district is administratively divided into 6 Tahsils, viz. Ratlam, Sailana, Jaora, Piploda, Bajna and Alot. The forests, which occupy a total area of 499.92 sq km or 10.26 %, are of the dry deciduous Vb type (Champion & Seth, 1968). The forests are dominated by Teak (*Tectona grandis*). Other notable woody species are Sajad *(Terminalia elliptica)*, Mahua *(Madhuca indica)*, Tendu *(Diospyros melanoxylon)*, Palas *(Butea monosperma)*, Kalam *(Mytragyna parviflora)*, Dhaora *(Anogeissus latifolia)*, Bel *(Aegle marmelos)*, Bija *(Pterocarpus marsupium)* and Bamboo *(Dendrocalamus strictus)* which constituted a significant chunk of the middle canopy of the forests. In Ratlam district there are two important rivers, namely, the Chambal flowing north along with its tributaries and the Mahi flowing to the north-west. The climate of the State is generally sub-humid with an annual rainfall ranging from 700-1200 mm. The minimum and maximum temperatures range between 6^0 C and 43 ^{0} C in the months of January and May, respectively. The black cotton soil predominate the entire district.

There are 1051 inhabited villages in the district. Bhils are the main inhabitants of different villages of the district. According to 2001 census, the population of Bhil tribe in the district was 2.26 lac, constituting about 23.27% of the total population.

Table 1. Area and tribal population of the Ratlam district (2001).

SN	Name of Study area / Tahsil	Area (sq. km.)	Geographical Status		Height above sea level (in Meter)	Tribal Population (2001)	% of Tribal Population
			Latitude	Longitude			
1	Ratlam	1331	23 05'-23 33'N	74 54' –75 20'E	488	66900	17.33
2	Alot	495	23 31'-23 55'N	75 16'-75 42'E	450	1046	0.69
3	Jaora	963	23 30'-23 51'N	74 49'-75 21'E	486	5644	3.82
4	Piploda	604	-	-	-	13478	13.74
5	Sailana	536	23 07'-23 33'N	74 30'-74 59'E	549	58335	75.41
6	Bajna	682	23 07'-23 28'N	74 30'-74 59'E	434	80753	91.04

Source: District Statistical Department, Ratlam.

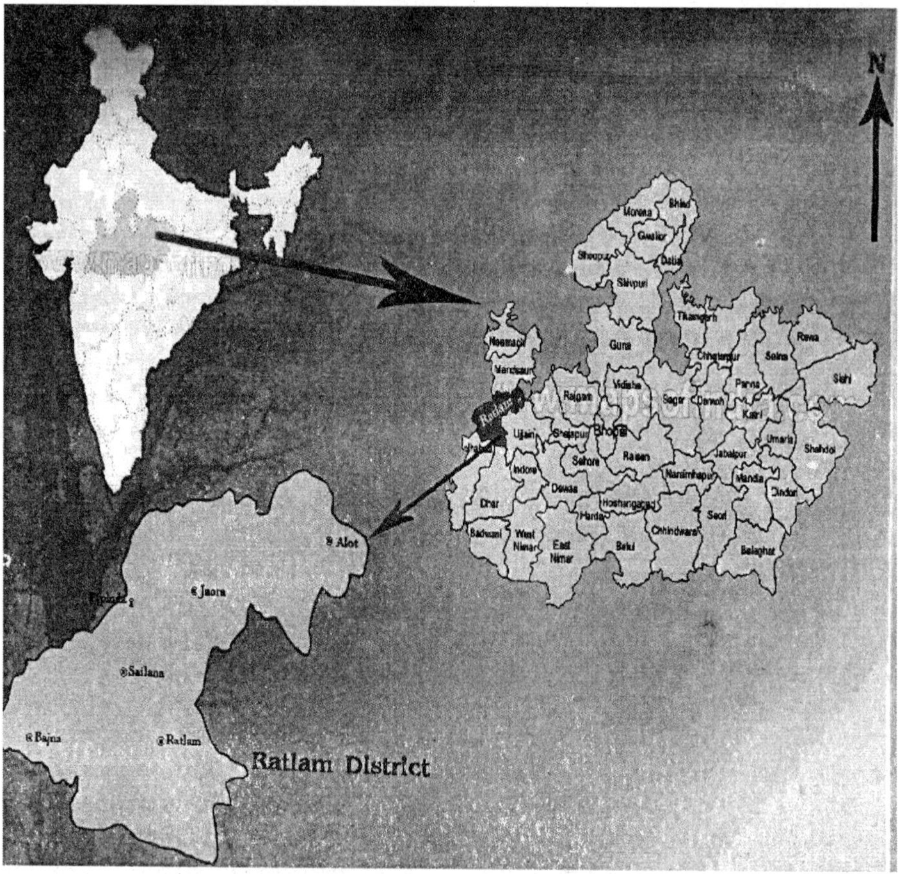

Fig. 2 Location of Ratlam district.

Table 2. Number of inhabited villages in the Ratlam district.

Tahsils	Number of Village	Tribal Population	Percentage of Tribal Population
Ratlam	170	66900	17.33
Alot	191	1046	0.69
Jaora	146	5644	3.32
Piploda	089	13478	13.74
Sailana	236	58335	75.41
Bajna	219	80753	91.04
Total	1051	226156	23.27

Source: District Statistical Department, Ratlam.

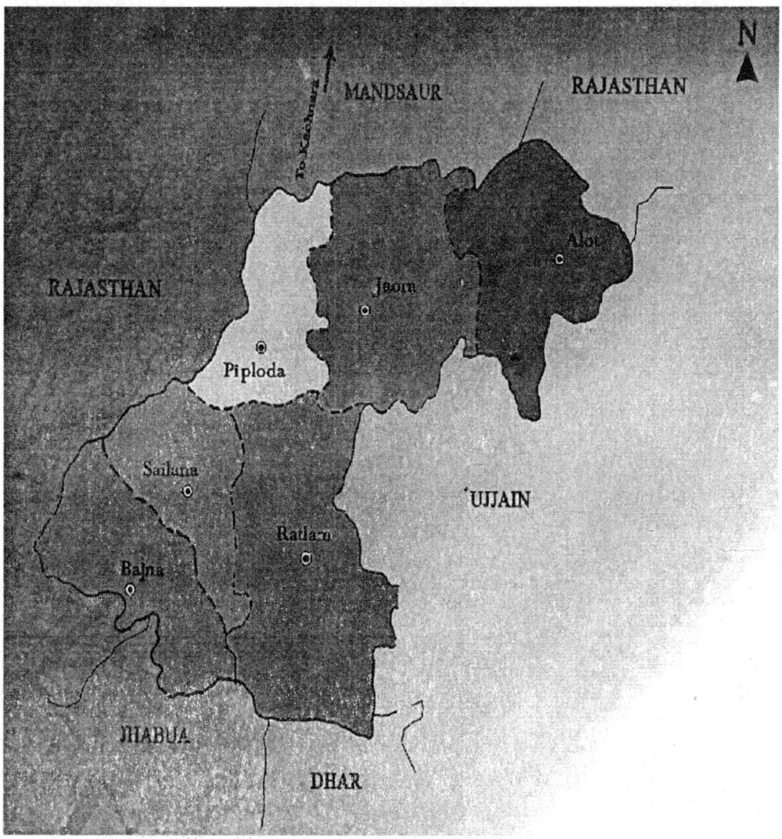

Fig. 3 Location of study area in Ratlam district.

Etymology

The district is known after the headquarters town, Ratlam. It was the head-quarters of Princely State of Ratlam which was merged in the Indian Union in June, 1948. The origin of the name of the town is obscure. Its popular derivation from Ratan Singh, the founder of the former Ratlam stands refuted as the town was already in existence before Ratan Singh was granted the territory. The town has been mentioned in the Ain-i-Akbari (Ratlam State Gazetteer, 1908).

Historical perspectives

The district of Ratlam was created in June 1948 and was reorganized in January 1949. It covers the area of the former Princely States of Ratlam, Jaora, Sailana, Piploda, Dewas Senior, Dewas Junior and parts of Mandsaur Tahsil of former Gwalior State of the Chief Commissioner's Province of Pant Piploda.

Ratlam was the head quarters of a *Mahal* in the Ujjain *Sarkar* of the *Subah* of Malwa in the Mughal Empire under Akbar. The town rose to importance when Ratan Singh was granted a jagir of 12 *Parganas,* viz., Dhar (Ratlam), Badnawar, Dagparawa, Alot, Titrod (Sitamau), Kotri, Gadgucha, Agar, Nahargarh, Kanar, Bhilara and Ramgharia, by Shah Jahan. Ratan Singh chose the place as his capital. The State was reduced to a very small size very soon after his death. After, in 1908, it was divided into two Tahsils of Ratlam and Bajna, each under a Tahsildar. Ratlam Tahsil was sub-divided in to the *Kamasdaris* of *Dhamnod, Dharar and Ringnia,* the *halka* of *Gird-Kasba* including the Town and *29 Jagirs.* Bajna was originally held by Mahida (on Mahi river) Bhils. It then passed to Bharmalot Rathors, and was made *Khalsa,* in lieu of Isarthuni Jagir given to them. Bajna was made a Tahsil and Ringnia was dropped.

Ghafur Khan, the Nawab of the Jaora was in the court of Holkars. Since he offered his submission to the British, he was guaranteed the possession of Sanjit, Malhargar, Tal, Mandawal, Jaora and Baranda Tahsils and the tribute of Piploda. In 1886 the State was divided into 4 tahsils of Jaora, Barauda, Tal-Barkhera and Sanjit-Malhargarh, each under a tahsildar and the *tappa* of Nawabganj under an *amil.*

Sailana State, guaranteed by the British in 1919, was divided into tahsil of Sailana town and Sailana district, each under a Tahsildar. This arrangement was made probably between 1850 to 1859. Only the district or tahsil was sub-divided into the *Kamasdaris* of Bilpank, *Bangrod* and Raoti. In 1919-20, the *Kamasdari* of Raoti was detached from the district tahsil and was counted as an independent unit, but the following year district tahsil was abolished and Sailana, Bangrod, Bilpank and Raoti were made four tahsils.

The Alot Tahsil of Dewas State (Senior Branch) and Ringnod Tahsil of Dewas State (Junior Branch) amalgamated in this district in January 1949, were kept as such even in the early years of the century.

The States of Central India signed a covenant on 22nd April, 1948, agreeing to unite and integrate into one State by the name of the United State of Gwalior, Indore and Malwa (Madhya Bharat). Accordingly the United State of Madhya Bharat was formally inaugurated on the 28th May, 1948. The Rulers of Gwalior and Indore handed over administration to the Raj Pramukh on the 16th June, while most others handed over on June 20 and 30. The signing of covenant by the Ruler of Jaora was delayed up to 12th August, 1948, pending a settlement of the dispute regarding the succession to the *Gaddi* of the State. While the districts of Indore and Gwalior States left undisturbed, the territories of the remaining States were grouped in to three charges of Chief Administrators and nine Administrator's districts. The Administrators of districts were appointed towards the end of June 1948. Ratlam was the head quarters of an Administrator as well as the Chief Administrator. The Chief Administrator of Ratlam controlled the Administrator's districts of Ratlam and Sitamau. The Administrator's district of Ratlam included the former States of Ratlam and Sailana, while Sitamau district included the former States of Sitamau and Piploda, and Ringod and Alot Tahsils of Dewas States.

The post of Chief Administrator was terminated in November, 1948 and that of the Administrators in January, 1949, giving place to the Suba, later called Collector.

The district was enlarged by the additions of former States of Jaora, and Piploda, Alot and Rignod Tahsils of Dewas States, Chief Commissioners territory of Pant Piploda and Mandsaur Tahsil of Gwalior State in January, 1949. During the decade 1951 to 1960 the only inter district adjustment was the transfer of 3 villages with an area of 4,885 acres form Alot Tahsil to Khachrod Tahsil of Ujjain district.

Present administrative aspects

The sistrict is administratively divided into 6 Tahsils, viz. Ratlam, Sailana, Jaora, Piploda, Bajna and Alot. These are grouped in to three revenue Sub-Divisions. Ratlam Sub-Divison covers the Tahsil of Ratlam. Sailana Sub-Division, with head quarters at Ratlam, is coextensive with Sailana & Bajna Tahsils. The third Sub-Division, namely, Jaora covers the remaining tahsils with Alot & Piploda head quarters at Jaora.

THE PEOPLE SETTING

The Bhils- Nature and Habits

Tribals are a distinct ethnic group who are usually confined to definite geographical areas, mainly forest area, speak a common dialect, are culturally homogeneous and unifying social organization. They form a sizeable percentage of total populational certain regions of India.

State Madhya Pradesh is the home of many tribals, scheduled castes, forest dwellers and aboriginals. Ratlam district is one of those regions where the rural population and forest dwellers form a considerable part of the population. The tribal population resides mostly in the interior areas. Like other communities, the original culture of the tribal communities is undergoing fast changes; it is due to urbanization, industrialization and acculturation with the consequent loss of knowledge and tradition.

Bhil is one of the oldest indigenous tribes of India. They are widely distributed throughout the study area. The word 'Bhil' is denoted to the ancient community expert in using bow and arrow and wearing leaves and toys of plants. Some Anthropologists consider them allied to Negroes, while others consider them Protoaustralloid. Bhils may be identified by their solid body, medium height, pointed nose and attractive eyes.

Bhil tribe lives in small houses called 'Tapri', made up of mud, timber, grasses, leaves etc on gentle slopes of hills near their agriculture fields. Several scattered houses ('Tapra') constitute a village called 'Phalia'. Some prosperous people, however, construct their houses with stones and unburnt bricks.

The tribal people are innocent, honest, unsophisticated, delightful, closely linked to nature and good looking with sharp features. They have great regard for truthfulness and generally depend on agriculture for their living. As regards social relations, they maintain strong family affinities. Women are more hardworking than the men. Besides their domestic duties, they also help their counterparts in agricultural operations. In their leisure time, men mostly enjoy tobacco smoking.

Bhils have faith in the nuclear family having husband, wife and unmarried children. It is also a male dominated community wherein girls have no claim in paternal property and boy are given a separate house and piece of land to lead their independent life after marriage.

Their main sources of income are agriculture, labour and forest products. Their names are based on the names of months, day, flowers, fruits and crops. The Bhil men & women are like to be called "Mama" and "Mami".

Tribal population

Bhil is the most numerous tribe scheduled throughout the district; they form 23.27% of the total population of the district (according to census 1991). They mainly distributed in all the villages of the district. The name Bhil is supposed to have been derived from the Dravidian word for a bow. According to Ratlam State Gazetter (1994), the Bhils have emigrated from Udaipur and Jodhpur (Rajasthan) to their present habitat. The following table gives the tahsil-wise tribal population of the district -

Table 3. Tribal population in the study area (1991).

Area / Tahsil	Tribal Population of the district			Rural tribal Population	Urban tribal Population	% in Total Population of district	Position in the district on the basis of ST Population
	Total	Male	Female				
Total	226156	114695	111461	216669	9487	23.27	-
Ratlam	66900	34143	32757	60033	6867	17.33	III
Alot	1046	537	473	890	156	0.69	VI
Jaora	5644	2957	2687	4885	759	3.32	V
Piploda	13478	6921	6557	13321	157	13.74	IV
Sailana	58335	29510	28825	56787	1548	75.41	II
Bajna	80753	40591	40162	80753	-	91.04	I

Source: District Statistical Department, Ratlam.

Literacy

They are educationally very backward. Very small percentage (12.66%) of the tribal people is formally educated and the rest are illiterate. Literacy rate among the females is negligible as shown in the Table 4. The table represents the status of literacy of ST in the district. This shows the miserable condition of literacy among the Bhils of the disrict. The condition has slightly improved in the last decade.

Table 4. Literacy among the Bhil tribe of Ratlam district.

Area	ST: Literacy			Percentage of Literacy to Total Population (1991)		
	Total	Male	Female	Total	Male	Female
Total	21914	18088	3826	12.66	20.53	4.50
Rural	18996	15991	3005	11.47	19.04	3.68
Urban	2918	2097	821	39.04	50.97	24.43

Source: District Statistical Department, Ratlam.

Occupation

The primary occupation of the Bhils in earlier times was hunting and fishing, however, now a days they are mainly engaged in agriculture and the great bulk of them are farm servants and labourers. For livelihood, they mainly depend upon physical labour, skilled and unskilled both. In search of work they migrate to neighbouring areas. Their main subsidiary occupation is cattle rearing and collection of forest products to supplement the food, and collection of firewood to sell in the market.

Agriculture

Most of the people live in the hilly areas of the forest and their lands are less fertile as compared to those cultivated by the non-tribal. Most of the lands cultivated by the tribal are devoid of irrigation facilities, the majority have only wells which rarely have water throughout the year. Thus tribal of the district are economically backwards. The villagers follow private and orthodox methods of agriculture. They use local, low productive seeds and chemical fertilizers are generally used to increase the production. Agriculture is mostly dependent on monsoon.

Since the land left with them is mostly infertile and unirrigated, therefore the agriculture is not their main sources of livelihood. Their capacity to invest in the field is also very limited with the result that agriculture is not an economic profession of tribal. The average land holding among them is marginal and uneconomic.

Minor Forest Products

The collection of minor forest products (NTFPs) by villagers is a profitable profession of Bhils; the minor forest products provide alternate food material to these people, they also earn some money by selling of surplus collections. Fruits of *Diospyros melonoxylon, Phyllanthus emblica, Madhuca indica, Annona squamosa, Manilkara hexandra, Phoenix sylvestris, Pithecellobium dulce, Tamarindus indica, Carissa carandas, Syzygium cumini, Limonia acidissima, Mangifera indica, Ziziphus mauritiana, Ziziphus nummularia* and *Aegle marmelos* are main produces which are collected and selled by them. Collection of minor forest produces is a sociocultural habit of the villagers.

Food

Main foods of Bhils are maize (*Zea mays*), onion, garlic and chilli which they cultivate in their small fields. The vegetables and fruits are collected by them from the forests. Wheat and rice are used at the time of festivals and other special occasions only. They keep self made bow and arrows, sword, knife, axe, pharsa etc. with them as weapons for self protection and hunting the wild fauna which also form the major part of their diet. They profusely use alcohol distilled by them from the flowers of Mahuwa (*Madhuca indica*). Meals are generally taken twice, at mid-day and evening. On festive occasions, various special preparations form the dish rich, i.e. makka (ghughri), genhu (ghughri, lapsi, Kasar and churma), chawal (kesariabad and kheer) etc. Bhils do not use milk, but new generation uses the milk in the form of tea. The Bhils are traditionally non-vegetarian. They eat the meat of Bakra (Bokro), Murga (Kukro), fish (Machli), Bull (Pado), Hiran, Titar, Bater etc.

Dialects

Bhili is the main dialect of the Bhil tribe. It forms a continuous chain between Rajasthani, Gujarati and Marathi and has some non-Sanskrit elements in it. Malwi and broken Hindi are also spoken and easily understood by the Bhils. Bhili has sufficient literature of its own. Bhili does not have script of their own and Nagri characters are frequently used while writing them.

Dress

The tribal people dress is the fashion common to the district. Males usually wear a 'Pagri' or 'Safa' (Turban) on the head, a 'angarkha' or 'Kurta', a 'dhoti' and 'pachedi' or 'gamcha' (Scarf). The males wear 'dhoti' up to the knee. All these are generally grey, white, except the 'pagri'/ 'safa'/'potya'/'phenti' (Turban) which are often coloured red, pink, purple, yellow etc. Except in the use of 'pagri', 'angarkha'/'kurta' and 'dhoti', little change is noticeable in the young generation, because educated Bhil boys do not wear 'pagri', 'kurta' and 'dhoti', but they wear modern dress i.e. trousers, shirt etc.

The tribal women wear small traditional dark red printed cloth 'sari' to cover the head ('Lugro'), a small bodice called a 'choli' or 'kanchli' and dark red printed cloth ‘ghagra’ (petticoat). The old aged women wear ‘Jakri’ (‘Angrakhi’) and the widow women wear black or brown coloured clothes.

Ornaments

Bhils, especially Bhil women, in general have much attraction towards the ornaments. Females adorn with a variety of gold and silver ornaments, but Bhil men-folk, being very poor, wear Pital, Gilat, Ranga, Kansa and Zusta ornaments. 'Kadas' on the wrist, 'bajuband' on the upper arm, locked or chain on the neck, 'balien' or ear rings on the ear-lobes and 'kardhani' on waist line form a complete set of ornaments which are ever now popular with men. Bhil men also wear black thread on the neck called ‘Karathi’. Their females adore theirself with a variety of ornaments i.e. hansli on the neck, 'Vithi' or rings (Fingers), 'Bor' on the head, 'zele-zumke' on the forehead, 'baliyan' on the ear-lobes, 'narniyan' (a type of Churiyan) on the wrist, 'nathni' on the nose, 'kandoro' on the waist line, 'pujeb' on the the lower legs and 'bichhudi' on the tees. Tattooing is very common in women folk belonging to Bhils. The period between engagement and marriage proper is regarded ideal for this purpose. It is regarded as a permanent ornamentation.

Social Customs and Beliefs

Marriage system

Among the Bhils of the district, marriage is performed for procretion purpose. Monogamy is the general rule among them. Bride’s cost called ‘Dhapa’ is the main base of marriage which is paid to the parents of bride by groom’s side. Generally young boys and girls (15 years or So) themselves select their partners or boy kidnaps the girl or married women and marry with her after paying Dhapa. In some subcastes of Bhils, exchange marriages are more popular in which Dhapa is not involved. In few clans, the parents put their boy on the service of girl’s parent to make them happy for marrying their daughter with their son (Sewa vivah). Widow marriage has found wide acceptance in the tribe, where female preferably remarry with the younger brother of her late husband and male with his sister-in-law. Divorce is also easy and common practice among Bhils. Natara is the local form of secondary marriage for women. They avoid marriages in their own clan.

Faith and Worship

Every village has its own local deity (Gramdev) and the families too have their Jatidev, Kuldev and Kul-Devi (house hold diety)which are symbolized by stones. 'Bhati dev' and 'Bhilat dev' are their sarpent god, 'Baba dev' is their village god, Harkulia dev is their crop god, Gopal dev is their gwala god, Bag dev is their lion god and Bhairav dev is their Dog god. Their some other gods are- Indel dev, Bada dev, Mahadevel, Tejaji, Lotha mai, Techma, Orka, Chichma, Kajal dev etc. A number of Hindu deities have now also been worshipped by them, viz. Hanuman, Shri Rama, Krishna, Shiva, Laxmi, Ganesh, Kuber, Indra, Durga, Kalika, Sitlamata, Hinglaj mata, Garkhanga mata etc. They are supposed to cure all ills, physical or economic, of the true worshipper.

On the whole, the Bhils are deeply religious, god fearing, superstitious and believes in ghosts, witches and evil spirits. Animal sacrifice is also prevalent.

Mostly Bhils have faith in Hindu mythology and, therefore, follow Hindu customs from birth to funeral. The dead bodies of Bhils are burnt except infants, which are buried. Sometimes commemorative stones are erected in honour of the deceased.

They have much faith in magico-religious beliefs and Bhopas for their physical, mental and psychological treatments.

Festivals

There are a number of festivals, viz. Rakhi, Navratri, Dasahra, Diwali, Holi which are celebrated by the Bhils. They also celebrate some tradi-tional festivals viz. Akhatij, Navmi, Howan Mata ki Chalavani, Sawan Mata ki jatar, Diwasa, Nawai, Bhagoria, Gal, Gar, Dhodi, Sanja, Indel, Doha etc with ceremonious zeal and enthusiasm.

During some festivals there are a number of tribal fairs (mela) held at different places of the district, viz. Navratri ka mela (garkhanga mataji ka mela), Bhagoria ka mela (during Holi festival) etc.

Communal dance, Festivities

The chief means of their recreation are folk songs and folk dances. Women-folk usually, dances at birth celebrations of a son, marriage functions and on a few festivals in traditional Bhili style to the accompaniment of the drum beat. Some of their dances are of Lathi dance, Dhol dance, Marriage dance, Holi dance, Battle dance, Bhagoria dance, Depawal dance, Hunting dance etc. Their musical instruments, viz. are Harmonium, Sarangi, Kundi, Bansuri, Apang, Khajria, Tabla, Jhanjh, Mandal, Thali etc are made by them from local products.

Leadership and Government

The Bhil tribe accept leadership in social and religious matters. Usually, the senior-most male member of the 'Phalia' (Village) acts as the official priest called "Gameti". The village Panchayat, called the 'Panch', advises the headman "Gameti" on all social and religious matters. Besides sarpanch (headman) there are 2 other members called 'Bhanjgaria' in the panch. Besides these, there are some other aged male persons also members of the panchayat. 'Sokara' is other institute organized by the adjoining 2 or more 'Phalia'. The 'Gameti' and 'bhanjgaria' of all the 'Phalia' are also the member of 'sokara'. The main concern of 'sokara' is to maintain security and solidarity of the 'Phalia'. The panchayat is the custodian of Bhil tradition and customs. Its main concern is to maintain the security and solidarity of the tribe.

METHODOLOGY

The ethnobotanical studies were carried out from January 2004 to September 2007 in different tribal inhabited villages of Ratlam district. Bhils are the main inhabitants of different villages of the district. These primitive people more or less use many common and uncommon local plants for the treatment of their routine ailments and various diseases. The interior localities provide a vast scope for the study of medicinal and economic plants.

The study is most effective and informative because information was recorded on the spot along with careful documentation of cultural and biological aspects of local knowledge, with voucher collections and transcribing local names etc. The study sites of the district were visited for the collection of plants in each season of the year.

Ethnobotany is a field – oriented study involving direct relationship of aboriginal with surrounding plants. Field study in the tribal areas gives first hand information. For this purpose, extensive field trips are organized for collecting the plant species and data using an integrated approach of botanical collections, group discussions, interviews and questionnaires. In Bhil tribe or in a village, normally there is a one elder person familiar with the surrounding plants; he is a local physician and a very resourceful person who is called as "Bhopa". "Bhopa" is a traditional rural medicine men or herbal practitioner of the tribe. This person is to be contacted for collecting information about ethnomedicinal plants. With an intension to give them the credit of their knowledge, a list of Bhopas is given below with their age and Village- 1. Vijay Dodiyar (30 yr, Dabar), 2. Khimaji Dodiyar (42 yr, Dabar), 3. Manglaji Hari (50 yr, Semaliamaal), 4. Vijay lal ji Rawat (45 yr, Khairda), 5. Lal chand Dodiyar (50 yr, Devla), 6. Chaitanji Ninama (40 yr, Naktipada), 7. Jotiji Ninama (40 yr, Ganawa), 8. Premchand ji Bhabhar (26 yr, Naktipada), 9. Manjiji Katara (40 yr, Chandragad Maal), 10. Punam Chand Bhabhar (32 yr, Chandragad Maal), 11. Mangal Singh Bhabhar (35 yr, Kangsi), 12. Virji Maida (34 yr, Mahapura), 13. Babuji Khadia (38 yr, Harthal), 14. Baluji Garwal (40 yr, Harthal), 15. Kodarji Garwal (42 yr, Harthal), 16. Mohan ji Maida (45 yr, Naharpura), 17. Bhagirath ji Katija (36 yr, Gayli), 18. Haroba (55 yr, Bibdod), 19. Phulji Bhil (45 yr, Matrunda), 20. Onkarnath ji (60 yr, Khankhai), 21. Kodar Ba (55 yr, Haldupada), 22. Lallu Maida (35 yr,

Rampuria), 23. Shambhu Singh Gamad (37 yr, Khejdipada), 24. Bahadur (38 yr, Vatpadi), 25. Bhairosingh Muniya (40 yr, Bhamat), 26. Lunja ji Charpota (55 yr, Kangsi), 27. Velji Ba (80 yr, Ranisingh), 28. Hurji Ba (45 yr, Ranisingh), 29. Lal Chand Ba (70 yr, Bashindra), 30. Babu (35 yr, Naharpura), 31. Lachman Ba (60 yr, Dholawad), 32. Premchand Charpota (28 yr, Molawa), 33. Bhonji Kodar (50 yr, Molawa), 34. Telarama Devda (79 yr, Molawa), 35. Ramu Muniya (60 yr, Bayedi), 36. Hakra lal ji (50 yr, Maliwada), 37. Naru Ninama (80 yr, Patakhari), 38. Ramesh Chatra (35 yr, Devda), 39. Galia Dharji (80 yr, Devda), 40. Raisingh Ninama (55 yr, Patakhari), 41. Harchand Katara (60 yr, Molawa), 42. Punja Bhuria (45 yr, Naharpura), 43. Nandu Avima (55 yr, Devda), 44. Babusingh Chouhan (65 yr, Rawti), 45. Babu Rangji (50 yr, Jaman), 46. Nandu Baria (55 yr, Ghatalia), 47. Nandu Khima (50 yr, Umar). The informations on other ethnobotanical plants i.e. food, fiber, wood, fuel, Shelter, fodder, narcotic, beverages, tannin, dye, gum, resin, cosmetics, crafts, fish poison and socio-religious beleifs etc. was collected from other knowledgeable people of the tribe. Sometimes the information was collected from them separately, but at other occasions, the information was collected having group discussions between Bhopa and other knowledgeable persons. The Bhopa and / or knowledgeable people were taken to the field to collect data, viz. local name, parts used, method of preparation of the drug and approximate dosage of administration. These data were recorded in the field book. Voucher specimens were collected for authentication of information and future reference. With the help of local name and specimen it was possible to verify the uses with other villagers and the data are considered valid if at least 2 informants provided similar uses about a medicinal plant, the data was considered valid. The plants have also been closely associated with many social customs and mythological rituals of the tribes; many flowers, fruits or whole plants are used for offering in worship and some plants are themselves worshipped or considered sacred. Some plants are believed to possess magical properties. The collection of information on these plants was also including in the proposed work. Herbarium specimens prepared following the standard method (Jain & Rao, 1978), have been deposited in the herbarium of IEMPS, Vikram University, Ujjain (Madhya Pradesh). Herbarium specimens were identified with the help of standard Floras (Duthie, 1903-29; Hooker, 1872-1894; Maheshwari, 1963; Mudgal *et al.*, 1997; Oomachan, 1977; Singh *et al.*, 2001; Verma *et al.*, 1993).

The present ethnobotanical work in different tribal inhabitated localities of Ratlam District is based on 210 plants used by the tribal for various ailments and for other purposes like food, fuel, fibers, house building, fodder, musical instruments and agriculture implements etc. Plants have been arranged in alphabetical order by botanical name. Only important synonyms, particularly basionyms are given. Name of family, locality, habit and habitat of plant, flowering and fruiting period, collection number, regeneration and distribution in the study area are also given.

The sequence of presentation of data in the book is: Plants are arranged alphabetically by their generic and specific names, followed by synonyms, name of family, vernacular and regional names, distribution, taxonomic description, flowering & fruiting, habitat ecology, regeneration, field no., parts used, chemical constituents, biological activity, folk uses and uses in literature.

ENUMERATION

(Alphabetical arrangement of species with attached ethnobotanical data)

Abelmoschus esculentus (L.) Moench. (Malvaceae)

Syn.: *Hibiscus esculentus* L.

Vernacular name - Bhindi, Bhinda

Regional Names - *Bengali* – Dherash; *English* - Lady's finger; *Gujarati* - Bhindu; *Hindi* - Bhindi; *Kannada* - Bhende; *Malayalam* - Vendakkai; *Marathi* –Bhendi; *Sanskrit* – Asrapatraka, Dindisa; *Tamil* - Vendakkai; *Telugu* - Bhenda-kaya.

Distribution - Cultivated throughout India.

Taxonomic description - Hispidly hairy herb, with 3 to 7-lobed leaves and yellow flowers with purple centre. Fruits cylendric, 5-angled, beaked.

Flowering & fruiting – Throughout the year.

Habitat ecology - Cultivated as a vegetable crop (Fruits) throughout the area.

Regeneration - By seeds.

Field no. - DJ-62

Parts used - Root, fruits, seeds.

Chemical constituents - Fruits contain 5-hydroxytryptamin, Oxalates, Iodine, Flourine, Calcium, Phosphorus, Magnesium, Vitamin-C, Fats, Carbohydrates, Fibres, Amino -acids, Polyphenols.

Biological activity -

Fruits - Aphrodisiac, constipating, cooling, demulcent, diuretic, emollient, stomachic.

Seeds - Antispasmodic, aphrodisiac, demulcent, emollient, stimulant.

Folk uses -

Food: The fruits are used as vegetable by the tribals.

Medicine:

- Extract of root is taken with water daily 3 times for a week to cure leucorrhoea.
- The root powder is taken orally with water to check undesirable discharge of semen with urine and to increase sexual potentiality.
- The root powder is soaked in water at night and taken orally in the morning to cure inflammation of genitals during mensus.
- Extract of root is taken with water daily 3 times for a week to check undesirable discharge of semen with urine.
- Extract of root is also useful in the cases of spermatorrhoea.

Miscellaneous: The dry seeds and fruits are sold in the local weekly market 'Haat'.

Uses in literature

Fruits – Abortifacient (Tarafder, 1983); Itchy eruptions, fever during teething, dysentery, diarrhoea (Jha *et al.,* 1997); Venereal diseases (Singh *et al.,* 1997); Diabetes (Ghosh, 2006); as vegetable (Shah, 1984).

Leaves - To make black and silky hair (External application) (Jadeja *et al.,* 2006).

Abrus precatorius L. (Papilionaceae)

Vernacular name - Lal Chermi, Jethimod.

Regional and other names – *Bengali* – Kooch; *English* - Crab's eye; *Gujarati* – Gunja; *Hindi* – Ratti; *Kannada* – Ganji; *Malayalam* – Kunni; *Marathi* – Gunchi; *Sanskrit* – Aruna; *Tamil* – Gundumani; *Telugu* – Guravindha.

Distribution - Throughout the tropics, often planted.

Taxonomic description - Twining shrubs, with 10-20 pairs of leaflets and pink or white flowers having 9 stamens. Pods 3 to 5 seeded.

Flowering & fruiting - August-February.

Habitat ecology - Common among the clumps of trees and shrubs throughout the area, particularly in hilly tracts.

Regeneration - By seeds.

Field no. - DJ-115

Parts used - Seeds.

Chemical constituents

Root contains Stigmasterol, Glycine, Glutamic acid, Aspartic acid, Tyrosin, Quercetin, Luteplin, Kaempferol, Taxalbumin; Alkaloid-abrin, Glucoside-abralin, Anthocyanin-abranin.

Seeds contain Lupeol acetate, Sitosterol, Sucrose, Gallic acid, Orietin, Isoorientin, Abrussic acid, Haemagglutinin.

Leaves contain Precol, Abrol, Abrasine, Precasine.

Biological activity

Root - Acrid, alexiteric, aphrodisiac, bitter, diuretic, emetic.

Leaves - Acrid, alexiteric, aphrodisiac, bitter, diuretic, emetic.

Seeds - Abortifacient, aphrodisiac, emetic, purgative.

Fruits - Acrid, aphrodisiac, bitter.

Folk uses

Magico-religious beliefs: Bhils believe that seeds can be used as an amulent to wardoff the evil spirits from their children.

Medicine:

- One teaspoonful powder of seeds is taken with water after menstruation period for 3 days early in the morning on empty stomach to promote conception.
- The leaves are chewed in mouth ulcer.
- The paste of leaves is mixed with jagerry and some small pills are prepared. One pill taken thrice a day to cure typhoid.

Miscellaneous: The leaves are boiled with tea in place of sugar for making sweet and tasty tea.

Uses in literature

Root - Abortifacient (Tarafder, 1983; Chevlan, 1998; Purkayastha & Nath, 2006); Snake bite (Chandra, 1995); Antiseptic, bronchitis, cold, blood purifier (Goel *et al.,* 1984); Leucoderma, promote conception (Pal & Jain, 1989); Anthelmintic (Gaur *et al.,* 1980); Sterility (Purkayastha &

Nath, 2006); Retention of urine (Kumar *et al.*, 2005); Syphlis (Prakash & Singh, 2006).

Leaves - Body pain (Sharma & Malhotra, 1984); Cough, sore throat, rheumatism (Bhalla *et al.*, 1982); Eye Complaint, skin diseases (Janardhanan, 1963; Singh *et al.*, 1998); Diabetes (Nagaraju & Rao, 1989); Scabies (Pal & Jain, 1989); Leucoderma (Saxena, 1983); Earache (Pereira, 2006); Thorat infection (Mokat & Deokule, 2006); Swelling (Bhalla *et al.*, 1982; Kumar *et al.*, 2006); Vetrinary medicine (Sebastian, 1984); As vegetable (Vartak, 1981); Masticatory (Shah *et al.*, 1981, Shah, 1984).

Seeds - Sterility (Jubilee Purkayastha & Nath, 2006); Abortifacient (Tarafder, 1983; Kapur, 1993); Antifertility (Bhalla *et al.*, 1982); Stomach complaint (Tarafder, 1983); Eczema (Purohit *et al.*, 1985; Upadhay *et al.*, 1998); Eye complaint (Bennet, 1985); Sciatica (Rajwar, 1983); Skin diseases (Janardhanan, 1963; Singh *et al.*, 1998); Fever (Srivastava & Rout, 1994); Paralysis, sciatica (Rajendran & Rengamani, 2006); Edible (Gunjatkar & Vartak, 1982).

Whole plant- Hepatitis (Dutta Choudhury & Shil, 2006).

Abutilon indicum (L.) Sweet. ssp. ***guineense*** (Schum.) Borss. (Malvaceae)

Syn.: *Abutilon asiaticum* (L.) Sweet.; *Sida guineense* Schum.

Vernacular name - Kangsi.

Regional & Other Names - *Bengali* - Petari; *English* - Indian mallow; *Gujarati* - Dabali; *Hindi* - Kanghi; *Kannada* -Tutti; *Malayalam* – Tutti; *Marathi* - Mudra; *Sanskrit* – Atibala; *Tamil* – Nallatutti; *Telugu* - Adavabenda.

Distribution - Throughout the tropics.

Taxonomic description - Undershrubs, with ovate – suborbicular leaves and yellow, axillary, solitary flowers. Fruits with 15 – 25 mericarps.

Flowering & fruiting - October-April.

Habitat ecology - Common in westlands, neglected corners of fields.

Regeneration - By seeds.

Field no. - DJ-65

Parts used - Root.

Chemical constituents

Plant contains Gossypetin, Glucosides, Cyanidin, Rutinosides, Essential oil.

Root contains Fatty acids, Aspargin, Rafinose.

Stem & Leaves contain Alkanol, Beta–Sitosterol, Vanillic acid, Caffeic acid, Fumaric acid, Fructose, Glucose, Galactose, Leucin, Histidine, Threonine, Serine, Glutamic acid and Aspartic acid.

Biological activity -

Root – Antipyretic.

Root bark - Astringent, diuretic.

Leaves - Demulcent, diuretic, emollient.

Seed - Aphrodisiac, demulcent, laxative.

Folk uses -

Medicine:

- Extract of root is taken with water daily 3 times for a week to cure leucorrhoea.

Uses in literature

Root - Paralysis (Jain & De, 1966); Cough (Singh & Pandey, 1980); Leprosy (Rajwar, 1983); Wound (Sudhakar & Chetty, 1998); Fever (Deokule, 2006).

Leaves - Boil, bronchitis, diarrhoea, ulcer (Bhalla *et al.,* 1982); Asthma, spermatorrhoea (Kumar & Chauhan, 2005); Headache (Kumar *et al.,* 1980); Wound (Rajwar, 1983; Tripathi *et al.,* 1996); Toothache (Aditya *et al.,* 2006); Bleeding piles (Sharma *et al.,* 2006); Gonorrhoea (Bhalla *et al.,* 1982; Sharma *et al.,* 2006); Jaundice, piles (Rajendran & Rengamani, 2006).

Fruits - Stomach Complaint (Tarafder, 1983).

Seeds - Piles (Shah *et al.,* 1983; Mishra & Verma, 1995; Tripathi *et al.,* 1996); Bronchitis (Mishra & Verma, 1995; Tripathi *et al.,* 1996); Arthritis (Katewa & Sharma, 1998); Gonorrhoea (Mishra & Verma, 1995; Tripathi *et al.,* 1996).

Acacia catechu (L. f.) Willd. (Mimosaceae)

Syn.: *Mimosa catechu* L. f.

Vernacular name - Kager; Khario.

Regional and other names – *Bengali* - Khayir; *English* - Cutch tree, Catechu; *Gujarati* - Kherio; *Hindi* -Khair; *Kannada* - Kaggli; *Malayalam* - Kadaram; *Marathi* - Khair; *Sanskrit* – Balapatra, Khadira; *Tamil* -Kadiram; *Telugu* - Khadiramu.

Distribution - Punjab, N.W. Himalaya, central India, Bihar, Burma, throughout the Konkan.

Taxonomic description - Trees, all parts pubescent; prickles hooked. Leaves 2-pinnate, with glands on the petiole and between the pinnae; leaflets 10-30 pairs. Spikes pale yellow. Pods flat.

Flowering & fruiting - May – September.

Habitat ecology - Rare in the forest of dry places & rocky wastelands.

Regeneration - By seeds.

Field no. - DJ-34.

Parts used - Stem bark.

Chemical constituents - Inner wood contains Catechin & Catechutannic acid (Catechutannic acid contain about 50% Tannin).

Biological activity

Stem bark - Acrid, anthelmintic, antidysenteric, antipyretic, astringent, bitter, cooling.

Resinous extract - Acrid, anthelmintic, aphrodisiac, astringent, bitter, hot, tonic.

Folk uses

Medicine: Gargles of stem bark decoction relieves the mouth ulcers.

Totem & taboo: This species is considered as a plant totem of Kharadi cloe of Bhils and, therefore, preserved and whorshipped by them.

Uses in literature

Stem bark - Abortifacient (Tarafder, 1983); Diarrhoea (Sharma *et al.*, 1979); Leucorrhoea (Tarafder, 1985); Gonorrhoea (Kapur, 1993); Stomachache (Jadeja *et al.*, 2006; Kumar *et al.*, 2006).

Resin - Diarrhoea (Borthakur *et al.*, 1996); Diabetes (Ghosh, 2006).

Wood - Skin diseases (Upadhay *et al.,* 1998).

Leaves - Dysentery (Maheshwari & Singh, 1987); Skin diseases (Gupta *et al.,* 1999); Pyorrhoea (Aditya *et al.,* 2006).

Flower - Gonorrhoea (Singh & Pandey, 1980).

Root - Rheumatism (Tarafder & Rai Chaudhuri, 1981); Toothache (Mishra & Billore, 1983).

Acacia leucophloea (Roxb.) Willd. (Mimosaceae)

Syn.: *Mimosa leucophloea* Roxb.

Vernacular name - Khejro

Regional and other names- *Bengali* – Safed babul; *English* – White babool; *Gujarati* –Bhabhal; *Hindi* – Safed babul; *Kannada* – Bilijali; *Malayalam* – Velvelam; *Marathi* –Hewar; *Sanskrit* – Arimedah; *Tamil* – Velvelam; *Telugu* – Tellatumma.

Distribution - India; Ceylon, Burma, Malaya.

Taxonomic description - Trees, with stipular-spines in pairs. Pinnae with concave glands in between and 5-30 pairs of leaflets. Heads pale-yellow, in terminal panicles. Pods flat, linear.

Flowering & fruiting - August-February.

Habitat ecology - Common in wastelands and roadsides.

Regeneration - By seeds.

Field no. - DJ-44.

Parts used - Root, Leaves, Stem-bark,Wood.

Chemical constituents - Stem Bark contain n-hexacosanol, β-amyrin, β-sitosterol, Tannin, Isookannin, Cyanin, Leuoxol.

Heartwood contain n-hexacosanol, β-amyrin.

Biological activity –

Bark - Alexeteric, anthelmintic, antipyretic, astringent, bitter, constipating, demulcent, expectorant, styptic, thermogenic, vulnerary.

Folk uses

Biofencing: The plant is used as bio-fences to protect the crops.

Fire-wood: Plant is a good source of firewood for the tribals.

Fodder: The leaves serve a good fodder for cattle.

Magico-religious beliefs: The leaves are used by the Bhopas in Jhada rituals to wardoff the effect of evil eyes.

Medicine:

- Extract of root bark is taken with water to cure diarrhoea, dysentery, headache, ascariasia and fever.
- Extract of inner-bark of the stem of *Acacia leucophloea* is mixed with the extract of inner-bark of the stem of *Butea monosperma* and *Azadirachta indica* and this mixture is taken with water to cure dysentery.
- The warm paste of stem bark is tied in piles.

Tannin: Bhils obtain tannin from the stem-bark which gives black to brown colour to the leather.

Timber: The wood is used for making the doors by the tribals.

Uses in literature

Root- Abortfiacient (Maheshwari & Singh, 1987).

Bark- Astringent (Sebastian & Bhandari, 1984); Ulcer (Apparanantham & Chelladurai, 1986); Diarrhoea, dysentery (Kataria & Gauchan, 2006); Astringent (Basak, 2006); Edible (Gunjatkar & Vartak, 1982; Bhandari, 1974).

Pod – Edible (Gunjatkar & Vartak, 1982).

Acacia nilotica (L.) Willd. ex Del. ssp. ***indica*** (Benth.) Brenan (Mimosaceae)

Syn.: *A. Arabica* (Lam.) Willd. Var. *indica* Benth.

Vernacular name - Bambul, Babul, Bawal, Bawar, Bawrio, Babool.

Regional and other names- *Bengali* - Babla; *English* - Indian Gum Arabic Tree; *Gujarati* - Babalia; *Hindi* -Babul; *Kannada* - Gobli; *Malayalam* - Karuvelam; *Marathi* - Babul; *Sanskrit* –Ajabaksha, Babbula; *Tamil* - Karuveli; *Telugu* - Nallatumma.

Distribution - Throughout the greater part of India.

Taxonomic description - Trees, with long, white stipular spines. Pinnae 2 –11 pairs and leaflets 7 – 25 pairs. Heads bright yellow, axillary, pedunculate. Pods moniliform.

Flowering & fruiting - August-March.

Habitat ecology - Common along edges of fields, roadsides.

Regeneration - By Seeds.

Field no. - DJ-01

Parts used- Root, Stem, Leaves, Stem bark, Fruits, Gum, Seeds.

Chemical constituents

Bark contains Beta-amyrin, Gallic acid, Tannins, Catechin, Quercetin, Leucocyanidin.

Fruit contains Gallic Acid, Catechin, Chlorogenic acid.

Gum contains Galactose, Arabic Acid, Calcium, Magnesium.

Seeds contains Ascorbic acid, Niacin, Thiamine, Amino acids.

Biological activity

Gum - Antidysenteric, astringent, styptic, tonic.

Stem bark - Acrid, alexipharmic, anthelmintic, aphrodisiac, astringent, demulcent, emetic.

Fruit - Acrid, astringent, cooling, expectorant, sweet.

Leaves - Antipyretic, astringent.

Folk uses-

Biofencing: The plant is used as bio-fences to protect the crops.

Fire-wood: This species is the good sources of fire-wood for the tribals.

Fodder: The leaves & fruits serve a good fodder for Cattle.

Gum & Resin: The gum obtained from the stem is eaten by the children.

Magico-religious beliefs:

- The root is tied on left hand with the help of sacred thread "Luccha". It is believed to cure Snake-bite.
- The leaves are used by the Bhopas in Jhada rituals to wardoff the effect of evil eyes.

Medicine:

- Paste of leaves is tied on anus to cure piles.
- Paste of leaves is applied on forehead to cure headache.
- Paste of leaves is applied over the soles of the feet to prevent cracking.
- Extract of root is effectively used as antidote of Snake-bite.

- A teaspoonful powder of seeds is taken early in the morning with water to cure lumbago.
- Extract of stem bark is taken with water to cure diarrhoea & dysentery.
- Paste of leaves is tied on eyes to cure conjunctivitis.
- Gum is useful in dysentery.
- Twigs used for brushing of teeth and in gum problem.

Miscellaneous:

- The dry pods are tied around feet as 'Paijeb' of metel by the children.
- The seeds and gum are sold in the market by the tribals.
- Tender branches are used as tooth-brush by the tribals.

Tannin: Bhil tribals use the stem bark for tanning the leather which gives black to brown colour.

Timber:

- The wood is used for making carts, agriculture implements, huts and household articles.
- The wood is also used in the formation of a musical instrument called 'Khanjri' or "Khurtal" or "Khajria".

Uses in literature

Gum - Burns (Janardhana, 1963); Dysentery (Singh & Pandey, 1980; Basak, 1997); Leucorrhoea (Kumar & Chauhan, 2005); Diarrhoea & diabetes (Basak, 1997); As masticatory (Bhandari, 1974).

Stem bark - Asthma & Bronchitis (Singh & Pandey, 1980; Sharma *et al.,* 2006); Diarrhoea, dysentery & diabetes (Bhalla *et al.,* 1982; Basak, 1997); Skin disease (Rajwar, 1983; Kumar *et al.,* 2006; Sharma *et al.,* 2006); bleeding piles (Kumar *et al.,* 2006; Sharma *et al.,* 2006); Toothache (Deokule, 2006); Edible (Bhandari, 1974).

Fruit - Dysentery (Singh & Pandey, 1980); Impotency & urino - genital disorders (Tripathi *et al.,* 1996).

Leaves - Eye complaint & throat infection (Singh & Pandey, 1980); Headache (Maheshwari & Singh, 1987); Stomatitis (Kumar & Chauhan, 2005); Diarrhoea (Mohanty *et al.,* 1996); Gout (Jadeja *et al.,* 2006).

Flowers - Diarrhoea & dystentery (Singh & Pandey, 1980); Ear complaint (Saxena & Vyas, 1983).

Roots - Liver complaint (Venugopal & Raghunathan, 1982).

Seeds- Spermatorrhoea (Prakash & Singh, 2006.).

Stem – Toothbrush (Shah, 1984; Shah *et al.,* 1981).

Acacia pennata (L.) Willd. (Mimosaceae)

Syn.: *Mimosa pennata* L.

Vernacular name - Khervel

Regional and other names- *Bengali* –Kuchui; *Hindi* -Aila; *Malayalam* -Kareencha; *Marathi* –Shembi *Telugu* –Karusikaya.

Distribution - India; Ceylon, Malaya, Tropical Africa.

Taxonomic description - Large, scandent bush, with polished bark. Prickles slightly recurved. Pinnae 8-15 pairs, with glands between the upper ones; leaflets 40-50 pairs. Flowers creamy-white or pale-yellow. Pods stalked, linear-oblong, glabrous.

Flowering & fruiting - July-March.

Habitat ecology - Ocassionally met in the forest areas on the slopes.

Regeneration - By seeds.

Field no. - DJ-113.

Parts used - Stem-bark

Chemical constituents

Bark contains Lupeol and a-spinasterol.

Stem contains Beta- Sitosterol.

Biological activity - Acrid, bitter, sour, hot.

Folk uses

Biofencing: The plant is used as bio-fences to protect the crops.

Poison: The stem bark is crushed and dropped in water as a fish poison by the Bhils.

Tannin: The tribals use the stem bark for tanning the leather.

Uses in literature -

Leaf- Headache, bodypain (Jain, 1965).

Bark- Snake bite (Das *et al.,* 1983); As fish poison (Das *et al.,* 1983); Tannin (Jain, 1963).

Achyranthes aspera L. (Amaranthaceae)

Vernacular name - Andhijharo, Ojijharo, Lutjiru, Lutjiro, Chirchito.

Regional and other names - *Bengali* - Apang; *English* -The prickly chaff- flower; *Gujarati* - Agheda; *Hindi* –Latjira; *Kannada* - Uttarane; *Malayalam* - Katalati; *Marathi* - Aghara; *Sanskrit* - Apamarg; *Tamil* - Nayurivi; *Telugu* – Apamargamu.

Distribution - Throughout India.

Taxonomic description - Erect hairy herbs, with oval–round, pubescent- villous leaves. Flowers rather plump in appearance, deflexed, spinescent, in lax spikes.

Flowering & fruiting - August- February.

Habitat ecology - Abundant in wastelands.

Regeneration - By seeds.

Field no. - DJ-02.

Parts used - Leaves, Roots, Seeds.

Chemical constituents

Seed contains Saponin A & B, Ecdy steron, Oligosaccharide.

Root contains Oleonilic acid, Maslinic acid.

Unripe fruit contains Saponin C & D.

Plant contains Nitrogenous compound- Achranthine & Alkaloids.

Biological activity - The plant is acrid, anodyne, anthelmintic, anti-inflammatory, bitter, carminative, demulcent, depurative, digestive, diuretic, stomachic, sudorific.

Folk uses

Magico-religious beliefs:

- Root is tied in hair of pregnant lady at the time of labour pain for normal delivery.
- Root is tied with the help of sacred thread 'Luccha' on left hand or left feet to cure fever.

Medicine:

- 10 leaves of the plant are crushed with 5 seeds of kali mirch (*Piper nigrum* L.) and a paste is prepared. Some small pills are prepared from this paste. One pill is taken with water for 4 days to cure piles.

- Extract of root is applied in the cases of scorpion sting.
- Extract of root is taken with water to cure diarrhoea.
- The twig is used as toothbrush in the cases of bad breath.
- Powder of roasted seeds is taken with honey to cure asthma.
- The paste of leaves and seeds is mixed with Haldi (*Curcuma longa*) and heated a bit with edible oil. This paste is tied in the case of corn.
- Extract of root is taken with water (Mishri may be added) in the case of burning in urethra during micturition.

Uses in literature

Roots - Whooping cough (Jain, 1984; Sharma *et al.,* 2006); Malaria (Singh, 1997); Abortifacient (Jain, 1965); Antifertility (Tarafder, 1983); Snake-bite (Nagendra Prasad & Abraham, 1984; Sharma *et al.,* 2006); Child birth (Tarafder, 1984); Pneumonia (Bennet, 1985); Dysentery (Saxena & Vyas. 1983); Wounds (Jain & De, 1966); Cough (Singh & Pandey, 1986; Sharma *et al.,* 2006); Toothache (Jain, 1965; Ganesan *et al.,* 2006; Sharma *et al.,* 2006); Asthma, bronchitis, cold, scorpion bite (Goel *et al.* 1984); Stomatitis (Malhotra & Moorthy, 1973); Spermato-rrhoea (Prakash & Singh, 2006.); Abdominal tumour (Sharma *et al.,* 2006).

Seed - Cough & asthma (Bhalla *et al.,* 1982); Headache (Mishra & Billore, 1983); Skin diseases (Rajwar, 1983); Pyorrhoea, toothache (Narain & Singh, 2006); As vegetable (King, 1969).

Stem - Snake-bite (Singh & Maheshwari, 1992); Toothache (Aditya *et al.,* 2006); Jaundice (Das *et al.,* 2006).

Inflorescence - Cold & cough (Lal & Yadav, 1983); Dog bite (Upadhyay & Jaswant Singh, 2005).

Leaves - Colic (Shah *et al.,* 1983); Debility (Maheshwari *et al.,* 1980); Piles (Baruah & Sharma, 1984; Baruah *et al.,* 2006); Ear complaint (Maheshwari *et al.,* 1980); Scorpion bite (Audichya *et al.,* 1980); Skin diseases (Rajwar, 1983); Leucoderma (Gupta, 1981; Sharma *et al.,* 2006); Sore & stomachic (Srivastava & Verma, 1981); Eye diseases (Kiruba *et al.,* 2006); Sexual debility, spermatorrhoea (Kumar & Chauhan, 2005); Asthma (Baruah *et al.,* 2006); Cold (Sharma *et al.,* 2006); Cough (Shah *et al.,* 1983; Sharma *et al.,* 2006); Jaundice (Das *et al.,* 2006); Menstruation trouble (Das *et al.,* 2006); As vegetable (Datar & Vartak, 1975).

Whole plant - Boils (Sudhakar & Rolla, 1985); Child birth, rheumatism, scabies, syphilis (Jain & Tarafder, 1970); Cough (Bedi, 1970); Liver complaint (Lal & Yadav, 1983); Skin diseases (Kiruba *et al.*, 2006); Dog bite (Jain *et al.*, 1973; Kadavul & Jisha, 2006); Piles (Sudhakar & Rolla, 1985; Chelvan, 1998; Tripathi *et al.*, 1996; Basak, 1997); Colic (Das, 1995); Renal disorders (Sudhakar & Chetty, 1998); Dropsy (Karnich *et al.*, 1981; Chelvan, 1998; Tripathi *et al.*, 1996; Basak, 1997; Das, 1995); Snake bite (Deokule, 2006); Diarrhoea, dysentery (Kataria & Gauchan, 2006).

Adansonia digitata L. (Bombacaceae)

Vernacular name - Gorakh-amli, Dakhi-amlo, Raja-rani.

Regional and other names - *Bengali* – Gorakamli; *English* - Monkey–bread tree; *Gujarati* - Gorak ambli; *Hindi* – Gorakhimli; *Kannada* – Magimavu; *Marathi* - Gorakh chinch; *Sanskrit* –Chitrala, Goraksi; *Tamil* – Anaipuliamaram; *Telugu* – Brahmaamlika.

Distribution - Grown in many places in India. Indigenous in tropical Africa.

Taxonomic description - Deciduous trees, with digitately 3 to 9-foliolate leaves and white, pendulous flowers opening during night. Fruits ellipsoid, woody.

Flowering & fruiting - March – May.

Habitat ecology - Rare, reported from Bibdod village.

Field no. - DJ-114.

Parts used - Fruits.

Chemical constituents

Pulp & fruit contains Free tarteric acid, Potassium Bitartrate.

Leaves contain Mucilage.

Stem bark contains Bitter crystalloid – Adansenin.

Biological activity

Leaves - Astringent, febrifuge, sudorific, tonic.

Stem bark - Astringent, cooling, stomachic.

Fruit pulp - Aperient, appetiser, astringent, cooling, demulcent.

Folk uses

Food: The mature fruits eaten by the tribals.

Uses in literature

Skin irritation & rheumatism (Shah & Gopal, 1992).

Aegle marmelos (L.) Corr. (Rutaceae)

Syn.: *Crateva marmelos* L.

Vernacular name - Billi, Billo, Bel.

Regional and other names : *Bengali* – Bela; *English* - Wood apple; *Gujarati* – Billy, Bil; *Hindi* – Bel; *Kannada* -- Bilpatre; *Malayalam* – Vilvam; *Marathi* – Baela; *Sanskrit* –Adhararuha, Bilva, Sri; *Tamil* – Bilwa; *Telugu* – Maredu.

Distribution - Wild in the sub-Himalayan tract, Central and S. India. Often planted all over India.

Taxonomic description - Deciduous trees, armed with spines. Leaves 3- foliolate and flowers creamy – white, in axillary panicles. Fruits globose, with stony rind and mucilaginous pulp.

Flowering & fruiting - March – April.

Habitat ecology - Found wild on dry slopes & also cultivated in garden.

Regeneration - By coppice, moderately by seeds.

Field no. - DJ-78.

Chemical constituents

Fruits, Leaves, Root bark.

Fruits contains Gum, Pectins, reducing Sugars, Sugars, Oil and Tannin.

Leaves & root contains Reducing sugars, Tannins, Iron, Calcium, Magnesium and Potassium.

Biological activity

Ripe fruit - Acrid, antipyretic, antiscorbutic, appetiser, aromatic, bitter, cooling, febrifuge, nutritive, stimulant, sweet, tonic.

Root - Sweet.

Leaves - Astringent, bitter, digestive, febrifuge, laxative.

Folk uses

Food: Mature fruits are eaten by the tribals.

Magico-religious beliefs: The leaves and unripe fruits are offered to Lord Shiva during worship.

Medicine:

- The juice is extracted from boiled ripe fruit and it is taken to cure cold.
- 15 leaflets are ground with 5 seeds of *Piper nigrum* and given with water in empty stomach to check undesirable discharge of semen with urine.
- Half of a ripe fruit is eaten thrice a day to cure dysentery.
- Dried leaves are boiled in edible oil and filtred. 2 drops of this filtered oil are dropped in the ear in the case of deafness.

Miscellaneous: Mature fruits are sold in local markets.

Poison: The powder of root bark is used as fish-poison for hunting them.

Uses in literature

Ripe fruit - Constipation (Singh & Pandey, 1980); Diarrhoea & dysentery (Shah, 1984; Jadeja *et al.,* 2006; Kataria & Gauchan, 2006); Stomach diseases (Jain & De, 1996); Fever (Singh, 1997; Baruah *et al.,* 2006); Dysentery (Purkayastha & Subhan, 2006; Pereira, 2006; Deokule, 2006; Baruah *et al.,* 2006); Asthma (Baruah *et al.,* 2006); diabetes (Mokat & Deokule, 2006; Ghosh, 2006); Eye diseases (Rajendran & Rengamani, 2006); Piles (Das *et al.,* 2006); Edible (Maheshwari & Singh, 1984; Maji & Sikdar, 1982; Jain & De, 1966; Goel *et al.,* 1984; Negi *et al.,* 1985; Bhargava, 1959; Nagar, 1985; Maheshwari *et al.,* 1981, 1986).

Root - Dogs bite (Saxena & Dutta, 1975); Hysteria (Kumar *et al.,* 2006).

Root bark- Fish poison (Joshi, 1986).

Leaves - Backache & vomiting (Bhattacharjee *et al.,* 1980); Jaundice (Goel *et al.,* 1984; Dutta Choudhury & Shil, 2006); Wounds (Singh & Maheshwari, 1983); Eye complaint (Singh & Pandey, 1980); Diabetes (Shah,1984; Singh & Prakash, 1994; Purkayastha & Subhan, 2006; Ghosh, 2006); Anti- inflammation (Katewa & Arora, 1997); Sun stroke (Singh & Prakash, 1994); Worm infestation (Pereira, 2006); Dysentery, fever, asthma (Baruah *et al.,* 2006); Abdominal pain (Kadavul & Jisha, 2006); Piles (Das *et al.,* 2006).

Bark - Snake bite (Singh & Prakash, 1994); Diarrhoea, dysentery & Stomach disorders (Saxena & Dutta, 1975); Intermittent fever (Kumar *et al.,* 2006).

Aerva lanata (L.) Juss. ex Schult. (Amaranthaceae)

Syn.: *Achyranthus lanata* L.

Vernacular name - Panruvo.

Regional and other names : *Bengali* – Chaya; *Gujarati* –Bur; *Hindi* – Gorakhganja; *Kannada* – Bilihindisoppu; *Malayalam* – Cerula; *Marathi* – Kapurmadhura; *Sanskrit* – Bhadra; *Tamil* – Cerupulai; *Telugu* – Pindicettu.

Distribution - Throughout India.

Taxonomic description - Perennial herbs, with suborbicular to lanceolate leaves. Flowers bisexual, in axillary spikes passing into leafy inflorescence. Fruits rotund, compressed.

Flowering & fruiting - Throughout the year.

Habitat ecology - Common in wastelands.

Regeneration - By seeds.

Field no. - DJ-197.

Parts used - Whole plant.

Chemical constituents

α-Amyrin, Campesterol, β-sitosterol, Palmitic acid, Betuline, Syringic acid, Vanllic acid, Feruloyltyramine, Aervine, Methylervine, Aervoside, Aervolanine.

Biological activity

Plant- Astringent, bitter, cooling, diuretic, emollient, suppurative, vermifuge.

Folk uses

Medicine: Decoction of whole plant is given by the tribals to their children in the case of dysentery.

Uses in literature

Whole plant- Anthelmintic (Tripathi *et al.,* 1996; Basak, 1997; Sudhakar & Rolla, 1985); Cuts, wounds (Goel *et al.,* 1984); Headache, skin diseases, swelling (Singh & Pandey, 1980); Malaria (Srivastava & Varma, 1981); Burn healing (Goel *et al.,* 1984; Upadhay *et al.,* 1998); Diuretic (Tripathi *et al.,* 1996; Basak, 1997); Dissolve kidney and gall bladder stones (Sudhakar & Chetty, 1998).

Leaf- Malaria fever (Ahmad, 1995; Singh & Singh, 1992); Ear complaint (Saxena *et al.,* 1931); As vegetable (Datar & Vartak, 1975).

Root- Boils (Ramachandran & Nair, 1981); Diarrhoea (Mohanty *et al.*, 1996); Kidney stone (Deokule, 2006).

Plant juice- Kidney stone (Kiruba *et al.*, 2006).

Ailanthus excelsa Roxb. (Simaroubaceae)

Vernacular name - Aydu, Aydu-Ba.

Regional and other names : *Bengali* – Maharukha; *English* -The tree of heaven; *Gujarati* – Motoaduso; *Hindi* – Maharuk; *Kannada* – Doddamara; *Malayalam* – Mattipongilyam; *Marathi* –Maharuk; *Sanskrit* – Aralu, Mahanimba; *Tamil* – Perumaran; *Telugu* – Peddamanu.

Distribution - Indigenous in the Indian peninsula, and often planted in various parts of India.

Taxonomic description - Large trees, with pinnate leaves crowed at the end of branches. Flowers yellow, in terminal and axillary panicles. Fruit a samara, spindle-shaped, 1-seeded.

Flowering & fruiting - December- April.

Habitat ecology - Usually planted along the roads, often found in wastelands.

Regeneration - By Seeds.

Field no. - DJ-04.

Parts used - Leaves, Stem bark.

Chemical constituents

Bark contains Glaucarubin, Glaucarubinone, Excelsin, Glaucarubol, Ailanthion, Ailantic acid, Malantahin, Tricontane, Hexatricontane.

Leaves contain Ailantic acid, Beta – sitosterol (vitexin).

Biological activity

Stem bark - Anthelmintic, antidysenteric, aromatic, astringent, bitter, diarrhoea, febrifuge.

Leaves - Bitter, carminative, febrifuge, tonic.

Folk uses

Magico-religious beliefs: The leaves are used by the Bhopas in Jhada rituals to wardoff the effect of evil eyes.

Medicine:

- Extract of stem bark is taken with water to cure typhoid and jaundice.
- The tribals give bath to the patients suffering from typhoid with the crushed stem bark.
- The decoction of stem bark is taken orally by the tribal people to cure abdominal swellings.
- The decoction of stem bark is given to the cattle suffering from abdominal swellings.

Miscellaneous: Leaves are used as Mosquito repellent.

Timber: Bhils make musical instrument called "Drum" (Dhol) from its wood.

Uses in literature

Bark - Cough & Cold (Joshi, 1982; Sharma *et al.,* 2006); Abortifacient (Tarafder & Chaudhuri, 1981); Bronchitis (Shah & Gopal, 1986); Asthma (Tripathi *et al.,* 1996); Rectal problem (Mohanty *et al.,* 1996); Diarrhoea (Siddiqui & Hussain, 1991; Deokule, 2006); Dysentery (Pandit *et al.,* 1996; Deokule, 2006; Jadeja *et al.,* 2006); Fever (Sharma *et al.,* 2006); veterinary medicine (Sebastian, 1984).

Leaves - Malaria (Aminduddin *et al.,* 1993); Rheumatism (Shah *et al.,* 1981); Diarrhoea, dysentery (Deokule, 2006); Veterinary medicine (Sebastian, 1984).

Albizia procera (Roxb.) Benth. (Mimosaceae)

Syn.: *Mimosa procera* Roxb.

Vernacular name : Guradyo, Gurad.

Regional Names - *Bengali* –Koroi; *English* –White sirish; *Hindi* – Safed siris; *Kannada* –Bellate; *Malayalam* –Karun thagara; *Marathi* – Kinhai; *Tamil* -Kondavagei; *Telugu* –Tella chinduga.

Distribution - India; Ceylon, Burma, Malaya, Tropical Africa.

Taxonomic description - Trees, with grey-white bark, 2-6 pairs of pinnae and 6-15 pairs of ovate-oblong leaflets. Flowers pale-white in panicles. Pods brown.

Flowering & fruiting - April-October.

Habitat ecology - Rare, in the deciduous forests.

Regeneration - By coppice moderately or by seeds.

Field no. - DJ-89.

Parts used - Root, Stem bark.

Biological activity

Leaves- Insecticidal.

Folk uses

Detergent: The root is used as a detergent by the tribals and substitute for soap for washing of cloths.

Medicine:

- Extract of stem bark is taken with water to cure kidney stone problem.
- Paste of stem bark is applied externally in the case of snake bite.
- 2 drops of extract of stem bark are dropped in ear to cure toothache.
- Extract of stem bark is taken with water to cure stomachache during menstrual period.

Poison: The root is crushed and dropped in water as a fish poison by the Bhils.

Uses in literature

Stem bark- Backache (Joshi *et al.*, 1980); Fish poison (Joshi, 1986).

Seed- Body pain (Sharma & Malhotra, 1984).

Allium cepa L. (Liliaceae)

Vernacular name - Piyaj, Kando, Kanda.

Regional and other names : *Bengali* – Pyanj; *English* – Onion; *Gujarati* – Dungari; *Hindi* – Piyaz; *Kannada* – Nirulli; *Malayalam* – Ulli; *Marathi* – Kanda; *Sanskrit* –Dirghapatra, Palandu; *Tamil* – Vengayam; *Telugu* – Nirulli.

Distribution - Cultivated throughout India.

Taxonomic description - Scapigerous, annual herbs, with bulbous stem, fistular, linear leaves and white flowers in umbels.

Flowering & fruiting - November-March.

Habitat ecology - Cultivated for its bulbs and leaves which are cooked as vegetable.

Regeneration - By seeds or by bulbs.

Field no. - DJ-127.

Parts used - Bulb, Leaves.

Chemical constituents - Bulb contains Glutamyl, Arginin, Cystein, Quercetin, Kaempferol, Methylcystein, Sulphoxide, Propylecystein, Saponins.

Biological activity - Aphrodisiac, diuretic, expectorant, stimulant.

Folk uses

Food: The bulbs are used as condiment. The leaves are used as vegetable.

Magico-religious beliefs: The tribals keep the bulb in the pocket during summer; it is believed to protect them from the effect of hot winds- "Loo".

Medicine:

- The paste of bulb is fried in pure ghee with jaggery. This paste is tied in the case of corn.
- The bulbs are largely eaten as such by the tribals to protect themselves from the effect of hot winds- "Loo" during summers.
- Bulb paste prepared in 250 ml mustard oil is given to animals to cure hoof disease.
- 2 drop of onion bulb are dropped in ear to check pus formation.

Uses in literature

Bulb - Anthelmintic (Gaur *et al.,* 1980); Asthma, eye complaint, bronchitis (Nagendra Prasad & Abraham, 1984); Boils (Malhotra & Moorthy, 1973); Ear complaint (Mahsehwari & Singh, 1984); Itching (Purohit *et al.,* 1985); Piles (Singh, 1988); Ringworm (Purohit *et al.,* 1985); Skin diseases (Shah, 1984); Stomach diseases (Bhalla *et al.,* 1982); Vomit (Rao & Jamir, 1982); Wound (Bhall *et al.,* 1982); Chicken pox, fever (Ganeshan *et al.,* 2006); Epilepsy (Deokule, 2006); Constipation (Jadeja *et al.,* 2006); Diabetes (Ghosh, 2006); As condiment (Abraham, 1981).

Leaves- As vegetable (Koelz, 1979).

Allium sativum L. (Liliaceae)

Vernacular name - Lasun, Lasan.

Regional and other names : *Bengali* – Lasan; *English* – Garlic; *Gujarati* – Lasan; *Hindi* – Lasun; *Kannada* – Bellulli; *Malayalam* – Vellulli; *Marathi* – Lasoon; *Sanskrit* – Rasona; *Tamil* – Ullipondu; *Telugu* – Vellulli..

Distribution - Cultivated throughout India.

Taxonomic description - Bulbous herbs, each bulb with several fleshy bulblets. Leaves linear. Flowers white, in umbels.

Flowering & fruiting - November-March.

Habitat ecology - Cultivated during winter for bulbs which are variously eaten or used as condiment.

Regeneration - By cloves (Segments of bulbs).

Field no. - DJ-67.

Parts used - Cloves, Clove oil.

Chemical constituents - Bulb contains A volatile oil which contains Allyl disulphide, Allyl-propyl disulphide and Poly sulphide; Allicin (antibacterial); Allicetion-I and Allicetion-II (Antibiotics).

Biological activity - Antibacterial, antiseptic, aphrodisiac, bone healing, digestive, diEphoretic, diuretic, expectorant, pungent, rejuvinator, stimulant, stomachic, tonic.

Folk uses

Food: The cloves are used as condiment.

Magico-religious beliefs: 7 to 21 cloves are tied in sacred thread "luccha" and a necklace is prepared. This necklace is worn around the neck for a week to cure typhoid and jaundice; it is believed that as the size of necklace increases automatically, the patients get cured.

Medicine:

- Clove oil is applied on ringworm.
- 2 drops of warmed clove oil are dropped in ear to cure toothache.
- 50 gm. cloves and 50 gm seeds of Rai (*Brassica nigra* L.) are ground together, this mixture is diluted with water and given to cattle in Afra disease.
- 2 drops of the juice of cloves are dropped in the nostrils to cure toothcahe.

Uses in literature

Bulb - Bronchitis, dropsy, wound, ringworm, tomour (Bhalla *et al.*, 1982); Fever (Nagendra Prasad & Abraham, 1984); Ear complaint, respiratory diseases (Rai, 1985); Eczema (Purohit *et al.*, 1985); Whooping cough (Lal & Yadav, 1983); Diarrhoea (Jadeja *et al.*, 2006); Epilepsy (Deokule, 2006); Bodyache (Mokat & Deokule, 2006); Diabetes (Ghosh, 2006); Cold (Gupta, 2006); Cough (Nagendra Prasad & Abraham, 1984; Gupta, 2006); Gout (Jadeja *et al.*, 2006); As condiment (Abraham, 1981).

Leaf - Piles (Bhalla *et al.*, 1982); Rheumatism (Saxena & Vyas, 1983).

Whole plant – Diabetes, hypertension, gastritis, amoebiosis, cancer, rheumatoid arthritis (Kumar *et al.*, 2006).

Aloe vera (L.) Burm. f. (Liliaceae)

Syn.: *A. Perfoliata* L. var. *vera* L.

Vernacular name - Guwarpatho.

Regional and other names : *Bengali* - Ghrita kumari; *English* - Indian aloe; *Gujarati* – Kadvikunvar; *Hindi* – Ghee - kanwar, gwarpatha; *Kannada* – Komarka; *Malayalam* –Kattavazha; *Marathi* – Korphad; *Sanskrit* – Kumari; *Tamil* - Bhottu – katrajhae; *Telugu* - China kata – banada.

Distribution - Planted in Indian gardens.

Taxonomic description - Perennial herbs, with fleshy musilagenous leaves in rosette having marginal spines. Flowers yellow, born in racemes on long scapes.

Flowering & fruiting - December-February.

Habitat ecology - Common in wastelands.

Regeneration - By suckers.

Field no. - DJ-56.

Parts used - Flowers, Leaf pulp and Leaf juice.

Chemical constituents

Leaves contain Carbohydrates, Tannins, Proteins, Chromones (Abesin & Aloesone), Malic acid, Citric acid, Tartaric acid.

Jelly from leaves contain Acetylated glucomannans, Iron, Calcium, Potassium & Magnesium.

Biological activity

Leaf pulp - Analgesic, anthelmintic, anti – inflammatory, carminative, digestive, purgative, stimulant, stomachic, tonic.

Folk uses

Food: They use the flowers of the plant for making a vegetable.

Medicine:

- The pulp of leaves is mixed with wheat flour and some small balls are prepared. These balls are roasted in dried cow dung. One ball is taken early in the morning with pure ghee in winter season for a period of 7 days for strength and vigour.
- The leaf pulp is heated with edible oil & turmeric powder. This paste is tied on anus to cure piles.
- Juice of leaves is applied externally to cure boils, minor burns & wounds, scratches.
- Juice of leaves is mixed with honey and taken in the cases of cough.

Uses in literature

Leaf pulp - Fever, liver complaint, constipation (Bhalla *et al.*, 1982); Skin diseases (Bhalla *et al.*, 1982; Singh *et al.*, 1998); Boils, gonorrhoea (Jain, 1985; Singh *et al.*, 1998); Rheumatism (Shah *et al.*, 1981); Wounds (Lakshamanan & Sankaranarayanan, 1998; Chatterjee & Bhatterjee, 2006); Piles (Bhalla *et al.*, 1982; Deokule, 2006); Dyspepsia (Jadeja *et al.*, 2006); Pimples (Mokat & Deokule, 2006); Boils, burns, swellings (Yesodharan & Sujana, 2006); Urinary complaints (Arinathan *et al.*, 2006).

Leaf- Veterinary medicine (Jain, 1984); Edible (Kanodia & Gupta, 1968).

Amaranthus spinosus L. (Amaranthaceae)

Vernacular name - Kantawari chauli.

Regional and other names - *Bengali* – Kantanotya; *English* - Prickly amaranth; *Gujarati* – Kantalodambho; *Hindi* - Kantewali chauli; *Kannada* – Mulladantu; *Malayalam* – Mullacherucheera; *Marathi* – Kantemath; *Sanskrit* –Alpamarisha, Tanduliya; *Tamil* – Mullukeeraj; *Telugu* –Nalladoggali.

Distribution - Throughout India; Ceylon, many tropical countries.

Taxonomic description - Monoecious herb, with ovate-rhomboid or elliptic leaves having axillary spines. Flowers in axillary clusters, passing to spikes. Fruits ovoid.

Flowering & fruiting - During & after rainy season.

Habitat ecology - Weed in wastelands and fields.

Regeneration - By seeds.

Field no. - DJ-201.

Parts used - Leaves.

Chemical constituents - Plant contains Carbohydrates, Proteins, Minerals, Iron, Vitamin – A and Calcium (Leaves).

Biological activity - Alexiteric, antipyretic, appetiser, cooling, digestive, diuretic, laxative, stomachic.

Folk uses

Food: Leaves are cooked as vegetable.

Uses in literature

Leaves - Jaundice (Tripathi *et al.,* 1996); Boils (Deokule, 2006); As vegetable (Maheshwari & Singh, 1984; Das *et al.,* 1984; Jain & De, 1966; Vartak, 1981; Abraham, 1981; Maheahwari *et al.,* 1986; Chaudhari *et al.,* 1985).

Seeds - Cough & cold (Joshi *et al.,* 1980).

Roots - Colic, eczema, gonorrhoea (Shah *et al.,* 1983); Snake bite (Goel *et al.,* 1984); Diarrhoea (Kataria & Gauchan, 2006); Dysentery (Kataria & Gauchan, 2006; Das *et al.,* 2006); Skin diseases (Das *et al.,* 2006).

Whole plant - Blood purifier, piles (Sharma *et al.,* 1979); Anthelmintic (Jain & Tarafder, 1970).

Amischophacelus axillaris (L.) Rolla Rao & Kammathy (Commelinaceae)

Syn.: *Commelina axillaries* L.

Vernacular name : Bokhli.

Regional and other names- *Hindi* -Baghanulla; *Tamil* -Nirpulli; *Telugu* –Golagandi.

Distribution - Throughout India.

Taxonomic description – Diffuse, fleshy herbs, rooting at lower nodes; leaves linear, usually glabrous, sheath short; flowers in sessile cymes enclosed in the leaf-sheaths; bracteoles linear- lanceolate, minutely ciliate; filaments purple-bearded, inflated at tips; capsules glabrous; seeds cylindric, pitted.

Flowering & fruiting - July-December.

Habitat ecology - Occasional in moist places.

Field no. - DJ-141.

Parts used - Leaf.

Folk uses

Medicine: A drop of leaf extract is dropped in eye twice a day to cure conjunctivitis.

Ammannia baccifera L. (Lythraceae)

Syn.: *A. salicifolia* Hiern.

Vernacular name - Akasia.

Regional and other names : *Bengali* – Dadmari; *English* – Blistering ammania; *Gujarati* – Jalaagiyo; *Hindi* – Dadamari; *Malayalam* – Nirummelneruppu; *Marathi* – Bhar jambhula; *Sanskrit* – Kurandika; *Tamil* – Kalluruvi; *Telugu* – Agnivendrapaku.

Distribution - Throughout India.

Taxonomic description - Erect or ascending herbs, with oblong or elliptic leaves and sessile flowers lacking petals and style. Capsules globose, reddish.

Flowering & fruiting - Throughout the year.

Habitat ecology - Common in water-logged habits.

Regeneration - By seeds.

Field no. - DJ-57.

Parts used - Whole plant.

Chemical constituents - Hentriaconite, Dotriacontanol, Betulinic acid, Lupeol, Ellagic acid, Quercetin, Lawsone.

Biological activity - *Plant* - Acrid, aphrodisiac, appetizer, bitter, cooling, diuretic, laxative, rubifacient, stomachic.

Folk uses

Medicine: The tribals use this plant species as fodder of cow in order to bring them to an early heat for breeding.

Uses in literature

Whole plant - Fever (Singh & Pandey, 1980; Joshi, 1982); Rheumatism (Singh & Pandey, 1980).

Leaf- Rheumatism (Shah *et al.,* 1981); Ringworm (Saxena & Vyas, 1983); Skin diseases (Goel *et al.,* 1984).

Ampelocissus latifolia (Roxb.) Planch. (Vitaceae)

Syn.: *Vitis latifolia* Roxb.

Vernacular name - Panivelo.

Regional and other names- *Bengali* –Govila; *Hindi* -Panibel; *Tamil* -Kattukkodimundivigui; *Telugu* –Bedasativa.

Distribution - Throughout India.

Taxonomic description - Large climbers, with hollow, striate branches. Leaves orbicular-cordate, lobed, serrate. Flowers reddish-brown, in pyramidal thyrsoid cymes. Berries globose, black, 2-seeded.

Flowering & fruiting - July-September.

Habitat ecology - Common among hedges and as forest undergrowth.

Regeneration - By seeds.

Field no. - DJ-199.

Parts used - Root, Stem.

Folk uses

Magico-religious beliefs: 7 to 21 pices of stem are tied in sacred thread "luccha" and a necklace is prepared. This necklace is worn around the neck for a week to cure typhoid; it is believed that as the size of necklace increases automatically, the patients get cured.

Medicine: Extract of root is given with water to cure jaundice.

Uses in literature

Root- Boils (Tarafder, 1984); Bone fracture (Joshi *et al.,* 1980; Saxena, 1986; Shah & Gopal, 1986); Dysentery (Malhotra & Moorthy,

1973); Inflammation (Trivedi *et al.,* 1985); Menstrual complaint (Shah *et al.,* 1981); Stomach pain (Joshi *et al.,* 1980); Skin diseases (Saxena, 1986); Snake bite (Mishra & Billore, 1983; Murthy *et al.,* 1986); Wounds (Shah *et al.,* 1983; Saxena & Dutta, 1975).

Fruit- Edible (Shah, 1984; Maji & Sikdar, 1982; Vartak, 1981; Shah *et al.,* 1981).

Anethum graveolens L. (Apiaceae)

Syn.: *Peucedanum graveolens* (L.) Hiern.

Vernacular name - Sowa

Regional and other names- *Bengali* – Sowa; *English* – Dill; *Hindi* – Soya; *Kannada* – Sabbasige; *Malayalam* – Satakuppa; *Marathi* – Balantashopa; *Sanskrit* – Satapuspa; *Tamil* – Satakuppi; *Telugu* – Satakuppivittulu.

Distribution - Cultivated throughout India.

Taxonomic description - A glabrous perennial herb. Leaves decompound. Flowers yellow, in umbels. Fruits dorsally compressed, conspicuously ridged & narrowly winged.

Flowering & fruiting - January-March.

Habitat ecology - Cultivated, also found as a weed of cultivation.

Regeneration - By Seeds.

Field no. - DJ-198.

Parts used - Seeds, Whole plant.

Chemical constituents

Fruits contain Vicenin, Phellandrene, Dillapiol, Carvone, Dihydro carvone, D-limonene, A new xanthone glycoside-dillanoside.

Seed oil contain Essential oil, Tripetroselinin, Petroselinicdi-olein, Dipetroselinicolein.

Biological activity - *Fruits* - Acrid, alexeteric, anodyne, anthelmintic, antidysenteric, anti-inflammatory, antispasmodic, bitter, carminative, digestive, diuretic, expectorant, febrifuge, stomachic, sudorific, thermogenic.

Folk uses

Food: Whole plant is used as vegetable.

Medicine:

- Seeds are taken in the form of tea to cure ammenorrhoea and dismennorrhoea.

- One teaspoonful powder of seeds with a pinch of common salt taken with water to check nausea.

Uses in literature

Fruits - Stomach complaints in children (Shah *et al.*, 1981); Flatulence (Vishwanathan, 1995); Dyspepsia (Jadeja *et al.*, 2006); Gout (Jadeja *et al.*, 2006); As vegetable (Shah *et al.*, 1981).

Leaves- As vegetable (Singh & Singh, 1981).

Anisomeles indica (L.) O. Kuntze (Lamiaceae)

Syn.: *Nepeta indica* L.

Vernacular name - Gobrio.

Distribution - Throughout India.

Taxonomic description - Suffruticose herbs, with ovate, acute or obtuse, crenate-serrate, softly pubescent leaves. Flowers bluish-purple. Nutlets broadly ovoid, with rounded ends, polished.

Flowering & fruiting - August-December.

Habitat ecology - Found in wastelands & also in open forests.

Regeneration - By seeds.

Field no. - DJ-116.

Parts used - Seeds, Leaves.

Chemical constituents - Plant contains essential oil.

Biological activity - Astringent, carminative, tonic.

Folk uses

Medicine:

- Powder of dried leaves (in shade) is given (7-10 gm.) in cow milk twice a day before sunrise and after sunset for a period of 2-3 weeks in the cases of Rickets in children.
- Powder of seeds is taken with water to cure loose motion and vomiting.
- Extract of root is given in intermittent fever.

Uses in literature

Whole plant- Antipyretic (Bhatt & Sabnis, 1987); Fever, nervous complaint (Jain & De, 1966); Muscular pain (Pal, 1984); Veterinary medicine (Sharma *et al.,* 1979; Ramachandran & Nair, 1981; Bhatt & Sabnis, 1987).

Annona squamosa L. (Annonaceae)

Vernacular name - Sitaphal.

Regional and other names- *Bengali* – Seetaphal; *English* - Custard apple; *Gujarati* – Anan; *Hindi* –Seetaphal; *Kannada* – Seethaphala; *Malayalam* - Seetha pazham; *Marathi* – Seetaaphala; *Sanskrit* – Seetaphala; *Tamil* - Sethapazham, atta; *Telugu* – Seethapalamu.

Distribution - A native of the W. Indies. Now cultivated throughout India.

Taxonomic description - Evergreen shrubs or small trees, with oblong-lanceolate leaves and yellowish green flowers. Fruits globose or ovoid with projecting areoles.

Flowering & fruiting - July – November.

Habitat ecology - It grows wild in open forest areas, also cultivated for edible fruits.

Regeneration - By seeds.

Field no. - DJ-88

Parts used - Fruits, Seeds, Leaves.

Chemical constituents - Leaves contain Anonaine, Roemerine, Corydine, Narcorydine, Norlaureline, Isocorydine, Noricocorydine, Glucine, Friedeline, Alkaloids, Essential oils, Volatile oil, Carvone, Michelabine, Liriodedine, Diazepine, Squamolone, Beta–Caryophylline, β-sitosterol, Camphor, Barmeol.

Biological activity

Root - Purgative.

Bark - Astringent.

Leaves - Anthelmintic, antispasmodic, stimulant, sudorific.

Fruits - Antiscorbutic, astringent, cooling, febrifuge, pectoral.

Seeds - Abortifacient, emetic.

Folk uses

Biofencing: Commonly planted along the boundaries of fields for fencing purposes.

Food: The mature fruits largely eaten by the tribals.

Medicine:

- Extract of seeds is taken with water to cure diarrhoea and dysentery.
- Extract of seeds is taken with water to prevent conception.
- The decoction of leaves is an antidote to vomiting caused by eaten the seeds of *Jatropha curcas.*
- The tribals give the extract of leaves with water to their cattle for removing the intestinal worms.

Miscellaneous: The mature fruits often sold in the market.

Uses in literature

Leaves - Dandruff (Tarafder, 1987; Basak, 1997); Cuts (Roy & Chaturvedi, 1986); Hysteria (Bhalla *et al.,* 1982); Ulcers (Bhattacharjee *et al.,* 1980); Worms (Roy & Chaturvedi, 1986).

Fruits - Skin diseases (Upadhay *et al.,* 1998); Tumour (Shah *et al.,* 1981); Diarrhoea, dysentery (Roy & Chaturvedi, 1986); Edible (Baruah & Sarma, 1984; Shah, 1984; Maji & Sikdar, 1982; Sudhakar & Rolla, 1985; Jain & De, 1966; Singh & Singh, 1981; Shah *et al.,* 1981; Jain, 1981; Goel *et al.,* 1984; Nagar, 1985).

Seed- Dandruff (Arinathan *et al.,* 2006); Abortification (Singh & Pandey, 1980; Shah *et al.,* 1981; Goel *et al.,* 1984; Roy & Chaturvedi, 1986); Menstrual complaint (Goel *et al.,* 1984).

Anogeissus latifolia (Roxb. ex DC.) Wall. ex Guill. & Perr. (Combretaceae)

Syn.: *Conocarpus latifolia* Roxb. ex DC.

Vernacular name - Dhawro.

Regional and other names- *Bengali* – Dhaoyo; *English* - Button tree; *Gujarati* – Dabria; *Hindi* –Dhawra; *Kannada* – Diadinduga; *Malayalam* – Vellanava; *Marathi* –Dhaura; *Sanskrit* – Dhava; *Tamil* – Vekkali; *Telugu* – Vellama.

Distribution - Throughout the greater part of India.

Taxonomic description - Tall, deciduous trees, with smooth, white-grey bark. Leaves silky tomentose when young. Flowers 2-5 in a head, yellow to pinkish- yellow. Fruits orbicular, glabrous, with entire wings.

Flowering & fruiting - October – June.

Habitat ecology - Found on hilly places.

Regeneration - By seeds.

Field no. - DJ-05

Parts used - Gum, Wood.

Chemical constituents

Seed contains Protein.

Leaves contain Phenol, Carboxylic acid, Saponin, Amino acids, Sugars.

Stem bark contains Shikimic acid, Quinic acid, Fructose, alanine, Phenylalanine, Tannins, Saponins.

Sap wood contains Quinic acid, Glucose, Phenylalanine, Ellagic acid.

Heart wood contain Quercetin, Myricetin, Glucose, Ellagic acid, Gallic acid, Phenylalanine.

Biological activity

Root - Acrid, pungent, stomachic.

Bark - Acrid, appetiser, cooling, pungent, sweet.

Folk uses

Gum : The gum collected from the stem is used as an ingredient of 'laddoos'- a sweet which forms the nourishing diet for ladies during pregnancy.

Medicine:

- Gum is fried in ghee and mixed with jaggery and safed musli (*Chlorophytum tuberosum* Baker). This mixture is warmed a bit with ghee and small balls are prepared. One ball is taken everyday after pregnancy for strength and vigour.
- Extract of stem bark is taken with water to cure dysentery.

Miscellaneous: Gum is sold in the local market by the tribals.

Tannin: Bhil tribals use the stem bark for tanning the leather.

Timber: The wood is used for making carts and cot (Khat).

Uses in literature

Root - Headache (Gopal & Shah, 1985); Stomach diseases (Shah *et al.*, 1981).

Gum - Dysentery (Roy & Chaturvedi, 1986); Edible (Vartak, 1959; Nagar, 1985; Maheshwari *et al.,* 1986; Datar & Vartak, 1975).

Bark - Headache (Shah *et al.,* 1981); Cough & stomachic (Ramachandran & Nair, 1981); Snake & scorpion bite (Mishra & Billore, 1983); Colic (Apparanantham & Chelladurai, 1986); Diarrhoea & dysentery (Maheshwari *et al.,* 1986); Skin diseases (Kumar *et al.,* 2006).

Argemone mexicana L. (Papaveraceae)

Vernacular name - Pilo dhaturo.

Regional and other names : *Bengali* – Shialkanta; *English* - Mexican poppy; *Gujarati* – Darudi; *Hindi* –Satyanashi; *Kannada* – Datturigidda; *Malayalam* – Bhrahmadande; *Marathi* – Pilvadhotra; *Sanskrit* –Svarnaksiri; *Tamil* – Bhrahmadandu; *Telugu* – Brahmadandi.

Distribution - Introduced and naturalized throughout India. Indigenous in tropical America.

Taxonomic description - Prickly herbs, with yellow sap and pinnatifid and spiny leaves. Flowers yellow, solitary terminal. Capsule prickly.

Flowering & fruiting - February-June

Habitat ecology - Comman throughout the area in wastelands, along road sides.

Regeneration - By seeds.

Field no. - DJ-06.

Parts used - Seeds, Leaf juice.

Chemical constituents

Root contains Berbetine, Protopine nitrate, Betasterol, Fatty acids, Ceryl alcohol, Allocryptopine.

Flowers contain Tannin, Resin, Argemone oil.

Leaves contain Glucosides, Kampferin, Anthraquinone, Essential oil & Chrysophanic acid.

Seeds contain tasteless oil of dark brown colour, Carbohydrates, Albumin, Phosphate, Sulphate.

Biological activity

Root - Alterative, anodyne, diuretic.

Seed - Demulcent, emetic, expectorant, laxative.

Folk uses

Medicine:

- Leaf juice is applied externally to cure ringworm.
- Paste of seeds is applied on boils.
- Yellow juice is applied to eyes twice a day to cure conjunctivitis.

Uses in literature

Root - Eye complaint (Bhalla *et al.,* 1982); Skin diseases (Singh & Maheshwari, 1983); Piles (Kamble & Pradhan, 1980); Dog bite (Pal, 1980); Scabies (Sudhakar & Chetty, 1998), Leucoderma (Singh *et al.,* 1997); Stomachache, vomiting (Subhangi & Patil, 2006); Fertility (Upadhyay & Jaswant Singh, 2005); Cuts & wounds (Das *et al.,* 2006).

Seed - Cough, asthma, whooping cough (Bhalla *et al.,* 1982); Skin diseases (Shekhawat & Anand, 1984; Kumar & Chauhan, 2005); Boils & ulcer (Sudhakar & Rolla, 1985); Headache (Singh & Pandey, 1980); Gum trouble (Bhatt & Sabnis, 1987).

Latex - Eye complaints (Maheshwari & Singh, 1984); Skin diseases (Shah, 1984; Pandit *et al.,* 1996; Gupta *et al.,* 1997; Deokule, 2006); Toothache (Singh *et al.,* 1987); Scabies (Jain *et al.,* 1973; Pandit *et al.,* 1996; Gupta *et al.,* 1997); Jaundice (Shah, 1984; Pandit *et al.,* 1996; Gupta *et al.,* 1997).

Leaves - Skin diseases, dropsy (Bhalla *et al.,* 1982; Dutta Choudhury & Shil, 2006); Boil, ulcer (Goel *et al.,* 1984); Eye complaint, Scabies (Jain *et al.,* 1973); Leprosy (Shah *et al.,* 1981); Syphilis (Bhatt & Sabnis, 1987); Ringworm (Jain *et al.,* 1973); Jaundice (Bhalla *et al.,* 1982; Dutta Choudhury & Shil, 2006; Das *et al.,* 2006).

Stem- As vegetable (Maji & Sikdar, 1982).

Argyreia nervosa (Burm. f.) Boj. (Convolvulaceae)

Syn.: *A. speciosa* Sweet

Vernacular name - Lambi-veldi.

Regional and other names - *Bengali* – Bichtarak; *English* - Elephant creeper; *Gujarati* - Samudra sosh; *Hindi* – Samunder kapat; *Kannada* – Chandrapada; *Malayalam* – Samudrapala; *Marathi* – Samudrasoka, Vriddhddaru; *Sanskrit* – Samudra sosha; *Tamil* – Samuthrappalai; *Telugu* – Chandrapada.

Distribution - Throughout India.

Taxonomic description - White tomentose twiners, with ovate-cordate leaves tomentose beneath. Flowers rose-purple and white.

Flowering & fruiting - August – December.

Habitat ecology - Often cultivated for its handsome flowers and foliage.

Regeneration - By seeds.

Field no. - DJ-64.

Parts used - Leaves.

Chemical constituents - Plant contains Arginine, Lycine, Valine, Behenic acid, Linolic acid, Linolenic acid, Oleic acid, Argoclavine, Clanoclavine–I & II, Elymoclavine, Ergine, Isoergine, Ergometrine, Fastuclavine, Isosetoclavin, Setoclavine, Lysergine, Lysergol, Lysergic acid, Caffeic acid, Anthocyanin, Cumarine, Flavonoids, Quercetine, Kaempferol, ß- Setosterol.

Biological activity

Leaf - Emollient.

Root - Alterative, bitter, diuretic.

Folk uses

Medicine: Leaves are heated a bit with edible oil and tied in the cases of rheumatism.

Uses in literature

Leaf - Boils (Pal, 1980); Headache (Saxena & Dutta, 1975); Swelling (Deokule, 2006); Gout (Jadeja *et al.,* 2006); As vegetable (Shah, 1984).

Root - Diabetes, gonorrhoea & ringworm (Tribedi *et al.,* 1982).

Artocarpus heterophyllus Lam. (Moraceae)

Vernacular name - Katar.

Regional and other names - *Bengali* – Kanthal; *English* – Jackfruit tree; *Gujarati* – Vanas; *Hindi* – Kathal; *Kannada* –Halasu; *Malayalam* – Plavu; *Marathi* – Phanas; *Sanskrit* – Panasah; *Tamil* – Palamaram; *Telugu* – Panasa.

Distribution - Throughout India.

Taxonomic description - Tree with dense, coriaceous elliptic leaves. Stipules glabrous. Male receptacle cylindric and female receptacle tubercled. The receptacles are lateral on trunk and branches, on special branchlets. The inflorescence is enclosed in 2, large, lathery, yellowish stipules belonging to the leaf.

Flowering & fruiting - November-June.

Habitat ecology - Cultivated occasionally in certain fruit orchards for its fruits.

Regeneration – By seeds.

Field no. - DJ-66.

Parts used - Fruits.

Chemical constituents

Leaves - Cycloartenone, Cycloartenol, B-sitosterol, Tannin.

Fruit - B-carotene, Amino acids, Aurantiamide acetate, Vitamin-A & C; Iron, Phosphorus.

Bark - Betulinic acid, Cycloheterophyllin, Iso-cycloheterophyllin, Cycloartenyl acetate, Cycloartenol, Cycloartenone.

Biological activity

Heart wood - Morin, dihydromorin, cyanonaclurin, artocarpin, iso-artocarpin, cyclo-artocarpin, artocarpesin, cycloartocarpesin, artocarpetin, artocarpanone, sucrose, ellagic acid.

Latex - Cycloartenon, cycloartenol, B-sitosterol, butyrospermol, leucine, isoleucine, tyrosine, valine, cycloartenol.

Unripe fruit- Acrid, astringent, carminative, tonic.

Ripe fruit - Aphrodisiac, cooling, latex, sweet, tonic.

Seeds - Aphrodisiac, constipating, diuretic, sweet.

Folk uses

Food: Fruit is used as vegetable.

Totem & Taboo:

- This species is considered as a plant totem of Katara clan of Bhils and, therefore, preserved and whorshipped by them.

Uses in literature

Leaves - Anthelmintic (Baruah & Sharma, 1984).

Root - Stomach complaint (Karnick *et al.,* 1981).

Latex - Skin diseases (Bhattacharjee *et al.*, 1980).

Fruit- Edible (Nagendra prasad & Abraham, 1984; Shah, 1984; Abraham, 1981; Datar & Vartak, 1975).

Wood- Furniture (Hajra & Baishya, 1981); Yellow dye (Vishnu mittre, 1981).

Asparagus racemosus Willd. (Liliaceae)

Vernacular name - Hemuri, Helio, Howansar, Motibapji, Arnapali.

Regional and other names- *Bengali* – Satmuli; *English* – Asparagus; *Gujarati* –Satavari; *Hindi* – Satawar; *Kannada* - Majjige-gadde aheruballi; *Malayalam* – Satavalli; *Marathi* -Satmuli; *Sanskrit* – Shatapadi, Satavari; *Tamil* – Ammaikodi; *Telugu* –Challagadda.

Distribution - Throughout India.

Taxonomic description - Scandent, spinous undershrubs, with faciculated tuberous roots. Leaves absent, branches modified into linear, leaf-like cladodes. Flowers in racemes. Fruits globose.

Flowering & fruiting - September – December.

Habitat ecology - Comman in the forests in hilly tracts, spreading on the ground or climbing on bushes.

Regeneration – Tubers roots.

Field no. - DJ-07

Parts used - Root.

Chemical constituents - Fleshy roots contain Sugars, Mucilage & Saponins.

Biological activity - *Roots* - Alterative, aphrodisiac, appetiser, astringent, bitter, cooling, indigestion, stomachic, sweet, tonic.

Folk uses

Magico-religious beliefs: 7 to 21 root pices are tied in sacred thread "luccha" and a necklace is prepared. This necklace is worn around the neck for a week to cure typhoid; it is believed that as the size of necklace increases automatically, the patient gets cured.

Medicine:

- Extract of root is taken with water to cure leucorrhoea and menorrhoea.
- Extract of root is taken with water to cure typhoid, gout and jaundice.

- Extract of root is given with water by the tribals to their cattle in the case of dysentery.

Uses in literature

Root - Rheumatism (Lakshamanan & Sankranarayanan, 1988); Cough (Sebastian & Bhandari, 1984); Fertility problem, To increase milk of mother (Mokat & Deokule, 2006); Dysentery & wound (Chaudhuri *et al.*, 1975); Diarrhoea (Sharma *et al.,* 1974); Gastric complaint (Joshi *et al.*, 1980); Gonorrohoea (Tarafder, 1984); Leucorrhoea (Hemadri & Rao, 1983); Impotency (Ramchandran & Nair, 1981); Stomach trouble (Hajra & Baishya, 1981); Dyspepsia & anaemia (Gopakumar, 1991); General debility (Singh & Aswal, 1992; Kumar & Chauhan, 2005); Typhoid (Jadhav, 2006); Improve sperm count (Udayan *et al.*, 2006 b); Retention of urine (Kumar & Chauhan, 2005); Tonic (Deokule, 2006); Fever (Baruah *et al.*, 2006); Dysentery (Rajendran & Rengamani, 2006); Edible (Vartak, 1981; Maheshwari *et al.*, 1981; Kanodia & Gupta, 1968).

Flower- Gout (Jadeja *et al.,* 2006).

Azadirachta indica A. Juss. (Meliaceae)

Vernacular name - Limdo, Limdi, Lim, Nim, Neem.

Regional and other names- *Bengali* – Nim; *English* - Margosa tree; *Gujarati* – Limba; *Hindi* –Neem; *Kannada* – Bevu; *Malayalam* – Veppa; *Marathi* – Limba; *Sanskrit* – Nimbaka; *Tamil* – Vembu; *Telugu* – Veepachettu.

Distribution - Throughout India.

Taxonomic description - Evergreen trees, with serrate or crenate, 7-17 leaflets and white flowers in panicles. Drupes ovoid, yellow when ripe.

Flowering & fruiting - March – May.

Habitat ecology - Abundant throughout the area.

Regeneration - By Seeds.

Field no. - DJ-08.

Parts used - Wood, Leaves, Seeds, Fruits, Bark.

Chemical constituents

Plant contains Margocin, Nimbadin, Sulphur, Nimban, Nimbosterol, Tannin, Essential oil, Resin, Glycosides, Fatty acids, Gum, Free amino acids.

Leaves contain Quercetin, β- Sitosterol (Nimbasterol).

Biological activity

Stem bark - Anthelmintic, astringent, bitter, demulcent, purgative, tonic.

Leaves - Anthelmintic, anti – inflammatory, antileprotic, stimulant.

Folk uses

Firewood: The wood is a good source for fuel.

Fodder: The leaves serve a good fodder for cattle.

Food: Ripe fruits are eaten by the children.

Magico-religious beliefs:

- The leaves are used by the Bhopas in Jhada rituals in the cases of a person suffering from typhoid.
- The leaves are used by the Bhopas in Jhada rituals to wardoff the effect of evil eyes.
- The tribal people place a bunch of leaves on the doors of their houses at the time of any person of the family suffering from smallpox.

Medicine:

- Juice of leaves is used as tonic to cure ascariasis and skin diseases (Itching, eczema, ringworm, boils, pimples etc.).
- Powder of seeds is taken with water to cure piles.
- Juice of leaves is taken with water to prevent miscarriages.
- Extract of leaves is taken in the cases of snake-bite.
- Extract of fruits is taken with water to cure bodyache and headache.
- Leaves are chewed in stomachic.
- Powder of stem bark is heated a bit with coconut oil and applied on burnt skin.
- Juice of leaves is taken with water to cure stomach trouble, typhoid and malaria fever.
- Extract of leaves is taken with water (sugar may be added) in the cases of vomiting.
- Juice of leaves is applied externally in case of eye diseases.

- Body of cattle is washed with decoction of leaves to remove the lice.
- Extract of inner-bark of the stem of *Acacia leucophloea* is mixed with the extract of inner-bark of the stem of *Butea monosperma* and *Azadirachta indica* and this mixture is taken with water to cure dysentery.

Miscellaneous:

- Seeds are sold in the market by the tribals.
- The tender twigs are used as tooth-brush.

Timber: The timber is used for making household articles and agricultural implements.

Uses in literature

Leaf - Boils & wounds (Shah, 1982); Cold & diabetes (Shah, 1984); Diarrhoea & dysentery (Kharkongor & Joseph, 1981); Blood purifier (Rao, 1981); Gout (Sebastian, 1984); Heart complaint (Rao, 1981); Scabies (Nagendra Prasad & Abraham, 1984); Skin diseases (Rai, 1985; Pereira, 2006; Deokule, 2006; Kumar *et al.,* 2006; Rajendran & Rengamani, 2006; Basak, 2006; Das *et al.,* 2006); Piles (Singh, 1988); Malaria (Shekhawat & Anand, 1984; Deka *et al.,* 2006); Snake bite (Lal & Yadav, 1983; Jain & Sahu, 1993); Mumps (Ganeshan *et al.,* 2006); Dyspepsia (Jadeja *et al.,* 2006); Itches (Sarma *et al.,* 2006); Hepatitis (Dutta Choudhury & Shil, 2006); Diabetes (Ghosh, 2006; Das *et al.,* 2006); Small pox (Rajendran & Rengamani, 2006); Eczema, ulcer (Basak, 2006; Das *et al.,* 2006).

Flower - Dyspepsia (Bhalla *et al.,* 1982); Wound (Maheshwari & Singh, 1984); Malaria (Deka *et al.,* 2006).

Fruits - Leprosy & skin diseases (Shah, 1982); Piles (Bhalla *et al.,* 1982); Edible (Singh & Singh, 1981; Ramachandran & Nair, 1981).

Seed - Antiseptic (Sudhakar & Rolla, 1985); Skin diseases (Singh & Pandey, 1980; Singh & Prakash, 1994); Malaria (Deka *et al.,* 2006).

Seed oil - Piles (Ved prakash & Jain, 1984); Skin diseases (Shah, 1984); Ulcer (Jain & De, 1966); For cooking (Vishnu Mittre, 1981).

Bark - Jaundice & liver complaint (Bhalla *et al.,* 1982); Wounds (Shekhawat & Anand, 1984; Deokule, 2006); Skin diseases (Shekhawat & Anand, 1984; Deokule, 2006; Kumar *et al.,* 2006; Basak, 2006); Fever (Kumar *et al.,* 2006; Arinathan *et al.,* 2006); Malaria (Deka *et al.,* 2006); Cold (Arinathan *et al.,* 2006); Swelling (Sharma *et al.,* 2006); Ulcer, eczema (Basak, 2006).

Gum - Snake bite & scorpion sting (Jain & Sahu, 1993); Diarrhoea (Mohanty *et al.,* 1996).

Balanites aegyptiaca (L.) Del. (Simaroubaceae)

Syn.: *B. rouxburghii* Planch.; *Ximenia aegyptiaca* L.

Vernacular name - Hingorio, Hingot.

Regional and other names : *Bengali* – Hingon; *Gujarati* - Hinger; *Hindi* - Hingot; *Malayalam* –Nanjunta; *Marathi* – Hingan; *Sanskrit* - Inguda; *Tamil* - Nanjundan; *Telugu* – Gara.

Distribution - Drier parts of India.

Taxonomic description - Spinous trees or shrubs, with 2-foliolate leaves. Flowers greenish-yellow, in axillary fascicled cymes. Fruits oblong, rind hard.

Flowering & fruiting - March – June.

Habitat ecology - Not very comman, found in Ghatla-patoli Village.

Regeneration - By seeds &root suckers.

Field no. - DJ-68.

Parts used - Fruit, Stem bark, Seed.

Chemical constituents

Fruits & seeds contain Oil of yellow colour and tasteless.

Pulp & fruit contains Saponin.

Biological activity

Flower - Bitter, heating, sweet smelling.

Fruits - Antidysenteric, alexipharmic, alterative, analgesic anthelmintic, bitter, sharp digestive, expectorant, laxative, taste.

Leaf - Anthelmintic, bitter, laxative.

Folk uses

Detergent: The tribals use the pulp of fruits as a detergent and subsitude for washing the cloths.

Medicine:

- Extract of root is taken with water during mensus to prevent conception.
- A drop extracted from fruit pulp is taken with a cup of water to cure typhoid, tetanus, hysteria and epilepsy.
- A syrup made up of extract of a seed with jaggery is given by the tribals to their children to cure pneumonia.
- Extract of fruit-pulp is applied in the cases of scorpion sting.

- Extract of seed (half seed only) is taken with water to cure menorrhoea.
- Extract of stem bark is useful in dog bite.
- The tribals give the extract of fruit pulp with water to their cattle for removing the intestinal worms.
- The decoction of stem bark is given orally to expel guinea-worms (Naru diseases).

Poison: The stem bark is used as a fish poison by Bhils.

Uses in literature

Seed oil - Burn & cough (Shah, 1981).

Bark - Cold (Bhalla *et al.*, 1982); Cough (Shekhawat & Anand, 1984); Swelling (Maheshwari *et al.*, 1986); Snake-bite (Deokule, 2006).

Fruits - Cold & cough (Bhalla *et al.*, 1982); Pneumonia (Saxena & Vyas, 1983); Snake bite (Singh, 1945); Skin diseases (Shekhawat, 1984; Pandit *et al.*, 1996; Sharma *et al.*, 2006); Dysentery (Jadeja *et al.*, 2006); Boils, leucoderma, whooping cough (Sharma *et al.*, 2006).

Fruit shell – Fireworks (Shekhawat & Anand, 1984).

Leaves- Snake bite (Deokule, 2006).

Root- Fish poison (Joshi, 1982).

Barleria prionitis L. (Acanthaceae)

Vernacular name : Vajradanti.

Regional and other names : *Bengali* - Kanta-jati; *Gujarati* - Pilikantashelio; *Hindi* - Vajradanti; *Kannada* – Kurantaka; *Malayalam* – Shemmuli; *Marathi* – Koranti; *Sanskrit* -Saireyaka; *Tamil* - Chemmuli; *Telugu* - Kanakam-baram.

Distribution - Throughout India.

Taxonomic description - Armed undershrubs, with elliptic or ovate-elliptic leaves and yellow flowers in dense spikes. Calyx-lobes glandular. Capsules 4-seeded.

Flowering & fruiting - October – January.

Habitat ecology - Common in shady & waste places.

Regeneration - By seeds.

Field no. - DJ-09

Parts used - Root, Leaves.

Chemical constituents - Plant contains Beta- Sitosterol, Scutellarein – 7 Rhamocyl Glucoside, Iridoids, Barlerin and Ac- Barlerin.

Biological activity - Acrid, anti-inflammatory, antiseptic, appetiser, bitter, diaphoretic, diuretic, expectorant, heating.

Folk uses

Medicine:

- Powder of root is taken with water to cure fever.
- Extract of root is taken with water to remove thorn from body.
- Boiled leaves are tied on boils.
- The paste of root is mixed with Jaggery and given in the case of corn.
- Extract of root is taken with water in the case of double pneummonia.

Uses in literature

Root - Boils (Bhalla *et al.,* 1982; Upadhay *et al.,* 1998); Toothache (Basak, 1997); Eczema (Deokule, 2006).

Bark - Dropsy (Bhalla, 1982); Gout (Jadeja *et al.,* 2006).

Leaf - Toothache (Sudhakar & Rolla, 1985; Ahmad, 1995); Wound (Ahmad, 1995); Cough (Baruah & Sarma, 1987; Ahmad, 1995); Rheumatism (Saxena & Vyas, 1983); Piles (Singh, 1988); Ear complaint (Mishra & Dixit, 1973).

Flower- Toothache (Aditya *et al.,* 2006).

Bauhinia racemosa Lam. (Caesalpiniaceae)

Vernacular name -Hetri, setri.

Distribution -Throughout India.

Regional and other names: *Bengali* –Vanraj; *English* –Downy mountain ebony; *Gujarati* –Asundro; *Hindi* –Jhinjhora; *Marathi* –Apta; *Sanskrit* –Ashmantak.

Taxonomic description - Deciduous trees; bark grey, rough. Leaves broader than long. Flowers in terminal, leaf-opposed racemes, creamy-white. Pods falcate, turgid. Seeds oval, black.

Flowering & fruiting - April-September.

Habitat ecology - Occasional along forest paths.

Regeneration - By seeds.

Field no. - DJ-79.

Chemical constituents

Stem bark contains Octacosane, B-amyrin, B-sitosterol.

Seeds- Protein, Pentosan, Mucilage.

Biological activity

Root bark- Astringent.

Stem bark, Leaves- Antipyretic, alexipharmic, refrigerent.

Parts used - Stem bark.

Folk uses -

Food: Pods are used as vegetable.

Medicine:

- Extract of stem bark is taken with water to cure loose motion.
- The necklace prepared from stem bark is worn around the neck for a week to cure conjunctivitis.

Tannin: Bhil tribals use the stem bark for tanning the leather.

Totem & taboo:

- This species is considered as a plant totem of Hari clan of Bhils and, therefore, preserved and whorshipped by them.

Uses in literature

Stem bark – Diarrhoea, dysentry (Singh & Pandey, 1980; Sinha *et al.,* 2006; Kumar & Lal, 2003); Swellings (Deshmukh *et al.,* 2003); Dysentery, malaria fever (Gogoi *et al.,* 2003).

Flower- Diarrhoea (Jadeja *et al.,* 2006).

Leaves- Wounds (Sivakumar *et al.,* 2003); Intestinal worms, fever, diabetes, piles, liver disorders (Kothari & Rao, 2003); Headache (Deshmukh *et al.,* 2003; Kumar & Lal, 2003); Dysentery, malaria fever (Gogoi *et al.,* 2003); Bidi making (Vishnu Mittre, 1981).

Gum- Diarrhoea (Deshmukh *et al.,* 2003); Tonic, soothing to intestines (Kumar & Lal, 2003); Malaria fever (Kumar & Lal, 2003).

Root bark- Wormicide (Deshmukh *et al.,* 2003).

Fruit- As vegetable (Singh & Singh, 1981).

Stem fibre- Basketry (Joshi, 1982).

Bauhinia variegata L. (Caesalpiniaceae)

Vernacular name - Khato hetro.

Regional and other names : *Bengali* - Raktakanchan; *English* – Camel's foot; *Hindi* - Kachnar; *Kannada* - Kanchavala; *Malayalam* - Mandarum; *Marathi* -Raktakanchan; *Sanskrit* - Kanchanara; *Tamil* - Taruvatti; *Telugu* –Adavimandara.

Distribution - Throughout India.

Taxonomic description - Medium- sized trees. Leaves orbicular, 2-lobed. Flowers white to pink. Pods flat, glabrous, brown.

Flowering & fruiting - April-June.

Habitat ecology - Found wild in the forest area.

Regeneration - By seeds.

Field no. - DJ-109.

Parts used - Stem bark.

Chemical constituents

Stem bark contains Sugar & Tannins.

Flowers contain Octacosanol, ß- stiosterol, Flavinoid.

Seeds & bark contain Myrcetol, Glycoside, Kaempferol, Glucoside.

Biological activity

Bark - Alterative, astringent, tonic.

Flowers - Acrid, astringent, cooling, sweet.

Folk uses

Medicine: Extract of stem bark is taken with water to cure dysentery.

Uses in literature

Bark - Leprosy, ulcer, skin diseases, syphilis (Bhalla *et al.,* 1982); Malaria, snake bite (Negi *et al.,* 1985); Dysentery (Megoneitso & Rao, 1983; Kataria & Gauchan, 2006); Piles (Rajwar, 1983); Tumour (Shah & Joshi, 1971); Diarrhoea (Negi *et al.,* 1985; Kataria & Gauchan, 2006); Wound, swelling (Kumar *et al.,* 2006).

Flower - Dysentery (Rajwar, 1987; Gopakumar, 1991); Worm & diarrhoea (Jain & De, 1966); Tumour (Shah & Joshi, 1971); Oedema (Gopakumar *et al.*, 1991); Piles (Gopakumar *et al.*, 1991; Kumar *et al.*, 2006).

Root – Piles (Kumar *et al.*, 2006).

Benincasa hispida (Thunb.) Cogn. (Cucurbitaceae)

Syn.: *Cucurbita hispida* Thunb.

Vernacular name : Bhuro khoro.

Regional and other names : *Bengali* – Kumra; *English* – White gourd melon; *Gujarati* – Koholu; *Hindi* – Petha; *Kannada* – Budikumbala; *Malayalam* – Kum palam; *Marathi* – Kohala; *Sanskrit* – Kusmandah; *Tamil* – Pusanikkai; *Telugu* – Budidagummadi.

Distribution - Cultivated throughout India.

Taxonomic description - A large climbing or trailing herb.

Flowering & fruiting - August-October.

Habitat ecology - Cultivated throughout the area.

Regeneration – By seeds.

Field no. - DJ-200.

Parts used - Fruits.

Chemical constituents : Fruit contains β-sitosterol, Asparagine, Luoeol, Mannitol, Proline, Arginine, Aspartic acid, Glutamic acid, Glutamine, Hydroxyproline, Isoleucine, Cysteine, L-leucine, n-triacontanol, Glucose, Rhamnose, Vitamins.

Biological activity

Fruits - Aphrodisiac, antiperiodic, cooling, diuretic, laxative, styptic, sweet.

Seeds - Anthelmintic, cooling, sweet.

Folk uses

Magico-religious beliefs: The fruits are offered to god and goddess by the tribals.

Totem & taboo: This species is considered as a plant totem of Bhuria clan of Bhils and, therefore, preserved and whorshipped by them.

Uses in literature

Fruits - Jaundice (Hemadri & Rao, 1984; Dutta *et al.,* 2006); Diabetes, epilepsy, nervous disorders (Jha *et al.,* 1997).

Boerhavia repens L. var. ***diffusa*** (L.) Hook. f. (Nyctaginaceae)

Syn.: *B. diffusa* L.

Vernacular name : Choti-dhankni.

Regional and other names : *Bengali*- Punarnava; *English* - Horse puruni spreading hog weed; *Gujarati* - Motosatodo; *Hindi* - Gadhapurna; *Kannada* - Balavadikae; *Malayalam* - Vasu; *Marathi* – Raktavasu; *Sanskrit* – Bhauma, Punarnava; *Tamil* - Kukaratte; *Telugu* – Keerai.

Distribution - Throughout India.

Taxonomic description - Decumbent or diffuse, perennial, extensive herbs. Leaves unequal, ovate-oblong. Flowers rose or pink, in panicles or subcapitate umbels. Fruits clavate, 5-ribbed, covered with stalked glandular hairs.

Flowering & fruiting - Throughout the year.

Habitat ecology - Common in the wastelands.

Regeneration - By seeds.

Field no. - DJ-117.

Parts used - Whole plant, root.

Chemical constituents- Plant contains Alkaloids – Punarnavan, Sterol, ß- Sitosterol, œ- Sitosterol, Elentine, Stearic acid, Palmitic acid, Minerals, Potassium nitrate, Sodium & Sulphate & Chloride.

Biological activity - Anti- bacterial, anti- inflammatory, antiviral, astringent, bitter, cooling, diaphoretic, laxative, stimulant, stomachic.

Folk uses

Magico-religious beliefs: 7 to 21 pieces of root are tied in sacred thread "luccha" and a necklace is prepared. This necklace is worn around the neck for a week to cure typhoid; it is believed that as the size of necklace increases automatically, the patients get cured.

Medicine:

- Extract of root is taken with water in the cases of typhoid, sexual impotency and general fever.

- Extract of root is applied in the cases of scorpion sting.
- Extract of whole plant is taken with water to cure bodyache.
- Root is tied in the hairs of pregnant lady at the time of labour pain for normal delivery.

Uses in literature

Whole plant - Kidney complaint, Gonorrhoea, dropsy (Bhalla *et al.*, 1982); Piles, cough (Rajwar, 1983); Fever (Bedi, 1978); Rheumatism & blood purifier (Janardhanan, 1963); Abortifacient (Maheshwari & Singh, 1987); Asthma (Sharma *et al.,* 1979); Jaundice (Das *et al.,* 1983); Dysentery (Maheshwari *et al.,* 1986); Stomach trouble (Tripathi *et al.*, 1996; Sudhakar & Chetty, 1998); Inflammation (Singh *et al.,* 1980).

Leaves- Urinary problems (Pereira, 2006); Spermatorrhoea, leucorrhoea (Prakash & Singh, 2006.); Asthma, worm affection, headache, liver and abdominal problem, cough, heart problems (Sharma *et al.*, 2006); As vegetable (Bhalla *et al.,* 1982; Jain & De, 1966; Ramachandran & Nair, 1981; Bedi, 1978; Saxena & Dutta, 1975; Maheshwari *et al.,* 1986; Chaudheri *et al.*, 1985).

Root - Eczema (Ganeshan *et al.*, 2006); Jaundice (Narain & Singh, 2006b; Deokule, 2006); Spermatorrhoea (Kumar & Chauhan, 2005); Conjuncitivis, blindness (Kumar *et al.,* 2006).

Seed- Jaundice (Jadeja *et al.,* 2006).

Bombax ceiba L. (Bombacaceae)

Syn.: *B. malabaricum* DC.; *Salmalia malabarica* (DC.) Schott & Endl.

Vernacular name : Hemlo, Semlo.

Regional and other names : *Bengali* – Shimool; *English* - Silk cotton; *Gujarati* - Shimul; *Hindi* -Semul; *Kannada* – Booruga; *Malayalam* – Poola; *Marathi* – Simalo; *Sanskrit* – Dirghayu, Salmali; *Tamil* - Mullilavau; *Telugu* – Buruga.

Distribution - Throughout the hotter parts of India.

Taxonomic description- Deciduous trees, with prickly trunk and digitately 5 to 7-foliolate leaves. Flowers crimson or yellow, crowded at the end of leafless branches. Capsules woody, 5-valved.

Flowering & fruiting - February – May.

Habitat ecology - Found in open forests.

Regeneration - By Seeds.

Field no. - DJ-10.

Parts used : Stem bark, Seed, Root, Wood.

Chemical constituents

Seed contains Hexacosanol, Tocopherol, Terenes.

Bark contains Lupeol, B- sitosterol.

Root contains Triacontanol, B- sitosterol.

Flowers (Petals) contain Essential oil, Hentriacontane, Hentriacontanol, Quercetin, Kaempferol, B- sitosterol.

Biological activity

Root - Astringent, cooling, demulcent, diuretic, sweet, tonic.

Bark - Acrid, astringent, demulcent, diuretic, tonic.

Flowers - Acrid, astringent, bitter, cooling.

Fruits - Alterative, aphrodisiac, cooling, digestive, diuretic, expectorant, stimulant, sweet, tonic.

Gum-Acrid, alterative, aphrodisiac, astringent, demulcent, tonic.

Folk uses

Food: The seeds are eaten by the tribals.

Medicine:

- Extract of stem bark is taken with water to cure diarrhoea and dysentery.
- Extract of roots or flowers is taken with water to promote conception. It is also useful to prevent miscarriages.
- Extract of roots or stem bark or flowers is taken with water to promote conception. It is also useful in leucorrhoea, menorrhoea and stomach pain during mensus.
- Extract of inner bark of young stem is taken with water to check undesirable discharge of semen with urine and to increase sexual potentiality.
- Paste of stem bark is tied on anus to cure piles.
- Paste of stem bark is applied on boils.
- The tribals give the extract of stem bark with water to their cattle for removing the intestinal worms.

Timber:

- The wood is used for making some musical instruments called 'Dholak' and 'Tambura'.
- The wood is also used for making fishing boats.

Uses in literature

Root - Abdominal pain & dysentery (Tarafder, 1983); Cough (Jain & Borthakur, 1980); Boils (Tarafder, 1984); Burn (Pal & Srivastava, 1976); diarrhoea (Rajwar, 1983); Spermatorrhoea, gonorrhoea (Prakash & Singh, 2006.).

Bark - Pimples & boils (Saxena, & Vyas, 1981); Dysentery (Malhotra & Moorthy, 1973); Bone fracture (Chaudhary *et al.,* 1982); Skin diseases (Rajwar, 1983); Sexual debility (Ved Prakash & Mehrotra, 1987).

Leaf - Anaemia (Gupta, 1981); Rheumatism (Chelvan, 1998).

Gum - Diarrhoea, dysentery (Sharma *et al.,* 1979); Asthma (Rajwar, 1983); Menorrhoea, leucorrhoea, (Singh & Aswal, 1992; Singh, 1993).

Seed - Small pox. (Saxena *et al.,* 1981); Liver trouble (Das *et al.,* 2006).

Fruits - Dysentery (Maheshwari & Singh, 1984); Edible (Vartak, 1981; 1959; Ved Prakash & Mehrotra, 1987; Datar & Vartak, 1975).

Stem - Gonorrohoea & syphilis (Tiwari *et al.,* 1979).

Flower - Sanke bite (Negi *et al.,* 1985); Edible (Purohit *et al.,* 1975).

Brassica rapa L. ssp. ***campestris*** (L.) Clapham (Brassicaceae)

Syn.: *Brassica campestris* L.

Vernacular name - Sarsu, Sarson.

Distribution - Throughout India.

Taxonomic description – Annual, erect herbs; leaves lyrate-sinuate to lanceolate. Flowers in oblong corymbs, elongating into racemes. Petals clawed, yellow. Fruits siliqua.

Flowering & fruiting - January-April.

Habitat ecology - Very largely grown during the cold season in rotation with other crops.

Regeneration - By seeds.

Field no. - DJ-139.

Parts used - Seed oil.

Folk uses -

Medicine:

- 250 gm. Mustard oil is mixed with 25 gm. Haldi (*Curcuma longa* L.) and given to cattles in the case of Afra disease.
- Oil is warmed a bit with 'deshi kapoor'. This oil is applied as ointment in the case of cervical pain.
- Oil is warmed a bit with cloves of Long (*Syzigium aromaticum*) and filtered. This oil is applied as ointment in the case of sprain.

Uses in literature

Seed- Muscular pain (Majumdar *et al.,* 1978); Skin diseases (Purohit *et al.,* 1985).

Seed oil – Boils (Dar *et al.,* 1984).

Brassica nigra (L.) Koch (Brassicaceae)

Syn.: *Sinapis nigra* L.

Vernacular name : Rai.

Regional and other names- *Bengali* –Raisarisha; *English* –Black mustard; *Gujarati* - Rai; *Hindi* -Rai; *Sanskrit* -Asuri; *Tamil* -Kadugu; *Telugu* –Avalu.

Distribution - Throughout India.

Taxonomic description - A rigid, Branching annual; upper leaves entire. Flowers bright yellow, in naked racemes. Fruits siliqua.

Flowering & fruiting - February-April.

Habitat ecology - Cultivated for its seeds throughout the area.

Regeneration - By seeds.

Field no. - DJ-11.

Parts used - Seeds.

Chemical constituents - Seed contains Glucoside sinigrin, Sinapine and a volatile Ssothiocynate and Essential oil.

Biological activity - *Seed*- Hot, sharp bitter taste.

Folk uses

Food: Leaves are used as vegetable.

Magico-religious beliefs: The tribals use the seeds of Rai, shivlingi and Urd as an amulet ('Tavij') which is tied around the neck of the children to protect them from evil spirits.

Medicine: The seeds are powdered and applied on forehead to cure headache.

Uses in literature

Seed oil- Rheumatism (Banerjee, 1977).

Seed- Edible (Banerjee, 1977).

Butea monosperma (Lamk.) Taub. (Papilionaceae)

Syn.: - *B. frondosa* Roxb. eX Willd.; *Erythrina monosperma* Lamk.

Vernacular name : Khankro, Khankru.

Regional and other names : *Bengali* – Palash; *English* - Flame of the forest; *Gujarati* - Khakhra; *Hindi* - Palash; *Kannada* – Mutthuga; *Malayalam* – Chamatha; *Marathi* – Palash; *Sanskrit* - Palasha; *Tamil* - Palasam; *Telugu* – Palasmu.

Distribution - Throughout the greater parts of India.

Taxonomic description - Medium-sized trees, with 3-foliolate large leaves and orange, showy flowers in racemes. Pods grey-silky.

Flowering & fruiting - March – June.

Habitat ecology - Extensively found in the hilly tracks of the area.

Regeneration - By seeds & vegetative method.

Field no. - DJ-13.

Parts used - Root, Leaves, Seeds, Gum, Flowers, Bark.

Chemical constituents

Plant contains Fats, Oils, Mucilage, Protein, Resin, Butin, Glucoside, Butein, Isomonospermoside, Coreopsin, Sulphruein, Palastirin, Butrin, Isobutrin.

Gum contains Thiamine, Riboflavin, Fatty acids, Myristi acid, Palmitic & Stearic acid.

Biological activity

Bark - Acrid, anthelmintic aphrodisiac, appetiser, bitter, hot, laxative.

Gum - Astringent.

Flower - Acrid, astringent, bitter, hot, sweet.

Fruit - Anthelmintic, aperient, digestive, dry, hot.

Seed - Anthelmintic, aperient, digestive, dry, hot.

Folk uses

Biofencing: Commonly planted along the boundaries of fields for fencing purposes.

Dye: The flowers yield a yellow dye which is used for colouring the clothS. The dye is commonly used by the tribals at Holi festival for colouring each other.

Fibres: The bast fibres extracted from the root is used for making ropes.

Fire-wood: This species is a good source of fire-wood for the tribals.

Gum: The gum obtained from the stem is used as medicine.

Medicine:

- Decoction of root is taken with water in cases of snake-bite.
- The extract of stem bark is taken with water to cure diarrhoea, dysentery, leucorrhoea and menorrhoea.
- The fresh leaves are heated a bit with edible oil and tied in cases of swelling & injury.
- The decoction of flowers is used by ladies for washing the organ in gonorrhoea.
- Boiled flowers are tied on stomach in the cases of burning in urethra during micturition.
- The powder of dried flowers is used as face-pack for the treatment of pimples.
- Powder of seeds is taken with water to cure bodyache, ascariasis & to promote conception.
- Paste of seeds is applied on injured body parts & boils.
- Extract of half seed is taken with water in cases of snake bite.
- The seed oil is used as massage oil in the case of arthritis.

- The crushed gum is put into water for whole night, next day its extract is taken with water thrice a day to cure diarrhoea, dysentery & backache.
- Gum is chewed in the cases of mouth ulcer and backache.
- The extract of stem bark is given with water to cattle to cure dysentery.
- Extract of inner-bark of the stem of *Acacia leucophloea* is mixed with the extract of inner-bark of the stem of *Butea monosperma* and *Azadirachta indica* and this mixture is taken with water to cure dysentery.

Miscellaneous:

- Seeds are sold in the market by the tribals.
- Leaves are used for making plates ('Baj') and bowls ('Dauna').

Oil: Oil extracted from seeds is used in medicine.

Poison: The stem bark is crushed and dropped in water as a fish poison by the Bhils.

Timber: The wood is used for making an musical instrument called 'Dhol'.

Uses in literature

Root - Dogs bite (Venugopal & Raghunathan, 1982).

Seed - Ascariasis (Jain & De, 1966).

Leaf - Boils (Singh & Maheshwari, 1983); Skin diseases (Malhotra & Moorthy, 1973); Diarrhoea (Mahanty *et al.,* 1996); Ophthalmia (Tripathi *et al.,* 1996); Liver problem (Purkyastha & Subhan, 2006); Cough & cold (Ganeshan *et al.,* 2006); Intestinal worms (Jadeja *et al.,* 2006); Plates & Bowls (Jain & De, 1966; Shah *et al.,* 1981).

Gum - Antifertilizer (Billore & Audichya, 1978); Piles (Singh, 1988); Diarrhoea (Sudhakar & Rolla, 1985; Kataria & Gauchan, 2006; Kumar *et al.,* 2006); Dysentery (Maheshwari *et al.,* 1986; Singh & Aswal, 1992; Kataria & Gauchan, 2006; Kumar *et al.,* 2006); Diabetes (Kumar *et al.,* 2006).

Bark - Bone fracture (Maheshwari & Singh, 1984; Tripathi *et al.,* 1996); Wound (Shah *et al.,* 1983; Yesodharan & Sujana, 2006); Dysentery (Maheshwari *et al.,* 1986; Tripathi *et al.,* 1996; Narain & Singh, 2006b); Diarrhoea (Joshi *et al.,* 1980); Fibre (Goel *et al.,* 1984; Joshi, 1982); Fish poison (Harja & Baishya, 1981).

Flower - Piles & diarrhoea (Bhalla *et al.,* 1982); Skin diseases (Tripathi *et al.,* 1996); Stomach disorders (Lal & Yadav, 1983); Pimples (Rajwar,1983); Eye complaint (Shah *et al.,* 1983); Heat stroke (Narain & Singh, 2006b); Dysentery (Kamble & Pradhan,1980); Antifertilizer (Ramachandran & Nair, 1981); As vegetable (Maheshwari *et al.,* 1986); Dye (Goel *et al.,* 1984; Agrawal, 1981).

Fruit- As vegetable (Nagar, 1985; Negi *et al.,* 1985).

Caesalpinia bonduc (L.) Roxb. (Caesalpiniaceae)

Syn.: *Guilandina bonduc* L.

Vernacular name : Ghatar.

Regional and other names : *Bengali* – Nata; *English* - Fever nut; *Gujarati* - Kakcha; *Hindi* - Katkaranj; *Malayalam* – Kalanji; *Marathi* – Kanchaki; *Sanskrit* –Kantakini, Kantaki karanja; *Tamil* - Kachukkai; *Telugu* – Gutsakai.

Distribution - Throughout India.

Taxonomic description - Armed shrubs; pinnae 6-8 pairs, with a pair of hooked stipular spines; leaflets 6-9 pairs, elliptic-oblong, obtuse. Racemes often panicled; pods shortly stalked, oblong, densely armed on the faces with wiry prickles, 1 to 2-seeded.

Flowering & fruiting - June – February.

Habitat ecology – Common, found in hedge around fields.

Regeneration - By Seeds.

Field no. - DJ-14.

Parts used - seeds.

Chemical constituents - Seed contains Fatty oils, Starch, Sucrose, Caesalpins, Bonducellin, Fatty acids, Phytosterinin, Saponins, Phytosterols, Amino acids, a bitter glucoside- Bonducin.

Biological activity

Leaf - Anthelmintic.

Flower - Bitter.

Fruits - Acrid, anthelmintic, aphrodisiac, astringent.

Seed - Anthelminitic, antiperiodic, styptic.

Folk uses

Medicine: A syrup made up of seed powder with jaggery is used to cure stomach trouble.

Uses in literature

Seed - Colic & fever (Shah *et al.,* 1983); Rheumatism, skin diseases (Singh & Pandey, 1980); Wounds (Saxena & Vyas, 1986); Swelling (Rajwar, 1983); Malaria (Singh & Pandey, 1980; Purkayastha & Subhan, 2006); Constipation (Jadeja *et al.,* 2006).

Leaf- Piles (Deokule, 2006).

Calotropis gigantea (L.) R. Br. (Asclepiadaceae)

Syn.: *Asclepias gigantea* L.

Vernacular name : Ankro, Ankru, Hafed ankro.

Regional and other names : *Bengali* – Akanda; *English* - Giant milk weed; *Gujarati* - Akru; *Hindi* -Madar; *Kannada* – Arka; *Malayalam* – Vellerukku; *Marathi* – Akand; *Sanskrit* – Shwetark, Alarka; *Tamil* - Verukku; *Telugu* – Mandaramu.

Distribution - Throughout India.

Taxonomic description - Shrubs or small trees, with elliptic-oblong or obovate-oblong leaves. Corolla-buds ovoid; corona-lobes hairy, shorter than staminal column.

Flowering & fruiting - Throughout the year.

Habitat ecology - Common in wastelands, road-sides.

Regeneration - By seeds.

Field no. - DJ-69.

Parts used - Leaves, Latex, Root, Flowers.

Chemical constituents

Plant contains Mudarol, Akundarol, Uscharidin, Calotropin, Frugoside, Calotroposides – A to G, Calotroplupenyl acetate- A, Leupeol acetate – B, Gigantursenyl acetate – A, Gigantursenyl acetate – B.

Biological activity

Plant - Alexipharmic, anthelmintic, Purgative.

Root bark - Diaphoretic.

Flower - Analgesic, anthelmintic, astringent, bitter, sweet.

Latex - Bitter, purgative.

Folk uses

Magico-religious beliefs: A bunch of the leaves of Ankra, Neem and Khajor is used by the Bhopas in Jhada rituals to wardoff the effect of evil eyes.

Medicine:

- 2 young leaves chewed with salt twice a day for 7 days to cure jaundice.
- Leaves are warmed with 'pure ghee' and tied on knee in case of painful inflammation of joints (gout) and swoellen part of the body.
- Leaves are warmed a bit with 'pure ghee and tied on boils.
- Latex of leaves is used as an ointment to cure backache and swelling.
- A flower is chewed daily early in the morning to cure asthma.
- 5 flowers of *Calotropis gigantea* and 5-10 cloves of *Syzygium aromaticum* are burnt with cotton fibres (*Gossypium herbaceum)* to prepare ash. About 10g. of ash is taken daily early in the morning with glassful of water to cure asthma.
- 2 drops of latex are dropped in ripe fruit of Banana *(Musa paradisiaca)* and eaten alternate day early in the morning to cure jaundice.
- 2 dried leaves (in shady condition) of *Calotropis gigantea* and 25 gm. fruits of *Foeniculum vulgare* are ground with 50 gm. mishri (crystalline sugar) and a teaspoonful powder is taken with water daily early in the morning to cure jaundice.
- Latex is applied to remove thorn from body.
- Latex is applied as ointment on boils, cuts/injury, wound and blisters in cattle.
- Extract of root is given with water to cattle in the case of Dog bite.
- Roasted gynostegium is eaten in cough.
- Latex is filled in teeth cavity with the help of cotton plug to cure toothache.

- Latex is applied as ointment in the case of corn.
- Latex is applied as ointment on blisters and wounds tc remove external worms in cattles.
- Paste of 3 flowers is fried in pure ghee and it is taken with wheat bread "Roti" to cure asthma.
- Latex obtained from leaves is filled in the endosperm of ripe, dried fruit of *Cocos nucifera* through hole. The hole is plugged with the removed piece of the fruit and the fruit is roasted in the burning cow-dung. After the fruit is roasted, it is made into a powder. A teaspoonful of powder with cow milk is taken in early morning to cure asthma.
- Leaves heated a bit with edible oil and tied on chest to cure cough.
- Latex is applied in the case of scorpion sting.
- Fomentation with poultice made from the leaves to cure rheumatism.
- The traditional ball 'Bati' are prepared from wheat flour and boiled with the leaves of *Calotropis gigantea.* These balls are roasted in burning cow-dung. These balls are taken with pure ghee in the case of tuberculosis.

Uses in literature

Flower - Cough (Kumar & Chauhan, 2005).

Leaf - Bodyache (Rao & Jamir, 1982); Burns (Harja & Baishya, 1981); Eye complaint (Sudhakar & Rolla, 1985); Sweeling (Singh & Pandey, 1980); Abdominal pain, joint pain, skin diseases (Baruah *et al.,* 2006); Chicken pox (Arinathan *et al.,* 2006); Rheumatism (Das *et al.,* 2006).

Seed - Antifertility (Tarafder, 1983).

Root - Leprosy (Bhalla *et al.,* 1982); Dysentery & elephantiasis (Sudhakar & Rolla, 1985); Scorpion bite & snake bite (Jain *et al.,* 1973); Leucoderma (Purkayastha & Subhan, 2006); Abdominal pain, joint pain, skin diseases (Baruah *et al.,* 2006).

Latex - Wound (Shah, 1984); Skin diseases, ringworm (Shah *et al.,* 1983); Abortifacient (Tarafder, 1983); Toothache (Jain *et al.,* 1973); Dog bite (Tiwari *et al.,* 1979); Asthma (Saxena, 1986); Swelling of joints (Deokule, 2006).

Calotropis procera (Ait.) Ait. f. ssp. hamiltonii (Wight) Ali (Asclepiadaceace)

Syn.: *C. hamiltonii* Wight

Vernacular name : Ankro, Ankru, Lal ankru.

Regional and other names : *Bengali* - Swet-Akanda; *English* - Milk weed; *Gujarati* - Ankru; *Hindi* -Safed-ak; *Malayalam* - Vellerukku; *Marathi* - Mandara; *Sanskrit* -Raktark; *Tamil* - Vellerukku; *Telugu* – Chinnajilledu

D stribution - Throughout India.

Taxonomic description - Shrubs or small trees, with ovate or ovate-oblong leaves. Corolla-buds hemispherical; corona-lobes glabrous, equalling or longer than staminal column.

Flowering & fruiting - Throughout the year.

Habitat ecology - Common in wastelands, road-sides.

Regeneration - By seeds.

Field no. - DJ-15.

Parts used - Latex, Flowers.

Chemical constituents - Plant contains Cardiac steroid glycoside-terpenoids, Resin, Glycosides, Carbohydrates, Aluminium, Iron, Magnesium, Sodium.

Biological activity - Analgesic, anthelmintic, anti- inflammatory, depurative, digestive, emetic, expectorant, laxative, resolvent, stomachic.

Folk uses

Midicine:

- Latex of leaves is used as ointment to cure backache and swelling.
- Latex is used as ointment to remove thorn from body.
- Latex is applied on wounds and boils.
- Latex is applied against scorpion sting.
- Latex is filled in teeth cavity with the help of cotton plug to cure toothache.
- Latex is applied as ointment in the case of corn.
- 2 flower are chewed daily early in the morning to cure asthma and T. B.
- Latex of leaves is mixed with wheat flour (*Triticum aestivum* L.) and small pills are prepared, these pills are taken in the case of dog bite.

Uses in literature

Flower - Cholera (Saxena & Vyas, 1981); Cough (Maheshwari & Singh, 1987; Kumar *et al.*, 2006); Leucorrhoea (Kumar & Chauhan, 2005).

Wood ash - Asthma (Shah, 1982).

Leaf - Boils, (Jain, 1985; Purkayastha & Subhan 2006); Skin diseases (Jain, 1985); Swelling & rheumatism (Bhalla *et al.*, 1982); Eczema (Shah, 1982); Cold (Maheshwari & Singh, 1987); Abdominal pain (Sharma *et al.*, 2006).

Root - Asthma, cough (Singh & Pandey, 1980); Elephantiasis (Bhalla *et al.*, 1982); Fever (Maheshwari *et al.*, 1986); Swelling (Shekhawat & Anand, 1984); Veterinary medicine (Jain & De, 1966).

Latex - Leprosy, dropsy, rheumatism (Bhalla *et al.*, 1982); Toothache (Maheshwari & Singh, 1984; Sharma *et al.*, 2006); Wound (Virjee *et al.*, 1984). Antiseptic (Lad & Yadav, 1983); Scorpion sting (Singh & Pandey, 1980); Swelling (Deokule, 2006; Sharma *et al.*, 2006); Cough (Kumar *et al.*, 2006); Gout (Jadeja *et al.*, 2006).

Stem - Piles (Mokat & Deokule, 2006).

Cannabis sativa L. (Cannabinaceae)

Vernacular name : Ganjo.

Regional and other names : *Bengali* – Bhang; *English* - True hemp; *Gujarati* - Bhang; *Hindi* - Bhang; *Kannada* – Bangi; *Malayalam* - Kanchavu; *Marathi* – Bhang-Bhanga; *Tamil* - Bhangi; *Telugu* –Ganjayee.

Distribution - This plant is not naturally growing in the area, but the plant products/organs are available in the market.

Taxonomic description - Dioecious, aromatic, woody herbs with 3 to 9- foliolate leaves. Flowers white or cream, males in panicels, females axillary, solitary, forming spikes.

Flowering & fruiting - January- April.

Field no. - DJ-118.

Parts used - Leaves.

Chemical constituents - Plant contains Cannabinoids, Cannabispirans, Alkaloids, Cannabig lendol, Cannabidiolic acid, Resin (Cannabinone), Gum, Sugars, Calcium, Phosphate and Volatile oil.

Biological activity - Analgesic, antispasmodic, aphrodisiac astringent, digestive, diuretic, narcotic, stiumulant, stomachic.

Folk uses

Medicine:

- One gram ash of 'Ganja' is mixed with 5 gram honey and this mixture is taken orally to cure asthma.
- The extract of leaves is taken with water to cure loose motion.

Miscellaneous: The plant has narcotic properties. Ganja, which is obtained from stigma of female flowers, is smoked with tobacco in the form of Chilam.

Uses in literature

Leaves - Piles (Lal & Yadav, 1983; Mishra & Billore, 1983; Shah & Joshi, 1971; Singh, 1988); Narcotic (Jain & De, 1966); Anthelmintic (Shah & Joshi, 1971); Bronchitis (Lal & Yadav, 1983); Dyspepsia, gonorrhoea (Bhalla *et al.,* 1982); Ear complaint, cuts (Shah & Joshi, 1971); Eye diseases (Shah, 1982); Skin diseases (Rao, 1981; Nautiyal, 1981); Pimples, boils, cuts and wounds (Jubilee *et al.,* 2006); Diarrhoea, dysentery (Kataria & Gauchan, 2006).

Fruits - Laxative (Rajwar, 1983).

Seed- Edible (Koelz, 1979).

Capparis zeylanica L. (Capparaceae)

Syn.: *C. horrida* L. f.

Vernacular name: Kantar, Kanthario.

Regional and other names- *Bengali* –Kalokera; *English* –Ceylon caper; *Hindi* –Ardanda; *Kannada* –Anatundikayee; *Marathi* –Govindi; *Tamil* -Kattotti; *Telugu* –Doddi.

Distribution - Throughout India.

Taxonomic description - Climbing shrubs; leaves ovate or oblong, entire; stipular spines hooked; petals densely hairy inside; berries subglobose, reddish-brown, with woody pericarp; seeds subglobose.

Flowering & fruiting - November-April.

Habitat ecology- Common among bushes and shrubs in wastelands; throughout the area.

Regeneration - By seeds.

Field no. - DJ-162.

Parts used : Seeds, Root.

Chemical constituents : An alkaloid, a phytosterol, a mucilaginous substance and water soluble acid- L-stachydrine, Rutin and Beta-sitosterol.

Seeds/Leaves contain Thioglucosie glucocapparin, n- tricontane, a- amyrine and Fixed oil.

Biological activity

Demulcent, sedative, stomachic, appetiser.

Folk uses

Medicine:

- Powder of seeds is taken with water to cure stomach trouble.
- Extract of root is mixed with gum powder of *Butea monosperma* and taken with water to cure dysentery and diarrhoea.
- Extract of root is taken with water in the case of bone fracture.

Uses in literature

Root bark- Anthelmintic (Sharma *et al.,* 1979); Boils (Maheshwari *et al.,* 1980; 1981); cholera (Sharma *et al.,* 1979).

Leaf- Boils (Sebastian & Bhandari, 1984); Piles (Sebastian & Bhandari, 1984; Sharma *et al.,* 1979); Swelling (Sharma *et al.,* 1979).

Fruit- Snake bite (Singh *et al.,* 1987); Pickle (Shah, 1984); As vegetable (Shrivastava & Varma, 1981; Vartak, 1959).

Capsicum annuum L. (Solanaceae)

Vernacular name : Mirach, Marach, Mirchi.

Regional and other names : *Bengali* – Morich; *English* – Chillies; *Gujarati* – Marcha; *Hindi* – Lalmirca; *Kannada* – Kempumenasu; *Malayalam* – Mulaku; *Marathi* – Mirchi; *Sanskrit* – Katuvirah; *Tamil* – Milagai; *Telugu* – Mivapakaya.

Distribution - Largely cultivated throughout India.

Taxonomic description - Perennial herbs, with lanceolate leaves, white flowers and slender, green, red or orange fruits.

Flowering & fruiting - Most part of the year.

Habitat ecology - Cultivated throughout the State for edible fruits.

Regeneration - By seeds.

Field no. - DJ-140.

Parts used - Fruits.

Chemical constituents - Fruits - A volatile alkaloid-Capsicin; a crystalline substance-Capsacin; a volatile oil-Solanine; Fatty acid, Resin.

Biological activity - *Fruits* - Acrid, antiperiodic, bitter, carminative, digestive, expectorant, laxative, rubefacient, stimulant, sudorific, thermogenic.

Folk uses

Magico-religious beliefs: Dried red 7 chillies are waved around the head of a person suffering from fever or other disease and placed at the crossing of roads with the belief that the patient will be cured and effect of evil spirits will be warded off.

Medicine: Chilli powder is smeared on wound/cuts for early healing.

Uses in literature

Fruits - Gout (Jadeja *et al.,* 2006); Condiment (Nagendra Prasad & Abraham, 1984).

Carica papaya L. (Caricaceae)

Vernacular name : Papito, Aren-kakari.

Regional and other names : *Bengali* – Papeya; *English* – Papaya; *Gujarati* –Chibda ; *Hindi* – Pappita; *Kannada* – Pappaya; *Malayalam* – Pappaya; *Marathi* – Papaya; *Sanskrit* – Erandakarkati; *Tamil* – Pappali; *Telugu* – Bobbasi.

Distribution -Widely cultivated throughout India.

Taxonomic description - A rapidly growing tree, upto 7 mm tall, with milky sap. Leaves palmetely lobed and flowers creamy yellow; female flowers in short clusters and males in long drooping panicles.

Flowering & fruiting - Greater part of the year.

Habitat ecology - Commonly cultivated for its edible fruits.

Regeneration - By seeds.

Field no. - DJ-119.

Parts used - Fruits, Roots, Leaves, Latex.

Chemical constituents

Fruits pulp contains Sucrose, invert Sugar, Papain, Malic acid, salts of Tartaric acid & Citric acid, Pectins, Carotenoid, Vitamins.

Latex contains Enzymes Papain & Chymopapain.

Biological activity

Fruits - Acrid, anodyne, anthelmintic, anti-inflammatory, antifungal, aphrodisiac, appetizer, bitter, carminative, demulcent, digestive, stomachic, thermogenic.

Latex - Anodyne, anthelmintic, febrifuge, sudorific.

Folk uses

Food: Unripe fruits are used as vegetable. Mature fruits are eaten by the tribals.

Medicine:

- Unripe fruit pulp is eaten to cure tumor in stomach.It is also useful in obesity.
- The powder of root is taken with water in the cases of kidney stone problem.
- The epicarp of unripe fruit is eaten in the cases of prolepsy of uterus and hydrocoel.
- 2-3 flowers chewed daily to cure jaundice.
- Latex is filled in teeth cavity with the help of cotton plug to cure toothache and swollen gums.
- Extract of leaves is taken with water (Mishri may be added) to cure menorrhoea.

Uses in literature

Leaf -Constipation (Megoneitso & Rao, 1983); Toothache (Das *et al.*, 2006).

Fruit - Gastric trouble (Rao & Jamir, 1982); Abortifacient (Tarafder, 1983); Gout (Jadeja *et al.*, 2006); Hepatitis (Dutta *et al.*, 2006); Dysentery (Das *et al.*, 2006); Constipation (Deokule, 2006); Edible (Shah, 1981); Pickle, vegetable (Shah, 1984).

Latex - Abortifacient (Ramachandran & Nair, 1981); Ringworm (Shah *et al.,* 1981; Maheshwari & Singh, 1984); Skin diseases (Megoneitso & Rao, 1983); Toothache (Singh *et al.,* 1985; Maheswhari & Singh, 1984).

Root - Abortifacient (Bhatacharjee *et al.,* 1980; Tiwari *et al.,* 1982).

Flower - Earache (Das *et al.,* 2006).

Carissa carandas L. (Apocynaceae)

Syn. : ***C. congesta*** Wt.

Vernacular name : Karamda, Karaunda, Karonda.

Regional and other names- *Bengali* –Karamcha; *English* –Christ's thorn; *Gujarati* -Karundo ; *Hindi* -Karaunda; *Kannada* –Karekai; *Malayalam* -Kala; *Marathi* –Karavanda; *Sanskrit* -Avighnah; *Tamil* - Karanta; *Telugu* –Kalivi.

Distribution - Throughout India.

Taxonomic description - Erect or scandent shrubs, with paired spines at each node. Leaves elliptic-oblong to ovate. Flowers white, in corymbose cymes. Berries elliptic-oblong, fleshy.

Flowering & fruiting - April-June.

Habitat ecology - Usually met with in the forests and wastelands.

Regeneration – By seeds.

Field no. - DJ-163.

Parts used : Fruits.

Chemical constituents

Root contains Beta-sitosterol, a Sesquiterpene as Carissone; a Lignancarinol, Odoroside-H, Digitoxigenin, Glucose, D-digitalose, Lupeol, Ursolic acid, Methyl ursolate.

Fruit contains Triterpene alcohol as Carissol, Glucose, Galactose, Serine, Glutamine, Alanine, Valine, Glycine, Carissic acid, Phenylalanine.

Biological activity

Leaves- Febrifuge.

Root- Anthelmintic, antiscorbutic, stomachic.

Unripe fruit- Aphrodisiac, appetiser, antipyretic, astringent, bitter, sour, constipating, thermogenic.

Ripe fruits- Sweet, cooling, appetiser, antiscorbutic.

Folk uses

Food: Unripe fruits are used as vegetable or pickled. Mature fruits are eaten by the tribals.

Miscellaneous: Fruits are sold in the market by the tribals.

Uses in literature

Leaf- Fever (Nautiyal, 1981).

Seed- Itching (Gupta, 1981).

Root- Wounds (Nautiyal, 1981).

Root bark- Veterinary diseases (Maheshwari *et al.*, 1986).

Cassia fistula L. (Caesalpiniaceae)

Vernacular name : Garmalo.

Regional and other names : *Bengali* - Bandarlate, Amultas; *English* - Indian laburnum; *Gujarati* - Garmalo; *Hindi* - Amaltas, Bandarlauri; *Kannada* – Kakke; *Malayalam* - Konna, saturangulam; *Marathi* - Bahava, bawa, boya; *Sanskrit* - Rajvriksha, dirdhaphaala, Chaturangula; *Tamil* - Konnai; *Telugu* – Kondrakaya.

Distribution - Throughout India.

Taxonomic description - Deciduous tree; branchlets glabrous; leaves eglandular; leaflets 4-8 pairs,ovate, silvery, pubescent beneath when young; flowers in lax racemes; pods long, cylindrical, nearly straight, smooth, shining, brown-black.

Flowering & fruiting - March – May

Habitat ecology - Found in mixed deciduous forest.

Regeneration - By seeds.

Field no. - DJ-70.

Parts used - Leaves, Flowers, Fruits, Seeds, stem bark.

Chemical constituents - Fruit contains Glycosides, Carbohydrates, Resin, Tannins, Magnessium, Calcium, Potassium, Iron & other compounds are Fistucacidin, Beta-sitosterol, n-tricontanol, Leucopelav-gonidin, Flavonoids, Glucosides, Gum, and Volatile oil.

Biological activity

Root – Astringent, cooling, febrifuge, purgative, tonic.

Bark - Anthelmintic, diuretic, emetic, febrifuge, laxative.

Leaves - Antiperiodic, depurative, laxative.

Flower - Acrid, bitter, cooling, demulcent, emollient, expectorant.

Fruit - Abortifacient, anti-inflammatory, antipyretic, cooling, diuretic, emollient, ophthalmic, purgative, sweet.

Folk uses

Fodder: Leaves serve a good fodder for cattle.

Food: Flowers are used as vegetable.

Magico-religious beliefs: Bhils protect their sleeping children from getting frightened by keeping the dry fruits under their pillows.

Medicine:

- Extract of seed is taken with water by children to cure pneumonia.
- Fruit extract is applied in the cases of scorpion sting.
- Extract of fruit is given by the tribals to their cattle in Afra disease.

Miscellaneous: Fruits are sold in the market by the tribals.

Tannin: The tribals use the bark as a tanning material.

Uses in literature

Root - Leprosy (Barua & Sarma, 1984).

Bark - Blindness (Joshi *et al.,* 1980); Dog bite (Tarafdar, 1987); Scorpion bite (Roy *et al.,* 1986).

Flowers - Antiseptic (Rajwar, 1983); Blood purifier (Srivastava & Verma, 1981); Rheumatism (Bhalla *et al.,* 1982); Ringworm (Bhalla *et al.,* 1982); As vegetable (Sharma & Malhotra, 1984; Vartak, 1981; Negi *et al,* 1985; Nagar, 1985; Maheshwari *et al,* 1986).

Fruits - Asthma (Tarafdar, 1987); Constipation (Singh & Maheshwari, 1983); Cough (Saxena & Vyas, 1981); Diarrhoea (Singh & Maheshwari, 1983; Kataria & Gauchan, 2006),; Dysentery (Maheshwari & Singh, 1984; Deokule, 2006; Kataria & Gauchan, 2006); Pimples (Bhalla *et al.,* 1982); Snake bite (Singh & Pandey, 1980); Stomachic (Saxena & Vyas, 1981); Toothache (Lal & Yadav, 1983); Diabetes (Ghosh, 2006); Edible (Ball, 1967).

Seeds - Jaundice (Maheshwari *et al*, 1981).

Leaf - Leprosy (Purkayastha & Subhan, 2006); Skin diseases (Mokat & Deokule, 2006; Kumar *et al.*, 2006).

Cayratia trifolia (L.) Domin (Vitaceae)

Syn.: *Vitis trifolia* L.

Vernacular name : Choti gurvel.

Regional and other names : *Bengali* - Amallata; *English* - Fox – grape; *Gujarati* - Khat; *Hindi* - Amalbel; *Kannada* – Heggoli; *Malayalam* – Amarcakkoti; *Marathi* – Amlatbel; *Sanskrit* -Gandirah; *Tamil* - Kattuppirandai; *Telugu* – Kurudinne.

Distribution - Throughout India.

Taxonomic description - Herbaceous climbers, with 3- foliolate leaves and greenish - white flowers in axillary or pseudo-terminal cymes. Berries obovate- globose, dark purple to black when ripe.

Flowering & fruiting - July – October.

Habitat ecology - Found in the damper jungles near water courses.

Regeneration - By seeds.

Field no. - DJ-146.

Parts used - Root.

Biological activity - Astringent, diuretic, sour.

Folk uses

Medicine:

- Extract of root is taken in the cases of snake bite and dog bite.
- Paste of root is applied on breast to cure tumor.
- Extract of root is taken to cure asthma.
- The powder of cloves of *Syzygium aromaticum* and the extract of root of *C. trifolia* are mixed with jaggery. This mixture is taken for 4 days early in the morning to cure T. B.
- Extract of root is given with water to cattle in the case of dog bite.

Uses in literature

Leaf - Boils (Singh, 1945); As vegetable (Vartak, 1981).

Root - Snake bite (Singh, 1945).

Whole plant - Piles (Kumar & Chauhan, 2005).

Fruit- Edible (Shah *et al.*, 1981).

Celosia argentea L. (Amaranthaceae)

Syn.: *C. cristata* L.

Vernacular name : Garko, Lambdi.

Regional and other names : *Bengali* – Swetmurge; *English* - Quail grass; *Gujarati* - Lambdi; *Hindi* - Safed murga; *Kannada* - Anne soppu; *Marathi* – Kurda; *Sanskrit* - Vitunna; *Tamil* -Pannakeerai; *Telugu* – Panchechettu

Distribution - Throughout India.

Taxonomic description - Erect annuals. Leaves linear-lanceolate or ovate-elliptic. Flowers white or pink. Seeds lenticular, black, shining.

Flowering & fruiting - October – March.

Habitat ecology - Common weed of the cultivated and fallow areas at the close of rains.

Regeneration - By Seeds.

Field no. - DJ-166.

Parts used - Leaves, Seeds.

Chemical constituents - Seed contain Saponin of Hyaluronic acid.

Biological activity - Aphrodisiac, astringent, cooling, nutritive, spasmolytic.

Folk uses

Food: Leaves are used as vegetable.

Medicine: Powder of seeds is mixed with jaggery and some pills are prepared. One pill is taken twice a day with water to cure asthma.

Uses in literature

Seeds - Diarrhoea (Shah *et al.,* 1983); Blood purifiers (Singh & Pandey, 1980); Sores in mouth (Sharma *et al.,* 1979).

Root- Kidney stone (Deokule, 2006).

Flower- Diarrhoea, dysentery (Kataria & Gauchan, 2006).

Leaf- As vegetable (Maheshwari & Singh, 1984; Singh & Singh, 1981; Ramachandran & Nair, 1981; Vartak, 1959; Nagar, 1985; Datar & Vartak, 1975).

Chenopodium album L. (Chenopodiaceae)

Vernacular name : Chil.

Regional and other names : *Bengali* - Batho shak; *English* - White goosefoot; *Gujarati* - Cheel; *Hindi* - Bathua sag; *Kannada* – Sakothinasoppu; *Malayalam* – Vastuccira; *Marathi* - Chandan bathua; *Sanskrit* - Vastukah; *Tamil* - Parappukeerai; *Telugu* – Pappukoora.

Distribution - Occuring chiefly in cultivated ground.

Taxonomic description - Herbs, with oblong-rhomboid, somewhate fleshy leaves. Flowers pale-green, born in clusters in paniculate, ebracteate spikes; perianth-segments without any swelling on the back. Seeds rugose.

Flowering & fruiting - November– February.

Habitat ecology - A weed of roadsides, wastelands and cultivated ground.

Regeneration - By seeds.

Field no. - DJ-157.

Parts used - Leaves.

Chemical constituents - Plant contains a Volatile oil which contains Ascaridol, Saponin, Vitamin-C, Magnessium, Phosphate, Carotene.

Biological activity - Acrid, anthelmintic, aphrodisiac, carminative, digestive, diuretic, laxative, sweet, tonic.

Folk uses

Food: Leaves are used as vegetable.

Uses in literature

Seeds - Skin diseases (Singh & Maheshwari, 1983).

Leaf - Urin complaint (Shah *et al.,* 1983); Anthelmintic (Deokule, 2006); Anaemia (Kumar *et al.,* 2006); As vegetable (Maheshwari & Singh, 1984; Vartak, 1981; Singh & Singh, 1981; Bhargava, 1959; Maheshwari *et al*, 1981, 1986; Chaudheri *et al.*, 1985).

Leafy twig - General debility & cardiac disorders (Chhetri, 2005).

Root - Jaundice (Shah, 2005); Anaemia (Kumar *et al.,* 2006).

Chlorophytum tuberosum Bak. (Liliaceae)

Syn.: *Anthericum tuberosum* Roxb.

Vernacular name : Dhori musri; Nari kuri.

Regional and other names : *English* - White musale; *Gujarati* - Dhori musli; *Hindi* - Safed mushli; *Sanskrit* –Mushli.

Distribution - E. Himalaya, Assam, Bihar.

Taxonomic description - Annual herbs, with many cylindric root – fibres ending in ellipsoid tubers. Leaves linear – lanceolate. Flowers white, in dense racemes. Capsules obovoid transversely veined.

Flowering & fruiting - July – August.

Habitat ecology - Rare, found on the hills.

Regeneration - Tuber.

Field no. - DJ-173.

Parts used - Tuber.

Chemical constituents - Root contains Steroids, Resin, Phenolics, Tannins, Carbohydrates, Calcium, Magnessium and Potassium.

Biological activity - *Root and tuber* - Aphrodisiac, cooling, diuretic, tonic.

Folk uses

Food: Leaves and tubers are used as vegetable.

Medicine:

- Dry root slices are mixed with jaggery and powder of dry rhizome of zinger (*Zingiber officinale* Rosc.). This mixture is warmed a bit with ghee and small balls (Laddoo) are prepared. One ball is taken every day after pregnancy for strength and vigour.
- Extract of root is taken with cow milk to cure leucorrhoea.

Uses in literature

Tuber - Fever (Roy & Chaturvedi, 1987); Leucorrhoea (Saxena & Vyas, 1983); Sex tonic (Roy & Chaturvedi, 1986); As vegetable (Bedi, 1978; Nagar, 1985).

Leaf- As vegetable (Vartak, 1959, 1981; Datar & Vartak, 1975).

Chrozophora rottleri (Geis.) A. Juss. ex Spreng. (Euphorbiaceae)

Syn.: *Croton rottleri* Geis.

Vernacular name : Jungli rukhdi.

Regional and other names- *English* -Turnsole; *Gujarati* - Kalookharad; *Hindi* –Sonballi; *Marathi* –Suravarta; *Sanskrit* – Suryavarta.

Distribution - Throughout India.

Taxonomic description - Stellately hairy herbs, branched in the upper region. Leaves entire or 3-lobed, ovate to suborbicular. Sepals of female flowers triangular; stigma red. Capsules purple.

Flowering & fruiting - August- may.

Habitat ecology - Common in wastelands.

Regeneration - By seeds.

Field no. - DJ-174.

Parts used- Leaves.

Biological activity : The plant has emetic, drastic, corrosive properties.

Folk uses

Medicine: Extract of leaves is taken with water in the case of dog bite.

Uses in literature

Root- Cough (Singh & Pandey, 1980).

Whole plant- Fish poison (Joshi, 1982; 1986).

Cicer arietinum L. (Papilionaceae)

Vernacular name : Channo, Chana, Chor.

Regional and other names : *Bengali* – Chola; *English* – Bengal gram; *Gujarati* –Chana ; *Hindi* – Cana; *Kannada* – Kadale; *Malayalam* – Katala; *Marathi* – Harbara; *Sanskrit* – Canakah; *Tamil* – Katalai; *Telugu* – Canakamula.

Distribution - Largely cultivated throughout India.

Taxonomic description - Much branched herbs, with pinnate leaves and turgid-oblong fruits.

Flowering & fruiting - December-March.

Habitat ecology - Commonly cultivated as a winter season crop in the area.

Regeneration - By seeds.

Field no. - DJ-16.

Parts used - Leaves, seeds, Husk.

Chemical constituents - Leaves contain Vanillic, P-hydroxybenzoic acid; Oxalic acid, Malic acid, Arginine, Tryosine, Lycine, Cystine, Tryptophane; Carotenoids & Vitamins.

Biological activity

Leaves - Abortifacient, purgative.

Seed - Aphrodisiac, cooling, expectorant.

Folk uses

Fodder: The husk (Chane ka kharia) serve a good fodder.

Food: Leaves are used for making chutney.

Medicine: Roasted seeds are taken with jaggery to cure jaundice.

Uses in literature

Seed - Jaundice (Lal & Yadav, 1983); To make silky hair (External application) (Jadeja *et al.,* 2006); Edible (Shah, 1984).

Cissus quadrangularis L. (Vitaceae)

Vernacular name : Haadjood, Arhand.

Regional and other names : *Bengali* - Harjora, harbhanga; *English* - Large granadilla, Bone – setter; *Gujarati* - Chodhari; *Hindi* - Hadjor; *Kannada* – Mangarol; *Malayalam* –Piranta; *Marathi* – Kandawela; *Sanskrit* - Asthisrnkhala; *Tamil* - Pirandai; *Telugu* – Nalleru

Distribution - Throughout India.

Taxonomic description - A large perennial climber. Stems fleshy, cactus – like jointed with quadrangular 4 – winged internodes and a tendril at some of the nodes; leaf – ovate or reniform and are opposite. Flowers greenish – white, in shortly peduncled umbellate cymes. Fruits berry, red when ripe.

Flowering & fruiting - July -August.

Habitat ecology - An extensive climber on hedges or trees.

Regeneration - By cuttings.

Field no. - DJ-52.

Parts used - Stem.

Chemical constituents - Plant contains Carotene, Calcium Oxalate, Vitamin- C, Alpha- amyrone and ß- sitostosterol.

Biological activity - *Whole plant* - Alterant, anodyne, anthelmintic, aphrodisiac, bitter, carminative, depurative, digestive, haemostatic, laxative, ophthalmic, sour, stomachic, sweet, thermogenic, union - promoting.

Folk uses

Medicine:

- Extract of a half of internode of the stem is given in the case of bone fractures in human being.
- A internode of the stem is given by the tribals to their cattle with Roti (Wheat bread) to cure fractures.

Uses in literature

Stem - Bone fracture (Shah *et al.,* 1983; Deokule, 2006; Sarma *et al.,* 2006; Basak, 2006); Jiont pain & Swelling (Shah *et al.,* 1983); Pickled (Shah *et al.,* 1981) .

Root - Dysentery & indigestion (Shah *et al.,* 1981); Wound & bone fracture (Gupta, 1981).

Whole plant - Bone fracture (Singh & Maheshwari, 1983; Nath *et al.,* 2006).

Fruit- Edible (Shah *et al.,* 1981).

Citrus limon (L.) Burm. f. (Rutaceae)

Syn.: *C. medica* L. var. *limon* L.

Vernacular name : Limbu.

Regional and other names : *Bengali* – Baranebu; *English* – Lime; *Hindi* – Jamiri nimbu; *Kannada* – Limbe; *Malayalam* – Cerunarakam; *Marathi* – Limbu; *Sanskrit* – Jambirah; *Tamil* – Elumiccai; *Telugu* – Nimma.

Distribution - Cultivated throughout India for its fruits.

Taxonomic description - Thorny trees, with ovate-elliptic leaves, white flowers tinged with pink-purple & oval or oblong fruits.

Flowering & fruiting - During winters.

Habitat ecology - Occasionally cultivated for edible fruits.

Regeneration - By seeds and by cuttings.

Field no. - DJ-18.

Parts used - Fruits, Leaves.

Chemical constituents

Oil from peel contains d-limonene, Camphene, Linalool.

Peel contains Bitter principle, Essential oil, Hesperidin, Citronellal, Bergamotine, Eriocitrin, Apigenin, Luteolin, Chrysoseriol, Quercetin, Isorhamnetin, Limocitin, Limocitriol.

Seeds contain Bitter principles, Obacunone, Limonin.

Roots contain Xanthyletin, Stigmasterol, ß-sitosterol.

Biological activity - *Fruits* - Anthelmintic, antiseptic. carminative, digestive, laxative, stimulant, sour, stomachic, thermogenic.

Folk uses

Medico-religious beliefs: 2 to 7 fruits are waved around the head of a person suffering from fever or other disease; than these fruits are split into 2 pieces; after adding sacred red powder 'Kumkum', all the pieces are placed in the sevage water with the belief that the patient will be cured and effect of evil spirits will be warded off.

Medicine:

- Leaves are warmed a bit with ghee and tied on eyes to cure swelling.
- Juice of fruits (Sugar may be added) is taken in the cases of burning in urethra during micturition.
- The fruit-juice is taken orally with salt/sugar by the tribals for relief in body-heat & diarrhoea.

Uses in literature

Fruit - Gout, dandruff (Jadeja *et al.*, 2006); Malaria (Deka *et al.*, 2006).

Cleome gynandra L. (Cleomaceae)

Vernacular name : Halul.

Regional and other names : *Bengali* –Hurhuria; *English*-Spider flower; *Gujarati*-Adiyakharan; *Hindi*-Hulhul; *Kannada* -Narun byalesoppu; *Malayalam* –Kathukatuku; *Marathi* –Satitalvani; *Sanskrit*-Suvarchala; *Tamil*-Naikadugu, Kattkadugu; *Telugu* -Vaminta, velakura.

Distribution - Throughout India.

Taxonomic description - Glandular hairy herbs, with 3 to 5-foliolate leaves. Flowers white, in terminal racemes. Capsules linear-cylindric.

Flowering & fruiting - June – November

Habitat ecology - Common weed in fields and wastelands during rainy season.

Regeneration - By seeds.

Field no. - DJ-120.

Parts used - Seeds.

Chemical constituents

Plant contains Essential oil.

Seed contains Cleomin.

Biological activity - *Root* – Hot, sharp taste, stomachic.

Folk uses

Medicine: Seeds are taken with water to promote conception.

Uses in literature

Leaf - Rheumatism, headache, ear complaint (Bhalla *et al.,* 1982); Boils (Shah, 1984); Headache, rheumatism (Kiruba *et al.,* 2006); Earache (Ganesan *et al.,* 2006).

Seed - Cough (Shekhawat & Anand, 1984); Headache (Baruah & Sarma, 1984); Skin diseases (Bedi, 1978).

Whole plant - Scorpion bites (Nagendra Pradsad & Abraham, 1984); Fever (Deokule, 2006).

Root - Fever (Kiruba *et al.,* 2006).

Cleome viscosa L. (Cleomaceae)

Vernacular name : Marchio, Marsio.

Regional and other names : *Bengali* – Hurhuria; *English* - Wild mustard; *Gujarati* - Talvani; *Hindi* - Hurhur; *Kannada* – Nayibelu; *Malayalam* – Atunarivela; *Marathi* – Kanphodi; *Sanskrit* -Ajagandha; *Tamil* - Naivelai; *Telugu* – Kukkavaminta.

Distribution - Throughout India.

Taxonomic description - Glandular haIry herbs, with 3 to 5-foliolate leaves. Flowers yellow, in leafy racemes; stamens more than 10. Capsules cylindric, beaked.

Flowering & fruiting - June – September

Habitat ecology - Common weed in fields and wastelands during rainy season.

Regeneration - By seeds.

Field no. - DJ-150.

Parts used - Seeds.

Chemical constituents - Seed contain Viscosic acid & Visosin.

Biological activity

Plant - Acrid, anthelmintic, sudorific, thermogenic.

Root - Antiscorbutic, stimulant, vermifuge.

Leaves - Digestive.

Seed - Anthelmintic, cardiac stimulant, carminative, constipating, febrifuge.

Folk uses

Medicine:

- Extract of root is taken with water to cure typhoid and fever.
- Juice of leaves is dropped in ear to cure pus formation.

Uses in literature

Leaf - Boils (Maheshwari *et al.,* 1980); Earache (Shah, 1984; Deokule, 2006); Headache (Ramchandran & Nair, 1981); Ulcer (Rajwar, 1983); Wound (Bedi, 1978); Boils (Kiruba *et al.,* 2006); Earache (Kiruba *et al.,* 2006; Kadavul & Jisha, 2006).

Seed - Diarrhoea & fever (Malhotra & Moorthy, 1973; Sharma *et al.*, 2006); Headache (Narain & Singh, 2006b); Boils & skin diseases (Purohit *et al.*, 1985); Intestinal worms (Jadeja *et al.*, 2006); Cardiac disorders, dyspepsia (Sharma *et al.*, 2006); Earache (Basak, 2006).

Fruit – Headache (Sarma *et al.*, 2006).

Stem- As vegetable (Vartak, 1981).

Clerodendrum multiflorum (Burm. f.) O. Ktze. (Verbinaceae)

Syn.: *Volkameria multiflorum* Burm. f.

Vernacular name : Arni.

Regional and other names : *Gujarati* - Arni; *Hindi* - Arni; *Kannada* - Taggi; *Malayalam* – Tirutali; *Marathi* – Arani; *Sanskrit* - Agnimantha; *Tamil* - Takkari; *Telugu* -Takkillaku.

Distribution - Throughout India.

Taxonomic description - Erect or scandent shrubs, with solid stem and opposite, rhomboid – ovate leaves. Flowers white, in panicles. Fruits obovoid.

Flowering & fruiting - October – February.

Habitat ecology - Common in wastelands.

Regeneration - By seeds and by stem cuttings.

Field no. - DJ-19.

Parts used - Flowers.

Chemical constituents - Plant contains Pectolinargenin, Scuttelarein, Apigenin, Clerosterol, Hispudilin, Luteolin, Clerodendrin, Cerolic acid, Cerylalcohol, Raffinose, Cholesterol, β – Sitosterol.

Biological activity - *Root* - Alexipharmic, bitter, heating, laxative, pungent, stomachic.

Folk uses

Medicine: Extract of flowers is taken with water for the prevention of miscarriages.

Uses in literature

Leaf - Headache (Goel *et al.*, 1984); Gout (Jadeja *et al.*, 2006); Veterinary medicine (Pal, 1980).

Root- Fever, cough, cold (Sharma *et al.*, 2006).

Clitoria ternatea L. (Papilionaceae)

Vernacular name : Visnukanta.

Regional and other names : *Bengali* – Aparajit; *English* - Butterfly pea; *Gujarati* - Garani; *Hindi* -Aparajita; *Malayalam* – Sankupushpam; *Marathi* – Gokurna; *Sanskrit* -Aparajita; *Tamil* - Kakkattan; *Telugu* – Sankupuspham.

Distribution - Cosmopolitan in the tropics.

Taxonomic description - Twinning herbs, with imparipinnate, 5 to 7-foliolate leaves and blue or white flowers. Pods spreading.

Flowering & fruiting - September – December.

Habitat ecology - Common in hedges.

Regeneration – By seeds.

Field no. - DJ-107.

Parts used - By Seeds.

Chemical constituents

Root bark contains Starch, Tannin, Resin.

Seeds contain Oil, bitter blue coloured Resin.

Biological activity

Root - Alexeteric, anthelmintic, aphrodisiac, bitter, depurative, diuretic, intellect promoting, laxative, ophthalmic, refrigerant, tonic.

Seed - Cathartic.

Folk uses

Magico-religious beliefs: The tribals use the seeds as an amulet (Tavij) which is tied around the neck of the children to protect them from evil spirits.

Medicine:

- Powder of seeds is taken with water to cure typhoid and fever.
- Paste of seeds is applied against snake bite.

Uses in literature

Root - Abortifacient (Ramachandranan & Nair, 1981); Goiter (Saxena & Vyas, 1983); Purgative (Malhotra & Moothy, 1973); Stomachache (Lakshamanan & Sankaranarayanan, 1988); Toothache (Subhangi & Patil, 2006); Impotency (Das *et al.*, 2006); Veterinary medicine (Ramachandranan & Nair, 1981).

Seed - Laxative (Shah *et al.,* 1983); Dropsy (Subhangi & Patil, 2006); stomach trouble (Das *et al.,* 2006).

Leaf - Swelling (Nagendra Prasad & Abraham, 1984); Dog bite (Subhangi & Patil, 2006).

Whole plant- Snake bite (Deokule, 2006).

Coccinia grandis (L.) Voigt (Cucurbitaceae)

Syn.: *Bryonia grandis* L.; *Coccinia. indica* W. & A.; *Cephalandra indica* (Wt. & Arn.) Naud.

Vernacular name : Tindori.

Regional and other names : *Bengali* – Telekucha; *English* - Little gourd, Ivy gourd; *Gujarati* - Galludda; *Hindi* - Kanduri; *Kannada* – Tondekayi; *Malayalam* – Kovakai; *Marathi* –Toondale; *Sanskrit* - Bimbi; *Tamil* - Kovakai; *Telugu* – Dondakaya.

Distribution - Throughout India.

Taxonomic description - Slender, dioecious climbers; nodes on the mature stem annulate; leaves entire or digitately 3 to-5 lobed, cordate; flowers white, usually solitary, axillary; fruits oblong-obtuse, green.

Flowering & fruiting - November- March.

Habitat ecology - Often noticed spreading or twining on trees and shrubs.

Regeneration - By seeds.

Field no. - DJ-179.

Parts used - Fruits.

Chemical constituents

Root contain β- Amyrin, Lupeol, β- sitosterol.

Fruits contain Beta-amyrin, Lupeol, Cucerbitacin, β- sitosterol, Taraxerol, Lycopene, β- carotene, Crypto xanthin.

Seeds contain a fat rich in Palmitic, Oleic, Linoleic acids.

Biological activity

Root - Aphrodisiac, cooling.

Leaf - Astringent, bitter, cooling, sweet.

Fruit - Antipyretic, astringent cooling, depurative, expectorant, sweet.

Folk uses

Food : Fruits are used as vegetable.

Medicine: Leaves chewed daily 2 times are very useful for diabetic patients.

Uses in literature

Root - Vomiting & headache (Kumar *et al.,* 1980).

Leaf - Diabetes (Bedi, 1978; Deokule, 2006); Earache (Pereira, 2006); Eye diseases (Rajendran & Rengamani, 2006).

Fruits - Throat affection (Sharma & Malhotra, 1984); Diabetes (Singh & Pandey, 1980); Cold & cough (Shah & Gopal, 1986); Edible (Maheshwari *et al.,* 1981, 1986; Chaudhari *et al.,* 1985); As vegetable (Shah, 1984; Maji & Sikdar, 1982; Singh & Singh, 1981).

Cocculus hirsutus (L.) Theob. (Menispermaceae)

Syn.: *Menispermum hirsutum* L.

Vernacular name : Dechri, Kankarvel, Kaliveldi.

Regional and other names : *Bengali* – Huyer; *English* - Broom creeper; *Gujarati* - Vevdi; *Hindi* - Jamitikibel, Jaljamni, Patalagarudi; *Kannada* - Dusari balli; *Malayalam* – Patalagarudakkoti; *Marathi* - Vasan vel; *Sanskrit* - Patalagarudah; *Tamil* - Kattukkoti; *Telugu* – Dusaratige.

Distribution - Tropical and sub-tropical India.

Taxonomic description - Scandent, dioecious shrubs, with cordate – lanceolate or subdeltoid leaves. Male flowers minute, yellowish – green, in panicles. Fruits reddish - purple, smooth. Seeds transversely rugose.

Flowering & fruiting - November – April.

Habitat ecology - Common throughout the area on the fringes of forests.

Regeneration - By seeds.

Field no. - DJ-121.

Parts used - Root, Leaves.

Chemical constituents

Leaves contain Trilobine & Isotrilobine.

Root contains D – Trilobine, DL – Coclaurine.

Stem contains β – Sitosterol, Ginnol & Mono methyl ether of Inositol.

Biological activity - *Root* - Acrid, alexeteric, alterant, anodyne, antipyretic, aphrodisiac, bitter, carminative, demulcent, depurative, digestive, diuretic, emollient, expectorant, laxative, tonic.

Folk uses

Medicine:

- Root extract is useful against snake-bite .
- Paste of fresh leaves is mixed with jaggery and about 7 small pills are prepared. One pill is taken every day early in the morning with curd for a period of 7 days to cure leucorrhoea.
- Root extract is taken with water to cure menorrhoea.
- Extract of leaves is taken with water to cure loose motion.
- Extact of root is mixed with jaggery and a syrup is prepared. It is taken in the case of stomach tumor.

Uses in literature

Leaf - Fever (Shah, 1984); Leucorrhoea (Chaudhari *et al.,* 1975); Cuts, wound (Johi, 1982); Rheumatism (Shah *et al.,* 1983); Eczema (Sudhakar & Rolla, 1985; Basak, 2006); Conjunctivitis (Audichya *et al.,* 1983); Eye ailments (Pereira, 2006); Gonorrhoea (Shah, 1984; Deokule, 2006); Skin diseases (Shah *et al.,* 1983; Basak, 2006); As vegetable (Vartak, 1981; Jain, 1964).

Root - Fever & rheumatism (Shah, 1984); Gout (Goel *et al.,* 1984); Leprosy (Basak, 2006).

Stem - Stomach diseases (Shah *et al.,* 1983).

Cocos nucifera L. (Arecaceae)

Vernacular name : Naral, Nariyal, Khopro.

Regional and other names : *Bengali* – Narikel; *English* – Coconut tree; *Gujarati* – Nariel; *Hindi* – Nariyal; *Kannada* – Tengu; *Malayalam* – Tennu; *Marathi* – Narel; *Sanskrit* – Narikelah; *Tamil* – Tenkai; *Telugu* – Tenkaya.

Distribution - Much cultivated in South India.

Taxonomic description - This plant is not naturally growing in the area, but the plant products/organs are available in the market.

Field no. - DJ-147.

Parts used - Fruits.

Chemical constituents : Kernel - Mannan, Galactomannan, Enzymes, Amino-acids, Proteins.

Biological activity

Root - Anthlemintic, astringent, diuretic.

Juice of the young spadix - Aperient, aphrodisiac, diuretic, intoxicating, refrigerant, sweet.

Shell - Cooling, diuretic.

Kernel - Aphrodisiac, appetizer, cooling, laxative, sweet.

Water - Anthlemintic, aperient, aphrodisiac, cooling, digestive, diuretic, sweet.

Oil - Aphrodisiac, appetiser, digestive, sweet.

Folk uses

Fibres: The fibers obtained from mesocarp are used for making rope.

Magico-religious beliefs: The fruits are offered to the god and goddess.

Medicine:

- Latex obtained from the leaves of *Calotropis gigantea* is filled in the endosperm of ripe, dried fruit of *Cocos nucifera* through hole. The hole is plugged with the removed piece of the fruit and the fruit is roasted in the burning cowdung. After the fruit is roasted, it is made into a powder. A teaspoonful of powder with cow milk is taken in early morning to cure asthma.
- The small pieces of dried seeds of *Trapa natans* are chewed with small pieces of dried endosperm of *Cocos nucifera* daily 3 times to cure lucorrhoea.

Timber: The fruit shell is used for making a musical instrument sarangi.

Uses in literature

Coconut water - Blood purifier, fever, urinary disorders (Bhalla *et al.*, 1982).

Root - Bronchitis, dysentery, liver complaint, uterine disorders (Bhalla *et al.*, 1982).

Coconut oil- Migraine (Upadhay *et al.,* 2006); Gout (Jadeja *et al.,* 2006); To make strong and healthy hair (External application) (Jadeja *et al.,* 2006).

Kernel - Diabetes (Ghosh, 2006).

Fruit flesh – Edible (Nagandra Prasad & Abraham, 1984).

Fibre- Cordage, fish trap (Agrawal, 1981).

Coix lacryma-jobi L. (Poaceae)

Vernacular name : Nathdi, Motida.

Regional and other names : *Bengali* - Kalo–kunch, Gugur; *English* - Job's tears; *Gujarati* - Kasal; *Hindi* -Sankru; *Kannada* – Manjutti; *Malayalam* – Kattugotampu; *Marathi* –Ranjondhala; *Sanskrit* - Gavedhukah; *Tamil* - Kattukuntumani; *Telugu* - Adavi Guruginja.

Distribution - Throughout India.

Taxonomic description- Robust, spongy, annual grass, often rooting at lower nodes. Leaves cordate at base, with a glabrous membranous ligule. Racemes consisting solitary basal female spikelet and several male spikelets upwards with glume hardly or narrowly winged.

Flowering & fruiting - September – November.

Habitat ecology - Common in semi-aquatic habitats throughout the area.

Regeneration – By seeds.

Field no. - DJ- 202.

Parts used - Seeds.

Biological activity - *Seeds* - Bitter, cathartic, depurative, diuretic, sweet, tonic.

Folk uses

Miscellaneous: The seeds are worn as necklace and ear-rings by the tribal children.

Uses in literature

Dysentery, diarrhoea, fever, (Jain & Tarafder, 1970).

Root- Menstrual disorders (Basak, 2006).

Seed- Edible (Singh & Singh, 1981; Jain, 1964); Ornaments-beads (Vishnu Mittre, 1981).

Colocasia esculenta (L.) Schott (Araceae)

Syn.: *Arum esculentum* L.

Vernacular name : Arvi.

Regional and other names : *Bengali* – Kachu; *English* - Cocoyam; *Hindi* - Arvi; *Kannada* – Samangadde; *Malayalam* – Shempu; *Marathi* - Kasalu, alu; *Sanskrit* - Alupam; *Tamil* -Samakilangu; *Telugu* – Chamadump.

Distribution - Wild and cultivated throughout the hotter parts of India.

Taxonomic description - Perennial herbs, with tuberous root – stock of various size, shape. Leaves cordate or sagittate at the base, entire, purple- blotched. Spadix with an apical barren appendage, shorter than spathes; spathes pale – yellow.

Flowering & fruiting - July – September.

Habitat ecology - Cultivated for edible tubers throughout the area.

Regeneration - By tubers.

Field no. - DJ-80.

Parts used - Tuber, Leaves.

Biological activity

Leaf - Rubefacient, stimulant, styptic.

Corm - Anodyne, demulcent, laxative.

Folk uses

Food: Tuber and leaves are used as vegetable.

Medicine: Tuber are used as in the form of vegetable curries to cure diarrhoea.

Uses in literature

Corm - Cuts, burns, stings of bees (Rao & Jamir, 1982); Wounds (Saxena, 1986); Edible (Nagendra Prasad & Abraham, 1981; Shah, 1984; Maji & Sikdar, 1982).

Leaf- As vegetable (Jain & De, 1966; Ball, 1967).

Commelina benghalensis L. (Commelinaceae)

Vernacular name : Bokhlo.

Regional and other names- *Bengali* –Kanisira; *English* –Day flower; *Hindi* -Kantauoa; *Marathi* –Kena; *Tamil* -Kanangakarai; *Telugu* – Nirukassuvu.

Distribution - Throughout India.

Taxonomic description - Diffusely branched, deumbent herbs, with elliptic-ovate leaves. Flowers blue; spathes sessile or subsessile. Capsules 3-gonous; seeds pitted.

Flowering & fruiting - July-October.

Habitat ecology - Common in moist & shady habitats throughout the area.

Field no. - DJ-180.

Parts used : Leaves.

Chemical constituents : The flower pigment contains Delphinidin diglucoside as the main constituent and p-Cumaric acid and Awobanol.

Folk uses

Food: Leaves are used as vegetable.

Uses in literature

Leaf- Diarrhoea (Saxena & Vyas, 1983); Scorpion bite (Tribedi *et al.,* 1985); Wounds (Kumar *et al.,* 1987); As vegetable (Maheshwari & Singh, 1984; Vartak, 1981; Singh & Singh, 1981; Ramachandran & Nair, 1981; Mahaehwari *et al.,* 1986).

Root- Fever, liver complaint (Rajwar, 1983).

Commiphora wightii (Arn.) Bhandari (Burseraceae)

Syn.: *Balsamodendron wightii* Arnott

Vernacular name : Guggal.

Regional and other names : *Bengali* – Gugal; *English* - Indian bdellium tree; *Gujarati* - Gugali; *Hindi* - Guggul; *Kannada* – Guggulu;

Malayalam – Gulgulu; *Marathi* – Guggala; *Sanskrit* - Gugguluh; *Tamil* - Gukkulu; *Telugu* – Guggulu.

Distribution - Bellary, Mysore, Kathiawar, Rajputana desert.

Taxonomic description - This plant is not naturally growing in the area, but the plant products/organs are available in the market.

Flowering & fruiting - March – April.

Habitat ecology - This plant is not naturally growing in the area, but the plant products/organs are available in the market.

Field no. - DJ-203.

Parts used - Resinous gum.

Chemical constituents

Gum-resin contains Sesamin, Steroids, Essential oil, Guggul, Sterol, Gugguls steron, Myrcene, Dimercene, Polymercene and Sugars.

Flowers contain Myrcyalcohol, ß- Sito sterol, Amino acids, Sugars, œ- Camphorene, Cembrene, Allysembrol, Flavonoids, & Ellagic acid.

Seed oil contains Lenoleic acid, Oleic acid, Palmitic acid, Stearic acid.

Biological activity- Acrid, alterative, anodyne, anthelmintic, anti-inflammatory, antiseptic, antispasmodic aphrodisiac, aromatic, astringent, bitter, depurative, digestive, diuretic, expectorant, rejuvenating, stimulant, thermogenic.

Folk uses

Medicine: The gum-resin is uesd for gargling against mouth ulcer.

Uses in literature

Resin - Asthma, boils, headache (Shah, 1982); Dysentery (Jadeja *et al.*, 2006).

Bark - Gum trouble & ulcer (Shekhawat & Anand, 1984).

Leaves- Gout (Jadeja *et al.*, 2006).

Twig- Toothbrush (Shekhawat & Sunil Anand, 1984).

Convolvulus arvensis L. (Convolvulaceae)

Vernacular name : Galobi.

Regional and other names: *Bengali* –Gondal; *English* –Dear's foot; *Gujarati* -Nari ; *Hindi* –Harin padi; *Marathi* –Chanduel; *Sanskrit* -Rajbali; *Tamil* –Hiranpadi.

Distribution - Throughout India.

Taxonomic description - Creeping or twining, perennial herbs, Leaves sagittate-hastate at base, Flowers pale pink,

Flowering & fruiting - November-March.

Habitat ecology - Common weed of fields and wastelands.

Regeneration - By seeds.

Field no. - DJ-204.

Parts used : Whole plant.

Chemical constituents

Root contains Cuscohygrine, Coumarins.

Aerial part contain n-Alkanes, n-Alkanols, a-Amyrin and Sterols.

Biological activity : Plant extract hypotensive, anticonvulsant, mascarinic and nicotinic.

Folk uses

Medicine:

- The plant is given to animals for increasing lactation.
- Boiled plant is tied in the cases of piles and injury.
- Boiled plant is tied in the case of breast tumour.

Uses in literature

Root- Purgative (Das *et al.,* 1983; Sharma *et al.,* 1979).

Whole plant- Detergent (Koelz, 1979).

Corchorus olitorius L. (Tiliaceae)

Vernacular name : Rajan.

Regional and other names: *Bengali* –Mithapat; *English* –Jew's mallow; *Hindi* -Koshta; *Marathi* –Motichunch; *Tamil* -Peratti; *Telugu* – Parinta.

Distribution - Throughout India.

Taxonomic description - Annual herbs, with ovate-lanceolate leaves & yellow flowers. Capsules 5-valved, 10-ribbed, with entire beak.

Flowering & fruiting - August-October.

Habitat ecology - Commonly in fields as a weed.

Regeneration - By seeds.

Field no. - DJ-20.

Parts used : Leaves, Seeds.

Chemical constituents

Fibres: Cellulose, Lignin, Nitrogen, Fat, Wax.

Seed: Crude protein, Crude fibre, Free invert sugar, Starch trace, Volatile oil, Minerals, Fixed oil, Fatty acids- Oleic and Linoleic acid, Arachidic acid, Palmitic, Stearic acid, Bitter crystalline substance-Corchoritin.

Biological activity

Leaves- Hot, acrid, astringent, alterative, alexiteric, demulcent, tonic, diuretic.

Seeds- Purgative.

Folk uses

Food: Leaves are used as vegetable.

Medicine:

- Seeds are chewed in the cases of dysentery and menorrhoea.
- Extract of seeds is given with water in the case of sexual impotency in male.

Uses in literature

Leaf - Piles (Sebastian & Bhandari, 1984); As vegetable (Maji & Sikdar, 1982).

Cordia dichotoma Forst. f. (Boraginaceae)

Vernacular name: Gundlo, Gundo.

Regional and other names : *Bengali* – Bahubara; *English* - Sebesten plum; *Gujarati* - Gundomoto; *Hindi* – Lasura; *Kannada* – Cikkacalli; *Malayalam* – Naruvari; *Marathi* – Bhokar; *Sanskrit* – Bahuvara, Slesmatakah; *Tamil* - Naruvilli; *Telugu* – Cinnanakkeru.

Distribution - Throughout India.

Taxonomic description - Deciduous trees, with 3 to 5-nerved, elliptic-ovate or obovate leaves longer than broad. Flowers white, in dichotomous cymes. Fruits globose, yellow to red when ripe.

Flowering & fruiting - March- July.

Habitat ecology - Found in wastelands and near villages.

Regeneration - By seeds.

Field no. - DJ-182.

Parts used - Leaves, Inflorescence, Fruits, Gum.

Chemical constituents

Bark contains Tannins, Allantonin, B-sitosterol, Flavanone, Rhamnopyranoside.

Fruits contain Mucilage & Sugars.

Seed oil contain B-sitosterol.

Seeds contain Flavonolglycosides, Taxifolin-3- rhamnoside.

Biological activity

Bark - Acrid, anthelmintic, astringent, bitter, constipating, cooling, depurative.

Leaf - Aphrodisiac.

Fruits - Anthelmintic, aphrodisiac, cooling, depurative, diuretic, emollient, expectorant, febrifuge, purgative, sweet.

Folk uses

Fodder: Leaves serve a good fodder for cattle.

Food: Inflorescence is used as vegetable. Mature fruits are eaten by the tribals.

Medicine:

- Gum is filled in teeth cavity to cure toothache.

- 2 drops of the extract of root are dropped in ear to cure toothache.

Uses in literature

Fruits - Chest inflammation, lung diseases (Bhalla *et al.,* 1982); Stomach diseases (Sudhakar & Rolla, 1985); Constipation (Roy & Chaturvedi, 1986); Urinary complaints (Deokule, 2006); Edible (Maheshwari & Singh, 1984; Das *et al.,* 1983; Jain, 1963; Singh & Maheshwari, 1985; Negi *et al.,* 1985; Vartak, 1959; Nagar, 1985; Maheshwari *et al.,* 1981; Datar & Vartak, 1985; Kanodia & Gupta, 1968); As vegetable (Shah, 1984; Singh & Singh, 1981).

Leaf - Chest infection, lung diseases, colic, stomach diseases (Sebastian & Bhandari, 1984); Cold & cough (Sexena & Vyas, 1983); Headache (Das *et al.,* 1983); Fever (Singh, & Maheshwari, 1983); Jaundice (Hemadri & Rao, 1984).

Bark- Toothache (Aditya *et al.,* 2006).

Flower- As vegetable (Singh & Maheshwari, 1985).

Cordia gharaf (Forssk.) Ehrenb. & Asch. (Boraginaceae)

Syn.: *Cornus gharaf* Forssk.

Vernacular name - Gundi, Gundli.

Distribution - Throughout India.

Taxonomic description – Medium-sized trees, with oblanceolate, feather-nerved leaves. Flowers white, in corymbose cymes. Drupes ovoid, yellow when ripe.

Flowering & fruiting - April-December.

Habitat ecology - Common in the forests and wastelands.

Regeneration - By seeds.

Field no. - DJ-138.

Parts used - Fruits.

Folk uses -

Food: The mature fruits are eaten by the tribals. Sometimes Bhils cook vegetable from the flowers and young fruits.

Uses in literature

Fruit- Edible (Singh & Singh, 1981; Kanodia & Gupta, 1968).

Coriandrum sativum L. (Apiaceae)

Vernacular name : Dhano, Dhanio.

Regional and other names : *Bengali* – Dhoney; *English* – Coriander; *Gujarati* –Konphir ; *Hindi* – Dhaniya; *Kannada* – Kottambri; *Malayalam* – Kottamalli; *Marathi* – Dhanya; *Sanskrit* – Dhanaka; *Tamil* – Kottamalli; *Telugu* – Kotimira.

Distribution - Extensive cultivated throughout the India.

Taxonomic description - A slender-branched, glabrous, annual herb with a strong smell. Leaves pinnately decompound. Flowers in compound umbel; inner actinomorphic and outer zygomorphic. Fruits ribbed. Seeds convexo-concave.

Flowering & fruiting - December-April.

Habitat ecology - Commonly cultivated throughout the area.

Regeneration - By seeds.

Field no. - DJ-21.

Parts used - Fruits.

Chemical constituents

Leaves - Oxalic acid, Calcium, Vitamin-C.

Fruits - β-sitosterol, Chlorogenic acid, Caffic acid, Umbelliferone, Scopoletin.

Seeds - Quercetin 3-0-caffeyl glucoside, Kaempferol 3-glucoside, Fatty oil, Essential oil.

Biological activity

Leaves - Acrid, anti-inflammatory, aromatic, astringent, styptic.

Fruits - Acrid, anit-inflammtory, appetizer, aromatic, astringent, bitter, carminative, digestive, diuretic, emollient, stimulant, sweet, thermogenic.

Folk uses

Food: Plant is used for making chuteney.

Medicine:

- Powder of roasted fruits is mixed with sugar; this mixture is called "Punjeri". It is taken daily 3 times to cure headache and stomach pain during mensis.
- Extract of seeds is mixed with water (Mishri may be added) and taken with water in chronic gastritis.

Uses in literature

Leaf - Colic, piles, flatulence (Bhalla *et al.,* 1982).

Seed - Stomachache, chicken pox (Negendra & Abraham, 1984); Ulcers, carbuncles (Amit Tomar & Singh, 2005).

Fruit - Flatulence (Jadeja *et al.,* 2006); Fever (Deokule, 2006); Condiments (Negendra & Abraham, 1984; Abraham, 1981).

Whole plant - Gout, dandruff (Jadeja *et al.,* 2006).

Crinum defixum Ker-Gawl. (Amaryllidaceae)

Vernacular name - Didigundo.

Regional and other names- *Bengali* –Sukdarshan; *Gujarati* - Nagrikand ; *Telugu* –Kesarchettu.

Distribution - Throughout India.

Taxonomic description - A perennial herb with avoid bulb; leaves linear, subfleshy, slightly concave, apex acute; scape 30-50 cm. long; bracts spathaceous, membranous; flowers white; fruit sub-globose, beaked; seeds 1-2, rogouse.

Flowering & fruiting - August-October.

Habitat ecology - Common along the banks of rivers.

Regeneration - By Bulb.

Field no. - DJ-22.

Parts used : Bulb.

Chemical constituents : Bulb/seed contain Lycorine.

Biological activity : *Bulb*- Emollient, emetic, diaphoretic, nauseant.

Folk uses

Medicine: Paste of bulb is heated a bit with edible oil and this paste is tied on anus to cure piles.

Uses in literature

Tuber- Abdominal disorders (Venugopal & Raghunathan, 1982); Bodyache (Singh & Maheshwari, 1983).

Leaf- Earache (Lal & Yadav, 1983).

Crotalaria hebecarpa (DC.) Rudd. (Papilionaceae)

Syn.: *Goniogyna hebecarpa* DC.

Vernacular name - Jungli rukhdi.

Distribution - Throughout tropical India & Sri Lanka.

Taxonomic description - A prostrate herb, clothed with long silky hairs. Leaves subsessile, cordate-ovate, hairy. Flowers yellow, solitary axillary; pods brown.

Flowering & fruiting - August-October.

Habitat ecology - A very common weed along roadsides & grassy fields.

Regeneration - By seeds.

Field no. - DJ-29.

Parts used - Whole plants.

Folk uses

Medicine: Extract of plant is used to cure small pox.

Cuminum cyminum L. (Apiaceae)

Vernacular name : Jiru.

Regional and other names : *Bengali* – Jira; *English* – Cumin; *Gujarati* – Jiru; *Hindi* – Jira; *Kannada* – Jirige; *Malayalam* – Jirakam; *Marathi* – Jira; *Sanskrit* – Jiraka; *Tamil* – Jirakam; *Telugu* – Jila karra.

Distribution - Cultivated throughout India.

Taxonomic description - Glabrous, annual herbs, with rose-coloured flowers. Fruits setulose, narrowly oblong.

Flowering & fruiting - Winter season.

Habitat ecology - Commonly cultivated throughout the area.

Regeneration - By seeds.

Field no. - DJ-148.

Parts used - Seeds.

Chemical constituents : Fruits contain Apigenin-7-O-glucoside, Luteolin-7-O-glucoside, Essential oil.

Biological activity : *Fruits* - Acrid, anaphrodisiac, anodyne, anthelmintic, anti-inflammatory, astringent, carminative, constipating, cooling, depurative, digestive, febrifuge, stomachic, sweet.

Folk uses

Medicine: The seed powder is mixed with curd & eaten to cure stomach pain & indigestion.

Uses in literature

Fruits - Diarrhoea, dyspepsia (Bhalla *et al.,* 1982); Stomachache (Nagendra Prasad & Abraham, 1984); Indigestion (Ganesan *et al.,* 2006); Flatulence (Jadeja *et al.,* 2006); Diabetes (Ghosh, 2006); Condiment (Negendra & Abraham, 1984).

Curcuma amada Roxb. (Zingiberaceae)

Vernacular name: Amba-haldi.

Regional and other names : *Bengali* – Amada; *English* - Mango ginger; *Hindi* - Ama–haldi; *Kannada* – Ambarasini; *Malayalam* – Mangainji; *Marathi* - Ambe haldi; *Sanskrit* - Amragandhiharidra, Karpurdharidra; *Tamil* - Mainji; *Telugu* - Mamide allam.

Distribution - Bengal, Malay peninsula, W. peninsula.

Taxonomic description - Stemless herbs, with ellipsoid, sessile tubers with the smell of unripe mango. Leaves oblong-lanceolate. Flowers white or pale -yellow, in spikes appearing with the leaves.

Flowering & fruiting - August-October.

Habitat ecology - Commonly cultivated throughout the area.

Regeneration - By rhizome.

Field no. - DJ-123.

Parts used - Rhizome.

Chemical constituents : Rhizome contains Essential oil (Pinene, ocimene, safrol), Resin, Sugars, Gum, Starch, Albuminoids, Crude fibres, Organic acid, Phytosterol, Zulenogeine, Camphor, Curcumene, Turmerone, Volatile oil.

Biological activity: Anti–inflammatiory, anti–pyretic. aphrodisiac, appetizer, aromatic, bitter, carminative, cooling, digestive, diuretic, expectorant, febrifuge, laxative, sour, stomachic, sweet.

Folk uses

Medicine: Powder of dried rhizome is mixed with honey and made into a paste, this paste is applied and bandaged in arthritis and fracture.

Uses in literature

Rhizome- Sprains (Deokule, 2006).

Root stock- As vegetable (Maji & Sikdar, 1982).

Curuma longa L. (Zingiberaceae)

Vernacular name : Haldi.

Regional and other names : *Bengali* – Halud; *English* - Turmeric; *Hindi* - Haldi; *Kannada* – Arisina; *Malayalam* – Mannal; *Marathi* – Halad; *Sanskrit* - Haridra; *Tamil* - Mancal; *Telugu* – Pasupu.

Distribution - Cultivated throughout the India.

Taxonomic description - Rhizomes large, yielding turmeric. Leaves oblong-elliptic. Flowers pale-yellow.

Flowering & fruiting - October-February.

Habitat ecology - Commonly cultivated throughout the area.

Regeneration - By rhizome.

Field no. - DJ-23.

Parts used - Rhizome.

Chemical constituents : Rhizome contains Active curcumin compound, Essential oils-Termeric oil, Terpenoids, A volatile oil-Curcumen (Terpene), Starch, Albumnoids.

Biological activity : *Rhizome* - Acrid, alterative, anodyne, anthelmintic, anti – inflammatory, antiperiodic, antiseptic, appetiser, bitter, carminative, depurative, diuretic, emollient, expectorant, febrifuge, laxative, stimulant, stomachic, thermogenic.

Folk uses

Dye: Yellow dye obtained from the rhizome is used for colouring the cloths.

Medicine:

- Powder of dried rhizome is taken with cow milk to cure arthritis and bodyache.

- Powder of rhizome is mixed with salt and ajwain (*Trachyspermum ammi*). This mixture is taken with water to cure vomiting.
- Powder of dried rhizome is mixed with salt, this mixture is heated on Iron frying pan. When it becomes red hot, a cup of water is added. Now filtered this mixture. This filtered hot water is given by the tribals to their children to cure cough.
- Powder of dried rhizome is warmed a bit with edible oil and applied in injury and swelling .
- Powder of dried rhizome is mixed with salt (3:1) and taken with water to cure stomach pain.

Uses in literature

Rhizome - Anthelmintic (Banerjee, 1977); Cough (Shah & Joshi, 1971); Headache (Lal & Yadav, 1983); Insect stings (Shah & Joshi, 1971); Swell body (Shah & Joshi, 1971; Rajendran & Rengamani, 2006); Wounds (Shah & Joshi, 1971; Bhattacahrjee *et al.,* 1980; Tarafder, 1984; Deokule, 2006); Leprosy (Purohit *et al.,* 1985); Migraine (Tiwari *et al.,* 1979); Pimples (Prohit *et al.,* 1985); Bodyache (Lal & Yadav, 1983); Cuts & wounds (Purkyastha & Subhan, 2006); Worm complaints (Udayan *et al.,* 2006; Gupta, 2006); Snake bite (Udayan *et al.,* 2006 b); Wounds (Deokule, 2006); Skin diseases, swelling (Baruah *et al.,* 2006); Diarrhoea, jaundice (Jadeja *et al.,* 2006); Hepatitis (Dutta Choudhury & Shil, 2006); Dysentery, stomach trouble (Gupta, 2006); Gout (Jadeja *et al.,* 2006); Constipation (Das *et al.,* 2006).

Cuscuta reflexa Roxb. (Cuscutaceae)

Vernacular name : Amarvel.

Regional and other names : *Bengali* – Swarnalata; *English* - Dodder; *Gujarati* - Amarabel; *Hindi* - Amarbel; *Kannada* – Amaraballi; *Malayalam* – Akasavalli; *Marathi* – Amarvel; *Sanskrit* -Akasavalli; *Tamil* - Kadiyagundal; *Telugu* – Savarapukada.

Distribution - Throughout India.

Taxonomic description - Leafless stem – parasites, with greenish – yellow stem. Flowers whitish, solitary or in clusters arranged in racemes; style 1; stigmas 2.

Flowering & fruiting - December- January

Habitat ecology - Commonly parasitising shrubs and undershrubs throughout the area.

Regeneration - By seeds and stem cuttings.

Field no. - DJ-24.

Parts used - Stem.

Chemical constituents

Plant contains Alkaloids, Proteins, Flavonoids, Resin, Tannins, Glycosides, Carbohydrates, Cuscutin, Cuscatalin, Reducing sugars.

Seeds contain Fixed oil, Amarbelin (Pigment), Bitter resin, Aluminum, Calcium, Sodium, Potassium & Reducing sugars.

Biological activity : *Whole plant* - Anthelmintic, aphrodisiac, astringent, bitter, carminative, diuretic, expectorant, sweet.

Folk uses

Medicine:

- Extract of stem is taken with water to cure epilepsy, bodyache, arthritis and jaundice.
- Extract of stem is used as antidote in the cases of snake bite.

Uses in literature

Seed - Stomach diseases (Shah & Joshi, 1971).

Stem - Jaundice (Joshi, 1982); Eye diseases & wound (Jain & De, 1966); Indigestion & liver complaint (Sharma *et al.,* 1979); Eczema (Deokule, 2006); Diarrhoea, dysentery (Kataria & Gauchan, 2006).

Whole plant - Skin diseases, burns, wound (Goel *et al.,* 1984; Sharma *et al.,* 2006); Fall of hair, diphtheria (Singh *et al.,* 1980); Bodyache (Kharkongor & Joseph, 1981; Sharma *et al.,* 2006); Bone fracture (Krishna & Singh, 1987; Sharma *et al.,* 2006); Cough (Kumar & Chauhan, 2005; Sharma *et al.,* 2006); Fever (Rajwar, 1983; Sharma *et al.,* 2006); Itching (Goel & Mudgal, 1988; Sharma *et al.,* 2006); Jaundice (Baruah *et al.,* 2006; Dutta Choudhury & Shil, 2006); Swelling of the testicle (Prakash & Singh, 2006.); Weakness, headache, indigestion, liver complaints, oedema (Sharma *et al.,* 2006); Veterinary medicine (Shrivastava & Sharma, 1981).

Cyperus rotundus L. (Cyperaceae)

Vernacular name : Hutla, Mothdo.

Regional and other names : *Bengali* – Mutha; *English* - Nut grass; *Gujarati* - Motha; *Hindi* -Nagavmotha; *Kannada* – Tungegadde; *Malayalam* – Muttanna; *Marathi* – Nagarmotha; *Sanskrit* - Musta; *Tamil* - Korai; *Telugu* -Tunga musta.

Distribution - Throughout India.

Taxonomic description - Sedges, with slender stolons terminated by ellipsoid or globose-ovoid tubers. Leaves not exceeding the culm. Spikes born in terminal corymbs; spikelets dusty brown, 8 to 28-flowered; rachilla winged. Achenes triquetrous, brown.

Flowering & fruiting - Most part of the year.

Habitat ecology - Common in all grassy places.

Regeneration - By seeds.

Field no. - DJ-154.

Parts used - Root.

Chemical constituents - Rhizome contains Essential oil which contains Penene, Cineol; Fatty oil which contains Linoleic acid, Oleic acid, Myristic acid, Stearic; Glycerol, Glycoside Cyperene, Cyperenone, Sugars, Gum, Carbohydrates.

Leaves contain Luteolin.

Biological activity

Tuber- Acrid, anthelmintic, anti–inflammatory, Antiperiodic astrin-gent, bitter, carminative, constipating, cooling, depurative, digestive, diuretic, expectorant, febrifuge, intellect promoting, stomachic.

Folk uses

Medicine: Decoction of root-tubers is useful in gout.

Uses in literature

Tuber - Dysentery (Rao, 1981; Kataria & Gauchan, 2006); Jaundice (Rao, 1981); Diarrhoea (Kataria & Gauchan, 2006); Malaria (Deka *et al.*, 2006); Edible (Shekhawat & Sunil Anand, 1984; Vartak, 1981).

Root - Stomach pain (Maheshwari & Singh, 1984; Basak, 2006); Snake bite (Mishra & Billore, 1983); Jaundice (Gogol & Boissya, 1984); edible (Ball, 1967).

Dalbergia sissoo Roxb. (Papilionaceae)

Vernacular name : Sisam.

Regional and other names : *Bengali* – Shisu; *English* - Indian red wood; *Gujarati* - Sissom; *Hindi* -Shisham; *Kannada* – Agaru; *Malayalam* – Shisavi; *Marathi* – Agaru; *Sanskrit* - Simsapa; *Tamil* - Sissu; *Telugu* – Nukku.

Distribution - Throughout India.

Taxonomic description - Deciduous tree, with 3-5, acuminate leaflets. Flowers yellowish, in axillary panicles. Stamens 9.

Flowering & fruiting - March – August.

Habitat ecology - Abundantly planted as a road side tree.

Regeneration - By seeds, by root suckers & by stem cutting.

Field no. - DJ-90.

Parts used - Leaves.

Chemical constituents - Plant contains Dalbergin, Dalbergenone, Methyl dalbergin, Nordalbergin, Isodalbergin, Dalbergiphenol, Tecterigenin, Isocaviudin, Glucoside, Isocaviunin, Essential oil which contains Bisboline & nerolidol.

Biological activity

Root - Astringent, constipating.

Leaf - Anthelmintic, bitter, constipating, digestive, diuretic, styptic.

Bark / heartwood - Abortifacient, acrid, anthelmintic, anti – inflammatory, antipyretic, appetiser, astringent, bitter, depurative, thermogenic.

Folk uses

Medicine:

- Five to eight leaves chewed daily 3 times for a week to cure peptic ulcer and mouth ulcer.
- Extract of leaves is taken daily early in the morning to cure leucorrhoea and menorrhoea.
- Decoction of leaves is useful in diarrhoea.

Uses in literature

Leaf - Gonorrhoea (Das *et al.,* 1983); Dysentery (Singh & Maheshwari, 1983; Kataria & Gauchan, 2006); Skin diseases (Rajwar, 1983); Spermatorrhoea (Kumar & Chauhan, 2005); Blood purifier (Shah, 2005); Diarrhoea (Kataria & Gauchan, 2006).

Stem - Dysentery & headache (Maheshwari, & Singh, 1984); Leprosy (Sharma, 1979).

Wood- Leprosy (Basak, 2006).

Datura metel L. (Solanaceae)

Syn.: *D. fastuosa* L. var. *alba* C. B. Clarke

Vernacular name - Karo dhaturo.

Regional and other names : *English* - Thorn apple; *Hindi* - Kaladhatura; *Kannada* – Unmatta; *Malayalam* –Unmattu; *Sanskrit* - Dhustura; *Tamil* - Ummattai; *Telugu* – Ummatta.

Distribution - Throughout India.

Taxonomic description - A shrubby herb with dichotomous branches. Leaves ovate, sinuate, unequal sided at the base. Flowers purple or white, large, funnel shaped. Persistent calyx. Capsule globose, prickled.

Flowering & fruiting - July – March.

Habitat ecology - An occasional shrubby weed in wasteplaces.

Regeneration - By seeds.

Field no. - DJ-110.

Parts used - Leaves, Seeds, Fruits, Root.

Chemical constituents - Seeds & flowers contain Daturadiol, Daturalone, Fastusine, β- sitosterol, Hyosine, Hyoscyamine, Atropine, Fastudin, Fastunine, Fastusic acid, Fastusidine, Fastusinine, Daluranlone, Citrostadienol, Obtusifoliol, Allantoin, Caproic, β- linolic, Linoleic acid, Oleic acid, Datumetine, Datumetelin, Vitamin- C, Chlorogenic acid, Fixed oil.

Biological activity

Root - Anodyne, antispasmodic, narcotic.

Seed - Antispasmodic, aphrodisiac, narcotic.

Folk uses

Magico-religious beliefs:

- The flowers & fruits are offered to 'Lord Shiva'.
- 7 to 21 pieces of root are tied in sacred thread "luccha" and a necklace is prepared. This necklace is worn around the neck for a week to cure typhoid; it is believed that as the size of necklace increases automatically, the patients get cured.

Medicine:

- Fomentation with poultice made from the leaves to cure rheumatism.
- Extract of seeds is taken with water in the cases of dog bite.
- The ash (1-gm.) of dried fruit wall (in Sunlight) is taken with honey to cure asthma.
- The ash (1-gm.) of dried fruit wall (in Sunlight) is taken with honey to cure cough, cold and pneumonia of children.The powder of dried fruits are boiled with edible oil and filtered. This oil is used as ointment in the cases of joint pain, sciatica and paralysis.

Miscellaneous: Seeds are sold in the market.

Uses in literature

Root - Leprosy (Jain & De, 1966); Boils, ringworm (Purohit *et al.*, 1985); Fever, skin diseases (Rajwar, 1983); Asthma (Singh *et al.*, 1980).

Fruit - Fever, skin diseases (Sudhakar & Rolla, 1985); Boils (Maheshwari & Singh, 1987).

Leaf - Whooping cough, gonorrhoea, bronchitis (Bhalla *et al.*, 1982); Boils (Maheswari & Singh, 1984); Wound (Lal & Yadav, 1983; Rajendran & Rengamani, 2006); Ringworm (Purohit *et al.*, 1985); Cough (Bedi, 1978); Tetanus, arthritis, menorrhagia, headache, asthma, paralysis, body pain (Subhangi & Patil, 2006); Asthma (Bhalla *et al.*, 1982; Rajendran & Rengamani, 2006).

Seed - Piles, skin diseases, tumour, urticaria, fistula (Bhalla *et al.*, 1982); Leprosy, wound (Jain & De, 1966); Ringworm (Purohit *et al.*, 1985); Cough (Singh & Pandey, 1980); Fever (Shah, 2005); Diabetes (Ghosh, 2006); Veterinary medicine (Bennet, 1985).

Datura stramonium L. (Solanaceae)

Vernacular name : Dhaturo.

Regional and other names : *Bengali* - Safeid–dhutura; *English* - Thorn apple, devil's apple; *Hindi* -Safed dhatura; *Kannada* – Ummatta; *Malayalam* - Ummata, matulam; *Marathi* – Dhotra; *Sanskrit* - Dhattura; *Tamil* - Ummattai, vellumattai; *Telugu* - Ummetta, dutturamu.

Distribution - Throughout India.

Taxonomic description - An erect, annual herb; leaves sinuate, lanceolate, acute at apex, narrowed at base, membranous, sparsely pubescent; flowers white; capsules erect, oblong, densely covered with long prickles.

Flowering & fruiting - September – December

Habitat ecology - An occasional shrubby weed in wasteplaces.

Regeneration - By seeds.

Field no. - DJ-71.

Parts used - Leaves, Root, Seeds.

Chemical constituents - Plant contains Hysocyamine, Hyoscine, Atropine, Scopalamine, Tigloidin, Tropine, Linoleic acid, Legnoceric, Myristic, Palmitic acid, Oleic, Stearic acid, β- carbolines, Compesterol, β-sitosterol, Stigmasterol, β- amyrin, Daturadiol, Gramisterol, Chlorogenic acid, Datugen, Datugenine, Daturalactone, Stramonolide.

Biological activity - *Seed* - Acrid, alexiteric, anthelmintic, bitter, emetic, febrifuge, heating, sharp taste, tonic.

Folk uses

Magico-religious beliefs: The flowers & fruits are offered to 'Lord Shiva'.

Medicine:

- Leaves are heated a bit and tied on boils.
- Extract of root is taken with water to cure pneumonia.
- Powder of seeds is taken with water to cure asthma.
- The extract of seeds is given by the tribals to their cattle in the case of cattle donot like to eat anything.

Uses in literature

Leaf - Toothache (Shah, 1982); Hydrocele (Chhetri, 2005); Asthma (Shah, 1982; Deokule, 2006); Earache, eye infection (Baruah *et al.,* 2006).

Seeds - Rheumatism (Dar *et al.,* 1984); Stomach pain (Shah, 1982); allergy (Das *et al.,* 2006).

Fruit- Veterinary medicine (Sebastian & Bhandari, 1984).

Dendrocalamus strictus (Roxb.) Nees. (Poaceae)

Syn.: *Bambos stricta* Roxb.

Vernacular name : Vans, Vansdo, Bans.

Regional and other names- *Bengali* –Halpa; *English* –Solid bamboo; *Hindi* -Bans; *Kannada* –Kiri bidiru; *Malayalam* -Arinkantam; *Marathi* –Bans; *Tamil* -Karanai; *Telugu* –Veduru.

Distribution - Throughout India.

Taxonomic description - Arborescent or shrubby bamboos, with smooth culms. Leaves with a petiole like base & articulated with sheath. Spikelets in dense clusters along the panicle branches. Rachilla internodes very short, not articulating. Fruits subglobose.

Flowering & fruiting - At the interval of many years.

Habitat ecology - Common in the hilly area forming small patches.

Field no. - DJ-58.

Parts used : Culm.

Biological activity : Tonic, astringent, cooling.

Folk uses

Fodder: Leaves serve a good fodder for cattle.

Magico-religious beliefs:

- The culms are used to make a ladder by Bhils for carrying dead body for funeral.
- 7 to 21 pieces of root are tied in sacred thread "luccha" and a necklace is prepared. This necklace is worn around the neck for a week to cure typhoid; it is believed that as the size of necklace increases automatically, the patients get cured.

Medicine:

- Extract of seeds is given with water to cure mental weakness.
- A solid round ball is obtained from the node of the stem. Extract of this ball is taken with water to cure stomachache.

Miscellaneous:

- Bamboo strips and culms are used in the formation of several musical instruments, viz. Sarangi (piece of culm and a slend stick), Vaansuri or bansuri (culm with 7 to 9 holes) and Kundi (2 sticks).
- A seed sowing device (seed-drill) called "Penna, Penni" is made from its culm.

Timber: The culms are used as poles in making huts and walking sticks (Lathi). The culm strips are used for making bows (Kamthi), baskets (Topli), brooms (Jharu), mat (Chatai), cradle (Jhalro), winnowing pan (Hoonpdo, Supro), rectangular shaped fans (Pankho) etc.

Wheather indicator: The tribal people consider the blooming as an indicator of famine.

Uses in literature

Root- Antifertility (Sharma, 1981); Fever (Mishra & Billore, 1983).

Leaf- Cough (Sebastian & Bhandari, 1984); Veterinary medicine (Sebastian, 1984).

Digera muricata (L.) Mart.(Amaranthaceae)

Syn.: *Achyranthes muricata* L.

Vernacular name : Karinjdo.

Regional and other names- *Bengali* –Gungatiya; *Gujarati* -Kanajaro; *Hindi* -Latmahuria; *Marathi* –Gitana; *Sanskrit* -Aranyavastuka; *Telugu* –Chanchali kura.

Distribution - Throughout India.

Taxonomic description - Erect or decumbent herbs; leaves ovate-lanceolate to oblong, entire or crenulate, acute, obtuse or abruptly acuminate. Flowers pink, in axillary, 10-30 cm. long racemes; bracts oblong, persistent, 2 outer tepals 3-4 mm long, 5 to 7-nerved; 3 inner tepals shorter, narrow, 1to 2- nerved; fruits ovate-globose, with 2-fid apex, 1-seeded.

Flowering & fruiting - Most part of the year.

Habitat ecology - Common weed in wastelands & cultivated fields throughout the area.

Regeneration - By Seeds.

Field no. - DJ-164.

Parts used : Leaves.

Biological activity *:* Sweetish, acrid, cooling, astringent, laxative.

Folk uses

Food: Leaves are used as vegetable.

Uses in literature

Leaf- Boils (Shekhawat & Sunil Anand, 1984); As vegetable (Singh & Singh, 1981; Singh & Maheshwari, 1985; Maheshwari *et al.,* 1986).

Dioscorea bulbifera L. (Dioscoreaceae)

Vernacular name : Veldi.

Regional and other names : *Bengali* – Banalu; *English* - Patato yam; *Gujarati* - Suariya; *Hindi* - Ratalu; *Kannada* – Heggenasu; *Malayalam* - Kaltu kanchil; *Marathi* – Manakund; *Sanskrit* -Varahi; *Tamil* - Kodikilangu; *Telugu* – Chedupaddudumpa.

Distribution - Throughout india.

Taxonomic description - Slender, glabrous, twining herbs with globose to pyriform tubers; leaves broadly ovate, glabrous, acuminate, with a deeply cordate base; male flowers in raceme like axillary spikes, greenish-purple; female flowers sessile, axillary or in fascicled spikes, purple; capsules quadrately oblong.

Flowering & fruiting - August – October.

Habitat ecology - Common among forest hedges near villages.

Regeneration - Tubers.

Field no. - DJ-168.

Parts used - Tuber.

Chemical constituents

Plant contains A poisonous glucoside (Dioscorine).

Tuber contains Starch, Vitamin – B, Calcium oxalate, Protein, Iron.

Biological activity : *Tuber* - Alternative, anthelmintic, appetiser, aphrodisiac, bitter, fattening, pungent, stomachic, tonic.

Folk uses

Food: Tubers and bulbils are used as vegetable.

Medicine:

- Extract of tuber is taken with water in the cases of prolepsy of uterus.
- Extract of tuber is taken with water in the cases of dog bite and small pox or chicken pox.

Uses in literature

Tuber - Abdomen pain, bone fracture (Joshi *et al.,* 1980); Boils (Jain & De, 1966); Dysentery, piles (Sharma *et al.,* 1979); Jaundice (Krishna & Singh, 1987); As vegetable (Datar & Vartak, 1975).

Diplocyclos palmatus (L.) Jeffrey (Cucurbitaceae)

Syn.: *Bryonia palmata* L.

Vernacular name : Shivlingi.

Regional and Other Names : *Bengali* – Mala; *Gujarati* - Shvalingi; *Hindi* - Shivlingi; *Kannada* – Lingatondi; *Malayalam* – Sivalingakkaya; *Marathi* – Shivalugi; *Sanskrit* - Lingini; *Tamil* -Sivalingakkaay; *Telugu* – Lingadauda.

Distribution - Throughout India.

Taxonomic description - Climbers, with 2-fid tendrils and 5-lobed leaves. Flowers greenish- yellow. Fruits globose, 6-striped; seeds with raised projections on both faces.

Flowering & fruiting - September- November.

Habitat ecology - Common among the hedges and bushes throughout the area.

Regeneration - By seeds.

Field no. - DJ-12.

Parts used - Seeds.

Chemical constituents - Seed oil- Punicic acid.

Biological activity - Acrid, alterant, anti – inflammatory, depurative, thermogenic.

Folk uses

Magico-religious beliefs: Bhils believe that seeds can be used as an amulent to wardoff the evil spirits from their children.

Medicine:

- Seeds are taken with water to promote conception.
- Powder of seeds is taken with water to cure typhoid.
- Seeds are taken with water to prevent miscarriages.

Uses in literature

Leaf - Eye infection (Goel *et al.,* 1984); Body pain (Arinathan *et al.,* 2006).

Seeds - Promote fertility in women (Jain, 1963).

Root- Dysentery (Deokule, 2006).

Diospyros melenoxylon Roxb.(Ebenaceae)

Vernacular name : Timru.

Regional and other names- *Bengali* –Bhum; *English* –Coromandel ebony, persimmon; *Hindi* -Tendu; *Kannada* –Abanasi; *Malayalam* -Kari; *Marathi* –Tendu; *Tamil* -Karundumbi; *Telugu* –Nallatumki.

Distribution - Throughout India.

Taxonomic description - Deciduous trees, with sub-opposite & alternate, ovate-elliptic leaves. Male flowers in drooping cymes; anthers awnless. Fruits globose or ovoid.

Flowering & fruiting - April-November.

Habitat ecology - Found in mixed forest.

Regeneration – Seeds, root suckers and coppices.

Field no. - DJ-91.

Parts used : Fruits, Stem bark, Leaves, Wood.

Chemical constituents

Bark-Tannins.

Biological activity

Leaves- Diuretic, carminative, laxative, styptic.

Bark- Astringent.

Fruits- Carminative, diuretic, laxative.

Folk uses

Food: Mature fruits are eaten by the tribals.

Medicine:

- Paste of stem-bark is applied over the soles of the feet to prevent cracking.
- Extract of stem bark is taken with water to cure stomach pain.

Miscellaneous:

- Fruits are sold in the market.
- Leaves are used to wrap tobacco for making bidis.

Poison: The stem bark is crushed and dropped in water as a fish poison by the Bhils.

Timber: The wood is used for making 'Khat' (cot) and cart.

Uses in literature

Bark- Edible (Bhatt & Sabnis, 1987).

Fruit- Urinary complaint (Bhatt & Sabnis, 1987); Edible (Sharma & Malhotra, 1984; Maji & Sikdar, 1982; Vartak, 1981; Singh & Singh, 1981; Jain, 1964; Vartak, 1959; Bhatt & Sabnis, 1987; Nagar, 1985).

Flower- Skin diseases (Malhotra & Moorthy, 1973).

Seed- Dysentery (Joshi, 1982).

Drimia indica (Roxb.) Jessop (Liliaceae)

Syn.: *Scilla indica* Roxb.

Vernacular name : Jangli Kando.

Regional and other names : *Bengali* – Banpiaaj; *English* - Indian squill; *Gujarati* - Rankando; *Hindi* - Jangli piyaz; *Malayalam* – Kattulli; *Marathi* – Janglipyajha; *Sanskrit* - Kolakanda; *Tamil* - Narivaengayam; *Telugu* - Adavi thellagadda.

Distribution - W. Himalaya, Bihar, Chota Nagpur.

Taxonomic description - A small geophyte with ovoid, tunicated, white bulb; leaves radical, linear-lanceolate; flowers pale-brown, on a leafless scape, in few flowered racemes; capsule long-ellipsoid; seeds elliptic, black.

Flowering & fruiting - March-May.

Habitat ecology - Found in gravelly soil.

Regeneration - By seed or by old bulbs.

Field no. - DJ-152.

Parts used - Bulb.

Chemical constituents : Bulb contain Glucosides, Scillian, Scillipicrin, Scillitin, Mucilage, Sugars, Calcium oxalate.

Biological activity : Anthelmintic, cardio tonic, diphoretic, diuretic, expectorant, laxative.

Folk uses

Medicine: Extract of bulb is taken with water in the cases of cholera.

Uses in literature

Bulb - Abortifacient (Tarafder, 1983); Skin dieseases (Malhotra & Moorthy, 1973; Sarma *et al.,* 1979); Fever (Tarafder, 1986); Wounds (Uniyal & Chauhan, 1973); Rheumatism (Saxena, 1986); Veterinary diseases (Tarafder, 1983).

Leaf - Bone fracture (Ganeshan *et al.,* 2006); As vegetable (Singh & Singh, 1981).

Echinops echinatus Roxb. (Asteraceae)

Vernacular name : Ontkato, Ontkataro.

Regional and other names : *Bengali* – Thakurkanta; *English* - Thistle; *Gujarati* - Shuliyo; *Hindi* - Ontkatara; *Marathi* – Kadechubak; *Sanskrit* – Utkantaka, Kantalu.

Distribution - Throughout India.

Taxonomic description - Thistle-like, annual herbs, with pinnatifid, spinescent leaves white woolly beneath. Heads white, blue or yellow, 1- flowered, clustered in spinous globose, compound heads.

Flowering & fruiting - October – March

Habitat ecology - Common in wastelands.

Regeneration - By seeds.

Field no. - DJ-25.

Parts used - Root.

Chemical constituents

Plant contain Hentriacontane, Hentriacontanol, B-amyrin and Leupeol.

Seeds contain Echinopsine.

Biological activity- Abortifacient, aphrodisiac, bitter, cooling, pungent, sweet.

Folk uses

Magico-religious beliefs:

- 7 to 21 pices of root are tied in sacred thread "luccha" and a necklace is prepared. This necklace is worn around the neck for a week to cure typhoid; it is believed that as the size of necklace increases automatically, the patients get cured.
- Root is tied in hair of pregnant lady at the time of labour pain for normal delivery.

Medicine:

- Powder of roots is mixed with clove (*Syzygium aromaticum*) and jaggery. It is taken orally with water to cure cold. Powder of root is also used to cure sore throat.
- Extract of root is used to cure cough, cold and fever.
- Extract of root is taken with water to cure hysteria.
- Decoction of root is useful in dysentery and typhoid.
- The root is tied in hair of pregnant lady for normal delivery.

Uses in literature

Whole plant - Fever, malaria (Saxena & Vyas, 1981).

Leaf - Fever (Sharma *et al.,* 1979).

Root - Skin diseases (Maheshwari *et al.,* 1986).

Eclipta prostrata (L.) L. (Asteraceae)

Syn.: *Verbesina prostrata* L.

Vernacular name : Pilvu, Kali-bhedo.

Regional and other names : *Bengali* – Keshori; *English* - Trailing eclipta; *Gujarati* - Bhangra; *Hindi* - Bhangra; *Kannada* - Garagada soppu; *Malayalam* – Kannunni; *Marathi* – Bhringuraja; *Sanskrit* - Bhringraj; *Tamil* - Kayyantakara; *Telugu* – Galagara.

Distribution - Bengal, Burma, Central India, Punjab, W. Rajputana, Peninsular India.

Taxonomic description - Erect or prostate herbs, often rooting at the lower nodes. Leaves elliptic-oblong. Heads white, solitary.

Flowering & fruiting - Throughout the year

Habitat ecology - Common in moist damp areas as a weed.

Regeneration - By seeds.

Field no. - DJ-111.

Parts used - Leaves.

Chemical constituents - Plant contains Glucosides – phytosterol – A, ß- Amyrin, Wedelolactone, Luteoline, Triterpenic acid, Fatty acids, Palmitic acid, Stearic acid, Oleic acid, Linolic acid, Nicotine, Ampliptene Alkaloid, Ecliptine & Resin.

Biological activity - Acrid, alterative, anodyne, anthelmintic, anti-inflammatory, aphrodisiac, bitter, carminative, depurative, digestive, diuretic, febrifuge, thermogenic.

Folk uses

Medicine:

- Paste of leaves is applied on wound & cuts.
- Decoction of whole plant is useful in fever.
- Paste of leaves is applied over the soles of the feet to prevent cracking.
- Juice of leaves is applied in the cases of scorpion sting.

Uses in literature

Root - Snake bite (Goel *et al.,* 1984); ulcer (Das *et al.,* 1983).

Whole plant - Elephantiasis, skin diseases, swelling, toothache (Bhalla *et al.,* 1982); Ulcer, Jaundice (Rajwar, 1983); Leucoderma (Shah *et al.,* 1983); Itching (Sexena & Dutta, 1975); Asthma, bronchitis (Shah *et al.,* 1981); Paralysis (Kumar & Chauhan, 2005); Diarrhoea, dysentery (Kataria & Gauchan, 2006).

Leaf - Fever, malaria (Goel *et al.,* 1984); Conjunctivitis (Banerjee, 1977); Scorpion sting (Das *et al.,* 1983); Gastric problem (kumar *et al.,* 1980); Skin dieases (Purohit *et al.,* 1985); Ulcer (Singh & Pandey, 1980); Dysentery (Purkayastha & Subhan, 2006); Blood purifier, diarrhoea, dysentery (Kiruba *et al.*, 2006); Wound (Yesodharan & Sujana, 2006); Diarrhoea (Purkayastha & Subhan, 2006; Arinathan *et al.*, 2006); Elephantiasis (Sharma *et al.*, 2006); As vegetable (Ramachandran & Nair, 1981).

Enicostema axillare (Lam.) Raynal. (Gentianaceae)

Syn.: *Gentiana axillaris* Lam.

Vernacular name : Nau, Nauli, Tawli, Nai, Nay.

Regional and other names : *Bengali* - Nagajivha; *Gujarati* - Mamejavi; *Hindi* - Chota chirayata; *Kannada* - Karibandita; *Malayalam* - Vellaruku; *Marathi* – Kadavinayi; *Sanskrit* - Nagajihva; *Tamil* - Vallari; *Telugu* - Chevvu – kurti.

Distribution - Throughout India.

Taxonomic description - Perennial herb, with 4-angular branches and linear-lanceolate leaves. Flowers white, in axillary clusters. Capsules 1-celled, 2-valved.

Flowering & fruiting - July – October.

Habitat ecology - Common in open grasslands.

Regeneration - By seeds.

Field no. - DJ-26.

Parts used : Whole plant.

Chemical constituents : Swertiamarin, Gentianine, Apigenin, Genkwanin, Isovitexin, Saponarin, Heptacosane, Nonacosane, Myristic acid, Oleic acid, Gentiocrucine, Alkaloids.

Biological activity : Acrid, anthelmintic, anti – inflammatory, anti-periodic, bitter, carminative, depurative, digestive, stomachic, thermogenic.

Folk uses

Medicine:

- Extract of leaves is useful in malaria fever.
- Decoction of whole plant is useful in gout.
- Extract of whole plant is taken with water to cure typhoid.
- Extract of leaves is applied on forehead to cure headache.
- Powder of dried leaves is taken with water to cure jaundice.

- Juice of leaves is used in curing the pus in ear.

Miscellaneous: Dried plants are sold in market by the tribal people.

Uses in literature

Leaf - Fever (Joshi, 1982); Worm infection (Pereira, 2006); Malaria (Narain & Singh, 2006).

Whole plant - Fever & snake bite (Singh & Pandey, 1980); Malaria (Saxena, 1986); Rheumatism, blood purifier (Kiruba *et al.*, 2006).

Erythrina suberosa Roxb.(Papilionaceae)

Vernacular name : Parabda, Paluda

Regional and other names : *Bengali* - Palita madar badisa; *English* - Indian coral tree; *Gujarati* – Bangaro; *Hindi* - Dadap, pankara; *Kannada* – Varjipe *Malayalam* – Murukku; *Marathi* - Pangara, harivana; *Sanskrit* – Paribhadrah; *Tamil* – Murukku; *Telugu* - Baridamu, bodita.

Distribution - Sea coast of India, Java.

Taxonomic description - Small trees, armed with prickles. Leaflets 3, pubescent beneath. Flowers scarlet, at the end of branches. Seeds 2 to 5.

Flowering & fruiting - March-June.

Habitat ecology - Rare, in wastelands & fringes of forests.

Regeneration - By seeds & by cuttings.

Field no. - DJ-92.

Parts used - Stem bark.

Chemical constituents

Bark contains a bitter Erytherine alkaloid & Resins.

Leaves contain Erytherine, Alkaloid.

Biological activity

Bark - Acrid, anthelmintic, anti – inflammatory, bitter, carminative, depurative, digestive, diuretic, expectorant, febrifuge, rejuvenating, sedative, stomachic, thermogenic.

Leaf - Anthelmintic, bitter, diuretic, laxative, stomachic.

Folk uses

Medicine:

- Ash of stem bark is mixed with edible oil and applied on boils.
- Ash of stem bark is mixed with edible oil and applied over the soles of the feet to prevent cracking.
- The fume, obtained from boiling of stem bark with water, is given through nasal tract to cure headache.
- Extract of stem bark is mixed with the extract of stem bark of Jambu (*Syzygium cumini*) and taken with water in the case of loose motion.

Uses in literature

Bark- Dysentery (Sharma *et al.,* 1979).

Eucalyptus globulus Labill.(Myrtaceae)

Vernacular name : Tejpan.

Regional and other names : *English* – Blue gum; *Hindi* – Yukeliptus; *Malayalam* – Yukkalimaram; *Sanskrit* – Tailpaparnah; *Tamil* – Yukkalimaram.

Distribution - Introduced from Australia.

Taxonomic description - A tall tree; bark peeling in long strips or sheets, greyish or bluish-white; leaves lanceolate. Flowers axillary, 2-3-nate or solitary, sessile. Fruits angular & warty.

Flowering & fruiting - Cold season.

Habitat ecology - Common along roadsides.

Regeneration - By seeds and coppices.

Field no. - DJ-155.

Parts used - Stem.

Chemical constituents : Leaves contain - Cis & Trans pinocarveol derivatives, α-pinene & Unsaturated α-ketone.

Biological activity

Fruits - Cystein, ornithine, asparagine, glutamic acid, threonine, α & β alanine, norvaline.

Oil - Acrid, anodyne, anthelmintic, antiseptic, astringent, bitter, carminative, digestive, diuretic, expectorant, rubefacient, stimulant, sudorific, thermogenic.

Folk uses

Timber: The stem is used in the huts like a beam & rafter.

Uses in literature

Leaf - Malaria, swelling (Megoneitso & Rao, 1983); Antiseptic (Basak, 2006).

Euphorbia hirta L. (Euphorbiaceae)

Vernacular name : Dudhi.

Regional and other names : *Bengali* - Baro kheruie; *English* - Australian asthma herb; *Gujarati* - Dudeli; *Hindi* - Dudhi; *Malayalam* - Nela palai; *Marathi* – Dudhi; *Sanskrit* -Dugdhika; *Tamil* - Pallavi; *Telugu* – Pacchaku.

Distribution -Throughout India.

Taxonomic description - Erect or procumbent herbs, with ovate-lanceolate or rhomboid-oblong leaves. Cyathia greenish, crowded in pedunculate cymes; glands stalked, pink. Capsules globosely 3-gonous.

Flowering & fruiting - Throughout the year.

Habitat ecology - Extensively found in waste places & also along the walls.

Regeneration - By seeds.

Field no. - DJ-59.

Parts used - Whole plant.

Chemical constituents : Plant contains Quercetin, Triacentane, Ellagic acid, Phenolics, Euphosterol, Phytosterol, Phytosterolin, Gallic, Mellistic, Palmitic, Oleic, Linoleic acid, Zanthorhamnin wax, Chlorophyll, Resin, Tannins, Sugar, Mucilage, Carbohydrates, Albuminoids, Catchin, Oxalate.

Biological activity : Astringent, diuretic, expectorant.

Folk uses

Medicine: Extract of whole plant is taken with water to cure leucorrhoea.

Uses in literature

Leaf - Ringworm, scabies (Shah *et al.,* 1981); Burn, pain in joints (Jain *et al.,* 1973); Gonorrhoea (Mishra & Sahu, 1984); Snake bite (Goel *et al.,* 1984); Dysentery (Ganeshan *et al.,* 2006); Wound (Yesodharan & Sujana, 2006); As vegetable (Datar & Vartak, 1975).

Latex - Eczema, scabies (Bhatt & Sabnis, 1987); Eye complaint (Bhalla *et al.,* 1982); Ringworm (Joshi, 1982); Scorpion bite (Maheshwari *et al.,* 1986); Wort (Mishra & Sahu, 1984); Pimples (Kiruba *et al.,* 2006); Skin irritation (Gupta, 2006).

Whole plant - Colic (Maheshwari & Singh, 1984); Dysentery, cough (Mishra & Sahu, 1984; Kataria & Gauchan, 2006); Cuts (Saxena & Dutta, 1975); Diarrhoea (Audichya *et al.,* 1983; Kataria & Gauchan, 2006); Toothache (Sebastian & Bhandari, 1984); Vomiting (Sexena & Vyas, 1983); Kidney diseases (Sudhakar & Rolla, 1985); Leucorrhoea (Kumar & Chauhan, 2005); Dysentery (Upadhyay & Jaswant Singh, 2005); Lucorrhoea & spermatorrhoea (Kumar & Chauhan, 2005); Cough (Deokule, 2006); Chronic bronchial affections, asthma (Mishra & Sahu, 1984; Kataria & Gauchan, 2006; Kumar *et al.,* 2006); Stomach trouble in children (Rajendran & Rengamani, 2006).

Euphorbia neriifolia L. (Euphorbiaceae)

Vernacular name : Gotla thuwar.

Regional and other names : *Bengali* – Monshasij; *Gujarati* - Thor; *Hindi* - Sij, shend; *Kannada* – Yalekalli; *Malayalam* – Illakkalli; *Marathi* – Nevagunda; *Sanskrit* -Sehund; *Tamil* - Illakkalli; *Telugu* – Akejemudu.

Distribution - W. Peninsula, cultivated elsewhere and in Shri Lanka.

Taxonomic description - Erect deciduous shrubs or small trees; stem 5-angled. Leaves obovate-oblong, crowded at the end of stem, subfleshy. Involucre 3-nate; glands without petaloid limb. Capsules obovoid-trigonous, deep red or pale brown.

Flowering & fruiting - February– March

Habitat ecology - Commonly planted along the boundaries of the field for fencing purposes.

Regeneration - By cuttings.

Field no. - DJ-27.

Parts used - Latex, Leaves.

Chemical constituents

Plant contains Euphol, Taraxerol, Friedelan, Resin, Gum.

Bark contains n- Hexacosonal, Euphorbol, Hexacos

Anoate, Pelargonine, Diglusoside, Euphorbin.

Root & bark contains Methylene Cycloartenol, Tulipanin 3-5 diglucoside.

Biological activity

Plant - Alexipharmic, bitter, carminative, laxative, pungent.

Milk - Appetiser, laxative, pungent.

Leaf - Appetiser, carminative.

Folk uses

Biofencing: Commonly planted along the boundaries of the field for fencing purposes.

Medicine:

- Latex is filled in teeth cavity with the help of cotton plug to cure toothache.
- Extract of leaves is used to cure cough and sore throat.
- Leaves are used as paste on boils.

Uses in literature

Milk - Earache (Deokule, 2006).

Leaf – Cough (Sarma *et al.,* 2006).

Stem – Fish poison (Joshi, 1986).

Euphorbia thymifolia L.(Euphorbiaceae)

Vernacular name : Dudhi.

Regional and other names- *Bengali* – Didiya; *Gujarati* - Nahanidudheli; *Hindi* -Dudhi; *Kannada* – Kempu nene hakki; *Malayalam* - Nilppala; *Marathi* – Choti dudhi; *Sanskrit* -Dugdhika; *Tamil* - Chinamampatchaiarisi; *Telugu* – Biduru mana siyyam.

Distribution - Throughout India.

Taxonomic description - An annual herb. Leaves very small and opposite. Involucers axillary, solitary or 2-3 in an axil. Capsules erect, obtusely keeled; styles short, 2-fid.

Flowering & fruiting - Throughout the year.

Habitat ecology - Common in wastelands.

Regeneration - By seeds.

Field no. - DJ-169.

Parts used : Whole plant.

Chemical constituents : Plant contains Epitaraxerol, n-Hexacosanol, Euphorbol, Esters, n-Alkanes, Sterols, Corilagin, Geranin, Bixanin, Casuarin, Desgalloylstachyurin.

Biological activity : Astringent, aphrodisiac, antidiarrhoeal, acrid, anthelmintic, alexipharmic, antibacterial, bitter, demulcent, depurative, diuretic, expectorant, laxative, purgative, vulnerary.

Folk uses

Medicine: Extract of whole plant is taken with water to cure small pox.

Uses in literature

Whole plant – Blood dysentery (Jain, 1984); Blood purifier, ringworm, skin diseases, snake bite (Mishra & Sahu, 1984); Spermatorrhoea (Maheshwari *et al.,* 1980, 1981); Veterinary diseases (Chaudhuri & Pal, 1976).

Leaf- Dysentery (Singh *et al.,* 1987).

Ferula asa-foetida L. (Apiaceae)

Vernacular name : Hing.

Regional and other names : *Bengali – Hing; English – Asafoetida; Hindi – Hing; Kannada – Hingu; Malayalam – Kāyam; Marathi – Hing; Sanskrit – Hingu; Tamil – Kayam; Telugu – Inguva.*

Distribution - Baltistan, Astor.

Taxonomic description - This plant is not cultivated in the study area, but the gum-resin, obtained from the roots, is commonly available in the market.

Flowering & fruiting -- June-December.

Field no. - DJ-28.

Parts used - Gum-resin.

Chemical constituents - Luteolin & its compound, Essential oil, Ferulic acid, Organic sulphur compound, Pinene, Isobutyl propanyl disulphide and Umbeliferone.

Biological activity - Acrid, anthelmintic, antispasmodic, bitter, carminative, digestive, diuretic, expectorant, laxative, sedative.

Folk uses

Medicine:

- Extract of roots is useful in toothache.
- The gum resin obtained from roots is dissolved in water. This water is applied as ointment on stomach to cure stomach pain & gastric trouble.

Uses in literature

Root extract (Gum-resin) - Flatulence, stomachache, dysentery (Jadeja *et al.*, 2006).

Ficus benghalensis L. (Moraceae)

Vernacular name : Vadlo, Vadli, Vad.

Regional and other names : *Bengali* – Bot; *English* - Banyan tree; *Gujarati* - Vad; *Hindi* - Bargad, Bar; *Kannada* - Mara, Ala; *Malayalam* - Ala, Vatam; *Marathi* - Vada, Wad; *Sanskrit* -Vata; *Tamil* - Alam, Vadam, Al; *Telugu* - Vati, Peddamarri.

Distribution -Sub-Himalayan tract and W. Peninsula. Planted elsewhere.

Taxonomic description - Large, evergreen trees, with prop roots and ovate-oblong leaves. Receptacles geminate, axillary, globose, red when ripe.

Flowering & fruiting - August – July.

Habitat ecology - Common in wastelands & along the roads.

Regeneration - By seeds & by cuttings.

Field no. - DJ-93.

Parts used - Leaves, Adventitious roots, Latex.

Chemical constituents : Plant contains Glucoside; Bengalenoside & Talbutamide, Leucoanthocyanidins, Phytosterolin, β- sitosterol, Freidelin, Quercetin, Galactoside, Ticlic acid, Tannin.

Biological activity : Acrid, astringent, sweet.

Folk uses

Fodder: Leaves serve a good fodder for the cattle.

Medicine:

- 2 drops of the latex obtained from leaves are dropped in a Batasa (Spherical spongy sugar cake) and eat it. It is useful in the cases of Ascariasis.
- Adventitious roots are chewed in the cases of toothache.
- Latex of leaves is mixed with wheat flour (*Triticum aestivum* L.) and small pills are prepared. One pill taken every day early in the morning for 7 days with cow milk to check undesirable discharge of semen with urine.
- 5 drops of the latex obtained from leaves are dropped in a Batasa (Spherical spongy sugar cake) and eat it. It is useful in the cases of premature ejaculation.
- Latex is applied over the soles of the feet to prevent cracking.
- A drop of latex is dropped in eye to cure conjuctivitis.
- The ash of leaves is given with water by the tribals to their cattle in Afra disease.

Uses in literature

Bark - Snake bite (Singh & Pandey, 1980); Diabetes (Negi *et al.*, 1985; Hembrom & Goel, 2005; Ghosh, 2006).

Aerial root - Diarrhoea (Jain & De, 1966); Menorrhoea (Joshi *et al.*, 1980); Leucorrhoea (Hemadri & Rao, 1983); As tooth brush (Nagendra Prasad & Abraham, 1984; Shah, 1984; Ramachandran & Nair, 1981).

Leaf - Blisters, boils (Tarafder, 1984); Cough (Singh & Pandey, 1980); Piles (Kumar & Chauhan, 2005); Cuts (Maheshwari & Singh, 1984).

Latex - Gum trouble & toothache (Nagendra Prasad & Abraham, 1984; Aditya *et al.*, 2006); Sex diseases (Rai, 1985); Eye complaint (Singh *et al.*, 1987); Blisters, boils (Singh & Maheshwari, 1983); Skin diseases (Maheshwari & Singh, 1987); Rheumatic pain (Deokule, 2006); Fever (Rajendran & Rengamani, 2006).

Fruit- Edible (Singh & Singh, 1981; Maheshwari *et al.*, 1986; Datar & Vartak, 1975; Kanodia & Gupta, 1968; Singh, 1945).

Ficus racemosa L. (Moraceae)

Syn.: *F. racemosa* L.

Vernacular name : Gulri, Umri, Umber.

Regional and other names : *Bengali* – Dumur; *Gujarati* – Umar; *Hindi* – Gular; *Malayalam* - Atti; *Marathi* – Gular; *Sanskrit* -Udumbara; *Tamil* - Adavam; *Telugu* – Atti.

Distribution - Throughout India.

Taxonomic description - A large tree; leaves alternate, ovate, elliptic-lanceolate or ovate-oblong, glabrous. Receptacles 2-4 cm across, clustered on short leafless branchlets, globose or pyriform, hairy, supported by 3 bracts at the base, green, turning reddish-purple when ripe.

Flowering & fruiting - February – June

Habitat ecology - Found in moist places & forest edges, along the nalas.

Regeneration - By seeds & vegetative method.

Field no. - DJ-81.

Parts used - Wood, Stem bark, Fruits, Latex, Tender twigs.

Chemical constituents : Bark contains Cerylbehenate, Lupeol, Alphaamyrin, Glaunol, ß- sitosterol, Keton, Hentriacontane, Tiglic acid, Glucose, Taraxeroll & Tannin.

Biological activity: *Plant* - Acrid, antidysenteric, cooling, sweet.

Folk uses

Fodder: Leaves serve a good fodder for cattle.

Food: Mature fruits are eaten by the tribals.

Medicine:

- Latex is filled in teeth cavity to cure toothache.
- Extract of stem bark is taken orally with water thrice a day for 3-4 days to cure dysentery.
- 1-2 drop of latex of stem is dropped in sugary sweet "batasha" and eaten to cure diarrhoea.
- Twigs are used as toothbrush to cure toothache.
- 1-2 drops of the latex obtained from leaves are dropped in a Batasa (Spherical spongy sugar cake) and eat it to cure mouth ulcer.

Timber: The wood is used for making a musical instrument called "Dhol".

Uses in literature

Latex - Diarrhoea (Kumar & Chauhan, 2005); Piles, diarrhoea (Deokule, 2006).

Bark- Piles (Pereira, 2006); Dysentery (Deokule, 2006); Diabetes (Kumar *et al.*, 2006).

Root- Diarrhoea, dysentery (Kataria & Gauchan, 2006); Diabetes (Kumar *et al.*, 2006).

Fruit – Diabetes (Kumar *et al.*, 2006); Edible (Maheshwari & Singh, 1984; Shah, 1984; Vartak, 1981; Singh & Singh, 1981; Ramachandran & Nair, 1981; Jain, 1964; Negi *et al.*, 1985; Nagar, 1985; Maheshwari *et al.*, 1981, 1986; Chaudhuri *et al.*, 1985; Datar & Vartak, 1975; Kanodia & Gupta, 1968; Singh, 1945); As vegetable (Vartak, 1959; Ramachandran & Nair, 1981).

Ficus religiosa L. (Moraceae)

Vernacular name : Pipali.

Regional and other names : *Bengali* – Ashathwa; *English* - Pipal tree; *Gujarati* - Pipro; *Hindi* - Pipal; *Kannada* - Ashwattha, arali; *Malayalam* – Ashvatham; *Marathi* – Ashvatha; *Sanskrit* -Pippala; *Tamil* - Aswattham; *Telugu* – Aswatthamu.

Distribution - Sub-Himalayan forests, Bengal, Central India.

Taxonomic description - Large, deciduous trees, with ovate to ovate-suborbicular leaves caudate-acuminate at apex and rounded at the base. Receptacles in axillary pairs, reddish-purple when ripe.

Flowering & fruiting - April - June

Habitat ecology - Commonly found along the roadsides.

Regeneration - By seeds & vegetative method.

Field no. - DJ-60.

Parts used - Leaves.

Chemical constituents - Bark contains Tannin.

Biological activity

Plant - Acrid, bitter, sweet.

Fruits - Alexipharmic, cooling.

Folk uses

Fodder: Leaves serve a good fodder for cattle.

Magico-religious beliefs: The young leaves are used in snake-bite. It is said that the leaf petiole when inserted in both the ears it is said that it sucks poison from the body.

Medicine: The sten bark is effectively used by tribals to check hair falling, for which the tribals apply the paste of stem bark on the head before taking bath.

Uses in literature

Fruits - Asthama (Bhalla *et al.,* 1982); Edible (Ball, 1967; Maheshwari *et al.,* 1986; Datar & Vartak, 1975; Kanodia & Gupta, 1968); As vegetable (Singh, 1945).

Bark - Ulcer, boils, blisters (Maheshwari & Singh, 1984); Scabies (Bhalla *et al.,* 1982); Snake bite (Murthy *et al.,* 1986); Whooping cough (Virjee *et al.,* 1984); Gonorrhoea (Bhalla *et al.,* 1982; Purkyastha & Subhan, 2006; Basak, 2006); Fibre (Jain & De, 1966).

Leaf - Ulcer, boils, blisters (Shah, 1982); Toothache (Shekhawt & Anand, 1984); Skin diseases (Malhotra & Moorthy, 1973); Snake bite (Maheshwari *et al.,* 1986); Scabies (Deokule, 2006).

Foeniculum vulgare Mill. (Apiaceae)

Vernacular name : Sonp.

Regional and other names : *Bengali* – Mauri; *English* – Fennel; *Gujarati* –Variari; *Hindi* – Sounph; *Kannada* – Badhesoppu; *Malayalam* – Perumjirakam; *Marathi* – Shepu; *Sanskrit* – Madhurika; *Tamil* – Sompu; *Telugu* – Peddajilakarra.

Distribution - Cultivated throughout the India.

Taxonomic description - Annual or perennial herb, with a characteristic aromatic smell. Leaves 2 to 4-pinnate. Leaf segments linear or filiform. Fruit oblong, with strong ridges, furrows 1-villate. Fruit variable in length. Flowers yellow.

Flowering & fruiting - January-March.

Habitat ecology - Often cultivated for its fruits and seeds.

Regeneration - By seeds.

Field no. - DJ-124.

Parts used - Seeds and Fruits.

Chemical constituents : Constituents of plant are Fatty oil, Protein, Anethose, Fenchone, Anisalehyde, Feniculin, Nelumboside, Flavons, Abscisic acid. Essential oil from fruit Anethole, Xanthotoxin, Bergapten, Psoralen, Scoparone, Vanillin, α-Pinene, Camphene, Limonene, α-Phellandrene, Fenchone, Methylchavicol.

Biological activity : *Fruits* - Acrid, alexipharmic, anthelmintic, antiemetic, aphrodisiac, bitter, carminative, digestive, diuretic, emollient, expectorant, febrifuge, haematinic, ophthalmic, refrigerant, stomachic, sudorific, sweet.

Folk uses

Medicine:

- Seeds are put into water for whole night, next day its extract is taken with water to cure menorrhoea.
- Extract of fruits is taken with water twice a day in the cases of burning in urethra during micturition.
- 2 dried leaves (in shady condition) of *Calotropis gigantea* and 25 gm. fruits of *Foeniculum vulgare* are ground with 50 gm. mishri (crystalline sugar) and a teaspoonful powder is taken with water daily early in the morning to cure jaundice

Uses in literature

Seed - Headache, kidney complaints, menstrual complaints, spleen complaints, dysentery, wounds, fever (Bhalla *et al.,* 1982); Toothache (Dar *et al.,* 1984).

Leaves - Colic, cough (Bhalla *et al.,* 1982).

Fruits - Flatulence, gout (Jadeja *et al.,* 2006).

Gloriosa superba L. (Liliaceae)

Vernacular name : Kaliali, Ragagaddi.

Regional and other names : *Bengali* – Ulatchanadal; *English* - Malabar glory lily; *Gujarati* - Dudhi ovachhnag; *Hindi* - Karihari, langali; *Kannada* – Agnisikha; *Malayalam* - Mallattamara, mettonni; *Marathi* - Kariannag, indai; *Sanskrit* -Kalihari; *Tamil* - Kalaippaikokishangu; *Telugu* - Adavi nabhi, deyyapaku, kalappagadda.

Distribution - India, Ceylon, Malaya, Tropical Africa.

Taxonomic description - Climbing herbs, with alternate, opposite or whorled leaves ending in a tendril. Flowers yellow, tinged with red in upper half; tepals undulating on the margins. Capsules oblong.

Flowering & fruiting - August-November.

Habitat ecology - Occasionally found in shruberies during rainy season .

Regeneration - By seeds & tubers.

Field no. - DJ-184.

Parts used - Root.

Chemical constituents : Rhizome tubers contain Alkaloid Colchicine; Superbine, Glorisine, Sitosterol, Glucosides, Benzoic acid & Resins.

Biological activity : Tuber - Abortifacient, acrid, alexiteric anthelmintic, bitter, heating, laxative, pungent.

Folk uses

Medicine:

- Paste of root is useful in the case of bone fracture.
- Extract of root is given in the case of dog-bite.

Uses in literature

Root - Anthelmintic (Sharma *et al.,* 1979).

Leaf - Asthma (Tarafder & Rai Chaudhuri, 1981); Gout (Jadeja *et al.,* 2006).

Rhizome - Scorpion bite, skin diseases & gonorrhoea (Singh *et al.,* 1980); Gout (Jain & Dam, 1979); Intermittent fever (Tarafder & Rai Choudhary, 1981); Abortifacient (Tarafder, 1983; Rajendran & Rengamani, 2006); Leprosy, piles (Singh & Pandey, 1980); Snake bite (Bedi, 1978); Stomach diseases (Baruah *et.al.,* 1987); Rheumatism (Chatterjee & Bhatterjee, 2006); Veterinary diseases (Jain & De, 1966).

Gossypium herbaceum L. (Malvaceae)

Vernacular name : Kapas.

Regional and other names : *Bengali* – Yula; *English* – Indian cotton; *Gujarati* – Kapas; *Hindi* – Kapas; *Kannada* – Janivara-hattimara; *Malayalam* – Parutti; *Marathi* –Kapas; *Sanskrit* – Karpasah; *Tamil* – Karpasam; *Telugu* – Pratti.

Distribution - A native of Asia and also of Egypt. Cultivated as a rainy season crop, often mixed with other crops.

Taxonomic description - An erect shrubby & hairy annual or perennial. Stem woody. Leaves divided to the middle, cordate, 3 to7-lobed. Flowers large, yellow with purple claw. Capsules ovoid-globose, mucronate; cells 6 to 7- seeded.

Flowering & fruiting - September-November.

Habitat ecology - Cultivated during rains throughout the area.

Regeneration - By seeds.

Field no. - DJ-61.

Parts used - Fibres, Seeds.

Chemical constituents : Betaline, Choline, Salicylic acid, Gossypol, Gossyfulvin, Gossypurpurin, Gossycaerulin, Gossypin, Gossipetin, Gossipitrin, Herbacitrin, Quercitin, Carotenoids, Vitamin- E, Methyl & higher Alcohols, Peptone.

Biological activity

Root - Abortiacient, acrid, antidysenteric, astringent, depurative, diuretic, emollient,

Leaves - Constipating, cooling, diuretic, haematinic.

Flowers - Diuretic, refrigerent, sweet.

Seeds - Anodyne, aphrodisiac, diuretic, emollient, expectorant, laxative, refrigerant, sweet.

Folk uses

Fibres: Textile fibres are obtained from the seed hair and sold in the market.

Medicine:

- Two drops obtained from unripe seed hair are dropped in eyes 2-3 times in a day to cure conjunctivitis.
- 5 flowers of *Calotropis gigantea* and 5-10 cloves of *Syzygium aromaticum* are burnt with cotton fibres (*Gossypium herbaceum)* to prepare ash. About 10g. of ash is taken daily early in the morning with glassful of water to cure asthma.
- The decoction of dried epicalyx is given at night for 3 days to regulate the menstration period.

Miscellaneous: Seeds are sold in market by the tribals.

Uses in literature-

Root - Abortifacient (Tarafder, 1983).

Bark - Abortifacient (Tarafder, 1983).

Epicalyx - Proper regulation of mensus (Jadhav, 2006).

Leaves - Snake bite (Basak, 2006); Headache (Deokule, 2006).

Grewia tiliifolia Vahl (Tiliaceae)

Vernacular name : Dhaman.

Regional and other names- *English* -Dhaman; *Hindi* -Dhaman; *Kannada* –Dadasalu; *Malayalam* -Catacci; *Sanskrit* -Dhanvangah; Catacci; *Telugu* –Peddatena.

Distribution - India, Ceylon, Malaya, Tropical Africa.

Taxonomic description - Small trees, with ovate to orbicular leaves and yellow flowers in axillary cymes. Drupes globose.

Flowering & fruiting - May-August.

Habitat ecology - Common in deciduous forests near the villages.

Regeneration - By seeds.

Field no. - DJ-170.

Parts used : Fruits.

Chemical constituents : Leaves, Tannin.

Biological activity : Astringent, sweet, acrid, refrigerant, expectorant, antipruritic, vulnerary, constipating, emetic, styptic, aphrodisiac, tonic.

Folk uses

Food: Fruits are eaten by the tribals.

Uses in literature

Root- Syphilis (Chaudhuri *et al.,* 1982).

Fruit- Edible (Vartak, 1981; Singh & Singh, 1981; Ramachandran & Nair, 1891; Nagar, 1985; Datar & Vartak, 1975).

Leaf- As vegetable (Sharma & Singh, 1988); Edible (Jain, 1963, 1964, 1981).

Haldina cordifolia (Roxb.) Ridsd. (Rubiaceae)

Syn.: *Adina cordifolia* (Roxb.) Hook. f. ex Brandis; *Nauclea cordifolia* Roxb.

Vernacular name : Haldu.

Regional and other names : *Bengali* - Dakom, Keli - Kadam; *English* – Haldu; *Gujarati* – Haldarvo; *Hindi* – Haldu; *Kannada* – Arsintega; *Malayalam* – Katampa; *Marathi* –Hedu; *Sanskrit* – Bahuphala, Haridru; *Tamil* – Kadambi; *Telugu* – Kamba.

Distribution - From the Himalayas to Myanmar and through Central and S. India to SriLanka.

Taxonomic description - Large tree with spreading branches; leaves elliptic-oblong; flowers sweet-scented; fruit fleshy.

Flowering & fruiting - July – March.

Habitat ecology - Not very comman, reported from Haldupada village.

Regeneration - By seeds & vegetative method.

Field no. - DJ-03

Parts used - Stem bark, Wood.

Chemical constituents

Fruit contains Flavonoids.

Leaves contain Ursolic acid, Quercetin 3–O– galactoside.

Heartwood contains Adinin pigment, Tannins.

Stem bark contains Alkaloids, Tannins, Pentosan, Lignins, Starch.

Root bark contains Coumarin Glycoside Adicardin.

Biological activity

Root - Astringent

Bark - Acrid, antiseptic, aphrodisiac, bitter, febrifuge, pungent, tonic, vulnerary.

Folk uses

Dye: Wood dust gives non poisonous yellow dye.

Medicine: The crushed stem bark is put into water for whole night, next day its extract is taken with water early in the morning by the women for prevention of miscarriage.

Tannin: Bhil tribals use the stem bark for tanning the leather.

Timber:

- The wood is used in the formation of musical instrument Drum (Dhol).
- The wood is also used in the formation of Comb ('Kangsi').

Uses in literature

Bark - Fever (Jain & De, 1996); Antiseptic (Mukherjee *et al.,* 1986; Basak, 2006); Dysmenorrhoea (Hemadri, 1992); Stomach disease (Singh & Prakash, 1994); Skin diseases (Jha *et al.,* 1997); Veterinary medicine (Roy & Chaturvedi, 1986).

Leaves - Cold & cough (Jain & De, 1966); Headache (Tribedi *et al.,* 1982); Veterinary medicine (Mukherjee *et al.,* 1986).

Stem - Eye complaint (Goel *et al.,* 1984); Veterinary medicine (Goel *et al.,* 1984).

Helicteres isora L. (Sterculiaceae)

Vernacular name : Atedo.

Regional and other names : *Bengali* – Atmora; *English* - East Indian screw tree; *Gujarati* - Murdasing; *Hindi* - Bhendu, jonkaphal; *Kannada* – Yedamuri; *Malayalam* – Valmpiri; *Marathi* - Kewan; *Sanskrit* - Avartani; *Tamil* – Valamabui; *Telugu* – Gubathada.

Distribution - From the Punjab and Bengal to SriLanka, Myanmar.

Taxonomic description - Shrubs, with ovate to orbicular leaves and red flowers. Fruits cylindric, woody, beaked, with spirally coiled follicles.

Flowering & fruiting - September- February.

Habitat ecology - Found in rocky places on the cool slopes of hills.

Regeneration - By seeds.

Field no. - DJ-94.

Parts used - Fruits.

Biological activity - Fruit - Astringent, demulcent, expectorant.

Folk uses

Medicine: Extract of fruit is useful in stomach trouble of children.

Uses in literature

Root - Asthma, cough (Singh & Pandey, 1980); Abortifacient (Tribedi *et al.,* 1985); Diabetes & snake bite (Roy & Chaturvedi, 1986); Sciatica (Sebastian & Bhandari, 1984); Cuts (Jain *et al.,* 1973); Stomach troubles (Sharma *et al.,* 1979).

Leaf - Eczema (Ramachandran & Nair, 1981).

Seed - Dysentery & stomach trouble (Singh *et al.,* 1985).

Stem bark - Diarrhoea, dysentery (Singh *et al.,* 1985); Ear complaint (Shah & Gopal, 1986).

Fruit - Colic & earache (Singh *et al.,* 1985); Cough, Cold, Fever (Mudgal & Pal, 1980); Dysentery (Sharma & Malhotra, 1984; Jadeja *et al.,* 2006); Stomach trouble (Maheshwari & Singh, 1984).

Flower- Edible (Ramachandran & Nair, 1981).

Bark fibre – Cordage (Jain & De, 1966; Shah *et al.,* 1981).

Hibiscus cannabinus L. (Malvaceae)

Vernacular name : Ambadi.

Regional and other names- *Bengali* –Machika; *English* -Mesta; *Gujarati* -Ambari ; *Hindi* -Pitwa; *Kannada* –Nalita; *Malayalam* - Kanchava; *Marathi* –Ambada; *Sanskrit* -Anvastha; *Tamil* -Kasini; *Telugu* –Gaynaru.

Distribution - Native of Africa; cultivated for its silky bast fibres.

Taxonomic description - Annual or perennial, prickly herbs. Corolla spreading, yellow with crimson centre.

Flowering & fruiting - August-December.

Habitat ecology - Cultivated for its silky bast fibres throughout the area.

Regeneration - By seeds.

Field no. - DJ-175.

Parts used : Bast fibres.

Chemical constituents

Seeds- Phospholipids, Palmitic acid, Stearic acid, Oleic acid, Linolic acid, Beta-sitosterol, Limonene, Citral.

Leaves- Aldobiouronic acid, Rutin, Isoquercitrin, Limonene, Phellandrene, Alpha-terpenyl acetate, Citral, Beta-sitosterol, Vanillic acid, Lignoceric acid, Cerotic acid, Montanic acid, Melissic acid.

Flowers- Anthocyanin, Fructose, Glucose, Mannose, Sucrose, Sterculic acid, Tridecanoic acid.

Biological activity

Seeds- Analgesic, aphrodisiac, appetising, stomachic.

Leaves- CNS-depresent, purgative.

Flowers- Antiulcer, antibilious, antipyretic, digestive.

Folk uses

Miscellaneous: The bast fibres are made into ropes & cordages which find a wide use in the area.

Uses in literature

Leaf- Purgetive (Janardhanan, 1963); As vegetable (Sharma & Malhotra, 1984; Shah, 1984).

Bark- Fibre (Vishnu Mittre, 1981).

Hibiscus rosa–sinensis L. (Malvaceae)

Vernacular name : Jasul.

Regional and other names : *Bengali* – Jaba; *English* - China rose; *Gujarati* - Jasuva; *Hindi* - Gurhal; *Kannada* – Dasavala; *Malayalam* - Chemparathy; *Marathi* – Jaswand; *Sanskrit* -Japa; *Tamil* - Chedi; *Telugu* - Java pushpamu.

Distribution - Cultivated in gardens throughout India.

Taxonomic description - Evergreen shrubs, with pink, red or rosy flowers.

Flowering & fruiting - November- February.

Habitat ecology - Cultivated ornamentally in the gardens.

Regeneration - By cuttings and by seeds.

Field no. - DJ-63.

Parts used - Buds, Flowers, Leaves.

Chemical constituents

Leaves contain Alkaloids, Glycosides, Reducing sugars, Sterols, Resins, Beta-sitosterol.

Root contain Mucilage, Fibres.

Flowers contain Fat, Calcium, Phosphorous, Thiamine, Riboflavin, Ascorbic acid, Cyanin chloride.

Biological activity

Analgesic, antipyretic, aperient, aphrodisiac, cardiac, contraceptive, demulcent, diuretic, emolient, laxative, refrigerant.

Folk uses

Medicine:

- 1 to 2 buds are chewed everyday early in the morning for a period of 7 days to cure leucorrhoea and menorrhoea.
- Extract of flowers is taken with water to cure menorrhoea, leucorrhoea and stomach pain during mensus.
- 1 to 2 leaves are chewed everyday early in the morning for a period of 7 days to check undesirable discharge of semen with urine and to increase sexual potentiality.

Uses in literature

Root - Cough (Shah, 1981; Basak, 2006); Gonorrhoea (Baruah & Sarma, 1984); Venereal diseases, menorrhagia (Basak, 2006).

Flowers - Skin diseases, fever (Rao & Jamir. 1982); Anti-fertility (Tarafder, 1983); Contraception (Bhattacharjee *et al.,* 1980); Inflammation (Baruah & Sarma, 1984); Menorrhoea (Shah *et al.,* 1981); Kidney diseases (Maheshwari *et al.,* 1980); Fever (Deokule, 2006); asthma (Arinathan *et al.,* 2006); Hair growth (Rajendran & Rengamani, 2006); Leprosy (Basak, 2006); Menstrual troubles (Das *et al.,* 2006).

Root - Gonorrhoea (Baruah & Sarma, 1984); Cough (Shah *et al.,* 1981).

Leaf - Headache (Purkyastha & Subhan, 2006); Wound, pimples (Pereira, 2006); Hair growth (Ganesan *et al.,* 2006; Rajendran & Rengamani, 2006); Dandruff (Gupta, 2006); Leprosy (Basak, 2006).

Holarrhena pubescens (Buch.-Ham.) Wall. ex G. Don (Apocynaceae)

Syn.: *Echites pubescens* Buch.-Ham.

Vernacular name : Dudhi.

Regional and other names : *Bengali* – Kurchi; *English* - Easter tree; *Gujarati* - Dhowda; *Hindi* - Kurchi; *Kannada* – Beppale; *Malayalam* – Kadakappala; *Marathi* – Kuda; *Sanskrit* -Kutaja; *Tamil* - Kulappalai, veppalai; *Telugu* - Amkudu, kadaga.

Distribution - Throughout India.

Taxonomic description - Small trees; leaves ovate to elliptic. Flowers white, in terminal corymbose cymes. Calyx-lobes oblong-lanceolate, ciliate; corolla puberulous; follicles cylindric, often dotted with white spots.

Flowering & fruiting - February – June.

Habitat ecology - Common on the cool places of hills but also elsewhere.

Regeneration - By seeds and root suckers.

Field no. - DJ-82.

Parts used : Stem bark, Root.

Chemical constituents

Root bark contains Holacetine, Connesine Alkaloids.

Seed contains Essential oil, Conessine & Kurchicine Alkaloids.

Stem bark contains Connesine, Holarrhenine, Holarrhimine, Kurchine, Kurchicine, Conamine & Conimine.

Biological activity

Bark - Acrid, anthelmintic, bitter, pungent.

Flowers - Acrid, anthelmintic, appetiser, bitter, cooling.

Seed - Anthelmintic, appetiser, astringent, bitter, cooling, pungent.

Folk uses

Fodder: Leaves and tender branches are fed to cattle.

Medicine: Decoction of stem bark is given orally with water to cure dysentery.

Miscellaneous: The latex of the plant is used for coagulating milk.

Poison: The root is crushed and dropped in water as a fish poison by the Bhils.

Uses in literature

Flower - Worms (Baruah & Sarma, 1984); Edible (Sharma & Malhotra, 1984); As vegetable (Maheshwari *et al.,* 1986).

Fruit - Stomach trouble (Maheswari *et al.,* 1986).

Leaf - Leucoderma (Baruh & Sarma, 1984); Dysentery (Baruah *et al.,* 2006); Gout (Jadeja *et al.,* 2006).

Root - Fever (Shah & Gopal, 1986); Stomach trouble (Sharma & Malhotra, 1984); Dysentery (Mokat & Deokule, 2006).

Seed - Asthma & colic (Shah *et al.,* 1983); Worms, diarrhoea & fever (Gupta, 1981); Leprosy (Baruah & Sarma, 1984); Jaundice (Sharma *et al.,* 1979); Malaria (Deka *et al.,* 2006); Diabetes (Ghosh, 2006).

Bark - Gout, snake bite, rheumatism, paralysis (Gupta, 1981); Piles, dysentery, diarrhoea (Baruah & Sarma, 1984); Cough, fever (Tarafder, 1986); Snake bite (Sebastian & Bhandari, 1984); Stomach diseases (Sharma & Malhotra, 1984); Malaria (Singh & Maheshwari, 1983); Cold (Srivatava & Varma, 1981); Gastric diseases (Tarafder, 1983); Headache (Joshi *et al.,* 1980).

Holoptelea integrifolia (Roxb.) Planch. (Ulmaceae)

Syn.: *Ulmus integrifolia* Roxb.

Vernacular name : Mojar.

Regional and other names : *English* - Indian elm; *Gujarati* - Kanjho; *Hindi* – Papri; *Kannada* - Thavasai, rasbija; *Malayalam* – Aval; *Marathi* – Papra; *Sanskrit* - Chirabilva; *Tamil* - Avali, kanji; *Telugu* - Nemali, tapasi.

Distribution - Sub-himalayas, Rajasthan, Bundelkhand, Bihar, Assam, W. Peninsula, SriLanka, Myanmar.

Taxonomic description - Large trees, with ovate, pinnately veined leaves and greenish-yellow flowers. Fruits broadly oblong to suborbicular, winged, flat, notched at apex.

Flowering & fruiting - February– May.

Habitat ecology - Common in wastelands & roadsides.

Regeneration - By seeds & vegetative method.

Field no. - DJ-95.

Parts used - Seeds, Leaves.

Biological activity - *Bark, Leaves* - Acrid, anti-inflammatory, anthelmintic, astringent, bitter, carminative, depurative, laxative, revulsive.

Folk uses

Food: Seeds are eaten by the tribals.

Medicine:

- Boiled leaves are tied on injured body part.
- The leaves of Mojar & Arni (*Clerodendrum multiflorum*) are crushed together and boiled in pure ghee and a paste is prepared. It is applied as ointement in the case of sprain.

Uses in literature

Leaf - Skin diseases (Singh & Maheshwari, 1983); Boils (Maheshwari *et al.*, 1986); Scorpion sting (Singh *et al.*, 1987); As fish poison (Joshi, 1986).

Stem bark - Ringworm (Maheshwari & Singh, 1984); Ulcer (Apparanantham & Chelladurai, 1986); Scabies (Mudgal & Pal, 1980); Hydrocele (Goel *et al.*, 1984); Bone fracture (Pal & Jain, 1989); Rheumatism, (Negi *et al.*, 1985; Kumar & Chauhan, 2005); As fish poison (Joshi, 1986).

Seed oil - Rheumatism (Pal & Jain, 1989).

Fruit- Edible (Singh & Singh, 1981).

Seed- Edible (Vartak, 1981).

Flower- As vegetable (Pal & Banerjee, 1971).

Hordeum vulgare L. (Poaceae)

Vernacular name : Jau.

Regional and other names : *Bengali* – Job; *English* – Barley; *Gujarati* – Jav; *Hindi* – Jau; *Kannada* – Yava; *Malayalam* – Yavam; *Marathi* – Jau; *Sanskrit* – Yavah; *Tamil* – Barliarisi; *Telugu* – Yava.

Distribution - Throughout India.

Taxonomic description - Erect, glabrous annuals. Spikes emerging from the leaf-sheaths; spikelets 2-ranked.

Flowering & fruiting - December-March.

Habitat ecology - Cultivated as a crop for grains and straw throughout the area.

Regeneration - By Seeds.

Field no. - DJ-205.

Parts used - seeds.

Chemical constituents : Fat, Starch, Protein. Alkaloid Hordenine, Luteolin,Glucoside, Orientoside, Orientin, Hordenine & its Methyl ether.

Biological activity: *Grains* - Acrid, aphrodisiac, astringent, digestive. diuretic, emollient, refrigerant, sweet.

Folk uses

Medicine: Flour is fed to the animals in the case of body weakness.

Uses in literature

Grains - Pimples (Purohit *et al.,* 1985); Urine complaint (Dar *et al.,* 1984); Diarrhoea, dysentery (Kataria & Gauchan, 2006); Edible (Koelz, 1979).

Hygrophila schulli (Buch.-Ham.) M. R. & S. M. Almeida (Acanthaceae)

Syn.: *Astercantha longifolia* Nees ; *Barleria longifolia* L. ; *Behel schulli* Buch.-Ham.

Vernacular name : Talmakhana.

Regional and other names : *Bengali* – Kuliakhara; *Gujarati* - Gokhru; *Hindi* - Gokulkanta; *Malayalam* – Vayalchulli; *Marathi* – Kolshinda; *Sanskrit* –Vajrakantaka, Kokilaksa; *Tamil* - Nirmulli; *Telugu* – Nirguviveru.

Distribution - Throughout India.

Taxonomic description - Hispid annuals, with thickened nodes. Leaves oblong or lanceolate, in verticils of 6 at a node. Axillary spines yellow. Flowers purple-blue, in axillary whorls. Capsules linear- oblong.

Flowering & fruiting - August– December.

Habitat ecology - Abundant in swampy habitats throughout the area.

Regeneration - By seeds & vegetative method.

Field no. - DJ-185.

Parts used - Seeds.

Chemical constituents - Seed contains Yellow coloured oil; Diastase, Lipase, Protease, salts of Potassium, Mucilage.

Biological activity

Latex - Aphrodisiac, bitter, sour, sweet.

Seed - Acrid, aphrodisiac, bitter, constipating, cooling, tonic.

Folk uses

Medicine: Powder of seeds is mixed with powder of tuber of Safed Musli (*Chlorophytum tuberosum*), one teaspoonful of this powder is taken daily on empty stomach to check undesirable discharge of semen with urine.

Uses in literature

Leaf - Bodyache, jaundice (Chaudhuri *et al.,* 1975); Joint pain (Ganeshan *et al.,* 2006); As vegetable (Arora, 1981).

Root - Dropsy, rheumatism, jaundice (Singh & Pandey, 1980).

Whole plant - Malaria (Sexena & Vyas, 1983); Fever (Hembrom & Goel, 2005); Diabetes (Kiruba *et al.,* 2006).

Indigofera cordifolia Heyne ex Roth (Papilionaceae)

Vernacular name - Baam.

Distribution - Throughout India.

Taxonomic description - Prostrate herbs, with simple, cordate leaves and red flowers in axillary racemes. Pods oblong, 1 or 2-seeded.

Flowering & fruiting - August-December.

Habitat ecology - Common in wastelands throughout the area.

Regeneration - Seeds.

Field no. - DJ- 210.

Parts used - Whole plant.

Folk uses

Medicine: The Bhils apply warmed paste of the plant on the abdomen to cure abdominal pain.

Use in Literature

Leaf- Swelling of mouth (Bhatt & Sabnis, 1987); Edible (Gunjatkar & Vartak, 1982; Bhandari, 1974).

Indigofera trita L. f. (Papilionaceae)

Vernacular name : Dantudi.

Regional and other names - *Gujarati* - Vekhario; *Tamil* - Kandaram; *Telugu* –Nakkanaru.

Distribution - Throughout India.

Taxonomic description - An undershrub, with trifoliolate leaves and raceme inflorescense.

Flowering & fruiting - July-August.

Habitat ecology - Common on field margins throughout the area.

Regeneration - Seeds.

Field no. - DJ-209.

Parts used - Roots.

Biological activity - *Seeds*- Nutritive tonic.

Folk uses

Medicine: Extract of roots is taken with water to cure fever.

Uses in literature

Twig- For strength of gum, teeth (Bhatt & Sabnis, 1987).

Ipomoea carnea Jacq. subsp. *fistulosa* (Mart. ex Choisy) Austin (Convolvulaceae)

Syn.: *I. Fistulosa* Mart. ex Choisy

Vernacular name : Nachedi, Nashedi, Besaram.

Distribution - Throughout India.

Taxonomic description - Erect, ascending or scandent, perennial undershrubs, with milky juice; stems fistular; leaves ovate-triangular, with a cordate base; flowers purple in many flowered cymes; capsules globose, glabrous; seeds black, villous.

Flowering & fruiting - Throughout the year.

Habitat ecology - Very common in hedges of fields, road-sides, banks of ponds, throughout the area.

Regeneration - By seeds.

Field no. - DJ-30.

Parts used : Leaves.

Chemical constituents : *Leaves*- Ipomose, an Anthracene glucoside, Gum,Jjalapin, Saponins.

Biological activity *:* Ethanol extract of the plant CNS depressant.

Folk uses

Medicine:

- Latex is applied against scorpion sting.
- Leaves are warmed a bit with edible oil and tied on wound & boils.

Uses in literature

Leaf- Cuts, wounds (Singh *et al.,* 1987); Sprain (Chattopadhyay, 1982); Ulcer (Sharma & Malhotra, 1984); As vegetable (Sharma & Malhotra, 1984).

Ipomoea batatas (L.) Lam. (Convolvulaceae)

Syn.: *Convolvulus batatas* L.

Vernacular name - Sakkarkand.

Regional and other names : *Bengali* – Ranga alu; *English* – Sweet potato; *Gujarati* –Sakaria; *Hindi* – Mitha alu, Sakkarkand; *Kannada* – Sihigenasu; *Malayalam* – Madhurakkilannu; *Marathi* – Ratalu; *Sanskrit* – Pindaluh; *Tamil* – Cakkaravallikkilanku; *Telugu* – Genusu.

Distribution - A native of America. Cultivated in India.

Taxonomic description – Prostrate, annual herbs. Tubers red, white or rarely yellow. Leaves ovate-cordate, acute, angular or more or less lobed. Peduncles long, many-flowered. Flowers purple. Seeds glabrous.

Flowering & fruiting - Cold season.

Habitat ecology - Cultivated for sweet edible tubers throughout the area.

Regeneration - By vegetative method.

Field no. - DJ-156.

Parts used - Tubers.

Chemical constituents - Tubers - Hydrocyanic acid, Oxalic acid, Phytic acid, Starch, Vitamins, Enzymes, ß-sitosterol, Oalmitic acid, NaCl.

Biological activity - *Roots* - Aphrodisiac, diuretic, laxative, refrigerant, sweet.

Folk uses

Food: Boiled tubers are eaten in the winter season by the tribals.

Use in Literature

Root- Edible (Nagendra Prasad & Abraham, 1984; Jain & De, 1966).

Jasminum officinale L. var. *grandiflorum* L. (Oleaceae)

Syn.: *J. grandiflorum* L.

Vernacular name : Chameli.

Regional and other names : *Bengali* – Chameli; *English* – Common Jasmine; *Gujarati* –Chambeli ; *Hindi* – Cameli; *Kannada* – Mallige; *Malayalam* – Piccakam; *Sanskrit* – Malati; *Tamil* – Kotimalligi; *Telugu* – Malati.

Distribution -Throughout India.

Taxonomic description-Scandent or climbing shrubs, with imparipinnate leaves and white flowers in lax panicles.

Flowering & fruiting - Throughout the year.

Habitat ecology - Cultivated ornamentally.

Field no. - DJ-186.

Parts used - Leaves.

Chemical constituents

Leaves - Salicylic acid.

Flowers - Essential oil in flowers contains Lactone, Pyridine & Nicotinate derivatives.

Biological activity

Root & Leaves - Acrid, anodyne, anthelmintic, antiseptic, aphrodisiac, astringent, bitter, depurative, diuretic, emollient.

Flowers - Ophthalmic, refrigerant.

Folk uses

Medicine:

- Leaves are chewed to cure mouth ulcer.
- Extract of root is used as 'Kajal' to cure conjuntivits.

Uses in literature

Flower- Stimulate hair growth (External application) (Jadeja *et al.*, 2006).

Bark - Burns, skin diseases (Sinha *et al.*, 2006).

Jasminum sambac (L.) Ait. (Oleaceae)

Syn.: *Nyctanthes sambac* L.

Vernacular name : Mogro.

Regional and other names : *Bengali* – Motia; *English* – Arabian Jasmine; *Gujarati* –Mogro; *Hindi* – Mogra, Bel; *Kannada* – Dundumallige; *Malayalam* – Mulla; *Marathi* –Mogra; *Sanskrit* – Mallika; *Tamil* – Malligai; *Telugu* – Malli.

Distribution - Throughout India.

Taxonomic description - Scandent or climbing shrubs, with simple leaves & white flowers.

Flowering & fruiting - January- December.

Habitat ecology - Commonly planted in gardens for ornamental purposes.

Regeneration – By vegetative method.

Field no. - DJ-32.

Parts used - Leaves.

Chemical constituents - Flowers contain Pyridine, Nicotinate derivatives.

Biological activity - *Flowers* - Acrid, alexipharmic, bitter, ophthalmic, purgative, refrigerant.

Folk uses

Medicine: Leaves are warmed a bit with ghee and tied on boils.

Jatropha curcas L. (Euphorbiaceae)

Vernacular name : Ratanjot.

Regional and other names : *Bengali* - Bag harenda; *English* - Purging nut; *Gujarati* - Jamalgota; *Hindi* - Jangli arand, ratanjot; *Kannada* – Adaluharalu; *Malayalam* – Kattalavanakka; *Marathi* – Moglieranda; *Sanskrit* – Suklaeranda, Akhuparnika; *Tamil* - Adalae, kattukkottai; *Telugu* – Adaviyamadamu.

Distribution - Grown in India. Native of tropical America.

Taxonomic description - Eglandular, soft wooded shrubs or small trees, with 3 to 5-angled ovate-cordate leaves. Flowers greenish-yellow, in corymbose cymes. Capsules globose-ovoid.

Flowering & fruiting - November – February.

Habitat ecology - Commonly planted along the boundaries of fields for fancing purposes.

Regeneration - By seeds and by cuttings.

Field no. - DJ-31.

Parts used - Leaves, Latex, Twigs.

Chemical constituents - Seed contains Amino acids, Arachid acid, Lionleic acid, Myrstic acid, Oleic acid, Stearic acid, Arabinose, Glucose, Rhamnose, Galactose, Xylose, Galactouronic acid, ß- sitoserol.

Stem bark contains B- Sitosterol, Taraxerol.

Leaves contain Sucrose, Vitexin, Compesterol, Triacontanol.

Biological activity

Fruit & Seed - Anthelmintic, puragative.

Root - Astringent, carminative, digestive, emetic.

Folk uses

Biofencing: Commonly planted along the boundaries of fields for fancing purposes.

Medicine:

- Leaves are heated a bit and tied on the boils.

- Latex filled in teeth cavity with the help of cotton plug to cur toothache.
- Twigs are used as toothbrush to cure toothache.
- A drop of latex is dropped in the ear to check pus formation.
- Latex is applied on cuts and injury to check bleeding.

Miscellaneous: Seeds are sold in the market.

Oil: Seed oil is used as fuel in lamps.

Poison: The stem-bark is crushed and dropped in water as a fish poison by the Bhils.

Uses in literature

Root - Abortifacient (Tarafder, 1983).

Seeds - Rheumatism, (Rajwar, 1983).

Stem - Tooth trouble (Gupta, 1981; Aditya *et al.,* 2006; Rajendran & Rengamani, 2006).

Latex - Burns & tooth trouble (Gupta, 1981); Cancer (Chaghatai & Javed, 1978).

Seed oil- For lighting (Rao & Neogi, 1980).

Twig- Toothbrush (Vishnu Mittre, 1981).

Justicia adhatoda L. (Acanthaceae)

Syn.: *Adhatoda vasica* Nees.; *A. zeylanica* Medic.

Vernacular name -Aruso.

Regional and other names - *Bengali* – Basak; *English* – Vasak; *Gujarati* – Aduso; *Hindi* – Adusa; *Kannada* –Adusoge; *Malayalam* – Adalodakam; *Marathi* – Adulsi; *Sanskrit* – Amalaka, Vasaka, Vansa; *Tamil* – Adatodi; *Telugu* – Atarushamu.

Distribution - Throughout India.

Taxonomic description - Shrubs, with elliptic leaves. Flowers white or creamish- white in elongate spikes; corolla – tube mottled. Capsules 4 – seeded.

Flowering & fruiting - January – April.

Habitat ecology - Common weed in shady habitats.

Regeneration - By stem cuttings.

Field no. - DJ-196.

Parts used - Leaves.

Chemical constituents

Leaves contains A bitter vasicine, Adhatodic acid, Volatile oil, Vasa, Resin, Gum, Sugar & Yellow pigment.

Plant contains Alkaloids (Vasicoline, Adhatodine, Vasicolinine, Anisolin), Vasicine & Vasicinone.

Biological activity - The plant is acrid, anti- inflammatory, anthelmintic, antileprotic, antiperiodic, antiseptic, antispasmodic bitter, bronchodilator, cooling, diuretic, expectorant, pungent, stimulant.

Folk uses

Medicine:

- The 2 leaves are boiled with a cup of tea and taken in cough.
- Extract of stem or leaves is given to cattles in the case of bronchial problem.

Uses in literature

Root - Anthelmintic, antipyretic, antiseptic (Bhalla *et al.,* 1982); Boils (Karnick *et al.,* 1981); Eczema (Singh & Pandey, 1980); Rheumatism, asthma (Tripathi *et al.,* 1996); Fever (Singh, 1993); Bronchitis (Singh, 1993; Tripathi *et al.,* 1996).

Seed - Boils (Singh & Pandey, 1980).

Leaves - Bronchitis (Bhalla *et al.,* 1982; Basak, 1997; Billore *et al.,* 1998); Asthma (Nagendra Prasad & Abraham, 1984; Basak, 1997; Billore *et al.,* 1998; Purkayastha & Subhan, 2005); Malaria (Bhalla *et al.,* 1982; Deka *et al.,* 2006); Swelling (Janardhanan, 1963); Rheumatism (Rao & Jamir, 1982; Basak, 1997; Billore *et al.,* 1998); Cough (Saxena & Vyas 1983; Basak, 1997; Billore *et al.,* 1998; Purkayastha & Subhan, 2005; Pereira, 2006); Cold (Basak, 1997; Billore *et al.,* 1998; Pereira, 2006; Kadavul & Jisha, 2006); Diarrhoea, dysentery (Jadeja *et al.,* 2006); Hepatitis (Dutta Choudhury & Shil, 2006); As vegetable (Bhargava, 1959).

Whole plant - Cough, asthma, bronchitis, anaemia (Kumar *et al.,* 2006).

Kalanchoe pinnata (Lam.) Pers. (Crassulaceae)

Syn.: *Cotyledon pinnata* Lam.

Vernacular name : Khatumdo.

Regional and other names : *Bengali* - Patharkuchi; *English* - Sprout leaf plant; *Hindi* - Zakhm-e-hayat; *Malayalam* - Elamarunga; *Marathi* - Panphuti; *Sanskrit* – Asthibhaksha, Asmabheda; *Tamil* – Runakalli; *Telugu* – Ranapala.

Distribution - A native of Tropical Africa, but naturalized throughout the tropics of the world.

Taxonomic description - A perennial, glabrous, succulent herb. Leaves opposite-decussate. Leaflets ovate or elliptic, crenate or serrate. Flowers in large spreading panicles, pendent, pale-greenish and reddish-purple.

Flowering & fruiting - December-february.

Habitat ecology - Spreading as an escape in rocky wastelands.

Regeneration - By leaf bud.

Field no. - DJ-125.

Parts used - Leaves.

Chemical constituents : Leaves contain Alkanes, Tricontane, Alpha- & Beta-Amyrin, Beta-sitosterol, Fumaric acid, Mallic acid, Citric acid, Quercetin, Kaempferol, Calcium sulphate, Calcium oxalate.

Biological activity : *Leaves* - Astringent, stimulant, styptic, wound healing.

Folk uses

Medicine: Leaf is heated a bit with edible oil and tied on boils.

Uses in literature

Leaf - Renal stone (Rao, 1981; Lal & Yadav, 1983; Hembrom & Goel, 2005; Mokat & Deokule, 2006); Influenza (Rao, 1981); Boils, wounds (Baruah & Sarma, 1984; Sudhakar & Rolla. 1985; Sebastian & Bhandari, 1984; Basak, 2006); Gastric complaint, insect bite (Baruah & Sarma, 1984); Cuts, wounds (Pereira, 2006; Das *et al.,* 2006); Muscular sprain, swelling (Deokule, 2006); Diarrhoea (Jadeja *et al.,* 2006); Headache (Rao, 1981; Das *et al.,* 2006); Dysentery (Jadeja *et al.,* 2006; Das *et al.,* 2006).

Lantana camara L. (Verbenaceae)

Vernacular name : Baramasi.

Regional and other names : *Bengali* – Chotra; *English* - Wild sage; *Hindi* - Aripple; *Kannada* - Nata hu gida, hesike; *Malayalam* - Arippu, puchedi; *Marathi* - Chadurang, ghaneri; *Tamil* - Unnichedi, arippu; *Telugu* - Pulikampa, lantana.

Distribution - Throughout India.

Taxonomic description - Prickly shrubs, prickles hooked. Leaves ovate or ovate-oblong. Flowers multi-coloured, orange-yellow, pink or red. Drupes globose, black when ripe.

Flowering & fruiting - Throughout the year

Habitat ecology - Common in wastelands and along roadsides.

Regeneration - By seeds.

Field no. - DJ-96.

Parts used - Flowers, Leaves, Stem.

Chemical constituents : Plant contains Essential oil.

Leaves contain Lantanine, Lantadene, Arachidic acid, Linoleic acid, Myristic acid, Palmitic acid, Citral, Eugenol, Furfural, Geraniol, Linalool, Alpha- amyrin, Beta- sitosterol, Lantanin acid, Lantadene-A & B, Verbascoses & Lantanane.

Biological activity

Antiseptic, antispasmodic, carminative, diaphoretic, laxative.

Folk uses

Biofencing: Commonly planted along the boundaries of field for fancing purposes.

Food: The ripe fruits are eaten by the children.

Medicine:

- Powder of dried flowers is taken with tea twice a day to cure jaundice.
- Extract of leaves is taken with water to cure dysentery and vomiting.

Timber: The stem-strips are used for making 'Tokri' (Basket) and 'Hoonpdo' or 'Supro' (Winnowing pan).

Uses in literature

Leaf - Itching, ringworm (Abraham, 1981); Tetanus (Purkayastha & Subhan, 2006); wounds (Deokule, 2006; Yesodharan & Sujana, 2006; Das *et al.*, 2006); Skin diseases (Baruah *et al.*, 2006); Cuts (Das *et al.*, 2006); Veterinary medicine (Pal, 1980).

Flower - Bleeding gums & decaying teeth (Upadhyaya, & Jaswant Singh, 2005); Hepatitis (Dutta Choudhury & Shil, 2006).

Fruit- Edible (Singh & Singh, 1981; Abraham, 1981).

Launaea procumbens (Roxb.) Ramayya & Rajagopal (Asteraceae)

Syn.: *Prenanthes procumbens* Roxb.

Vernacular name - Karvi galobi.

Distribution - Throughout India.

Taxonomic description - Annual or perennial herbs, with mostly radical, rosette leaves. Heads yellow, with white pappus.

Flowering & fruiting - September-April.

Habitat ecology - Commonly found as a weed in cultivated field.

Regeneration - By seeds.

Field no. - DJ-206.

Parts used - Whole plants.

Folk uses

Medicine: The plant is given to animals for increasing lactation.

Uses in literature

Leaf- Fever (Malhotra & Moorthy, 1973); As vegetable (Vartak, 1959).

Lawsonia inermis L. (Lythraceae)

Syn.: *L. alba* Lamk.

Vernacular name : Mendi.

Regional and other names : *Bengali* – Gotanti; *English* - Henna; *Gujarati* - Mendi; *Hindi* - Mehendi; *Kannada* – Mayilanchi; *Malayalam* – Mayilanji; *Marathi* – Mendi; *Sanskrit* - Medika; *Tamil* - Marudondri, maruthondi; *Telugu* - Gorintaku, aivanary.

Distribution - Drier parts of India.

Taxonomic description - Much branched shrubs, with broadly lanceolate leaves and creamish, fragrant flowers. Capsules globose.

Flowering & fruiting - May – November.

Habitat ecology - Cultivated as a hedge plant in the gardens.

Regeneration - By seeds & vegetative method.

Field no. - DJ-33.

Parts used - Leaves.

Chemical constituents : Leaves contain Heena dye, Hanno – tannic acid, Tannin, Olive green resin.

Biological activity : Anti- inflammatory, antipyretic, burning sensation, diuretic, emetic, expectorant, prophylactic.

Folk uses

Biofencing: Commonly planted along the bounderies of fields for fencing purposes.

Dye: The leaves yield a dye which is widely used as a colouring material for hand, feet & hairs, especially by ladies.

Medicine:

- Paste of leaves is applied on anus to cure piles.
- Paste of leaves is applied over the soles of the feet to prevent cracking.
- Extract of leaves is taken with water to cure diarrhoea and vomiting.
- The juice of leaves (Sugar may be added) is taken orally with water to check undesirable discharge of semen with urine.

Uses in literature

Leaf - Burning sensation (Nagendra Prasad & Abraham, 1984; Deokule, 2006); Falling of hair (kumar & Chauhan, 2005); Bodyache, Muscular pain (Pereira, 2006); Hair growth (Rajendran & Rengamani, 2006).

Root - Jaundice (Hemadri & Rao, 1984).

Seed - Weakness (Gupta, 1981).

Lepidagathis cristata Willd. (Acanthaceae)

Vernacular name : Hiyar-gatto.

Regional and other names- *Marathi* –Bhuyaterada.

Distribution - Throughout India.

Taxonomic description - Procumbent, suffruticose perennials, with spreading, winged branches. Leaves linear-oblong, margins spinous-serrulate. Flowers in subradical, globose heads. Seeds ovoid-oblong, hairy.

Flowering & fruiting - October-March.

Habitat ecology - Commonly found in rocky to sandy habitats throughout the area.

Regeneration - By seeds.

Field no. - DJ-176.

Parts used : Inflorescence.

Chemical constituents : Leaves- Protein, Fat, Carbohydrates, Fibres, Iron, Calcium, Phosphorous, Silica.

Folk uses

Magico-religious belief: The tribal people place a bunch of this plant on the doors of their houses to keep away Jackal.

Medicine: Dried inflorescence is burnt to prepare ash. This ash is mixed with coconut oil and a paste is prepared. This paste is applied as ointment on burns & eczema.

Uses in literature

Whole plant- Antiallergic (Sudhakar & Rolla, 1985); Wounds (Singh & Pandey, 1980).

Lepidium sativum L. (Brassicaceae)

Vernacular name : Asalio.

Regional and other names : *Bengali* – Halim; *English* – Water cress; *Gujarati* – Asahio; *Hindi* – Halim, Candsur; *Kannada* – Alavibija;

Malayalam – Asali; *Marathi* –Ahliva; *Sanskrit* – Asalika; *Tamil* – Aliviral; *Telugu* – Aditya.

Distribution - Cultivated in the major part of India.

Taxonomic description - Erect annual herb. Leaves pinnatifid, sessile. Racemes corymbose. Flowers white. Fruits pod. Seeds brown.

Flowering & fruiting - February-May.

Habitat ecology - The cultivated fields & roadsides.

Regeneration - By seeds.

Field no. - DJ-126.

Parts used - Seeds.

Chemical constituents : Glucotropoeoline, Sinapine, Sinapic acid, ß-sitosterol, Benzylcyanide, Vitamins, Essential oil.

Biological activity

Root - Acrid, bitter.

Leaves - Antibacterial, diuretic, stimulant.

Seeds - Aphrodisiac, bitter, depurative, diuretic, rubefacient.

Folk uses

Medicine: The seeds are cooked (~50 gm) in pure cow milk (250 gm) to prepare 'Kheer"(Sugar may be added) and taken by the children (10-18 years old) for 3 days for improvement in their height.

Uses in literature

Plant - Asthma, bronchitis, colic, dysentery, pain, stomachache (Jain & Tarafder, 1970).

Fruit - Gout (Jadeja *et al.,* 2006).

Leaf- As vegetable (Jain & De, 1966).

Leucas cephalotes (Koen. ex Roth.) Spreng. (Lamiaceae)

Syn.: *Phlomis cephalotes* Koen. ex Roth.

Vernacular name : Kubhi.

Regional and other names : *Bengali* – Barahalkasa; *Gujarati* - Doshinokubo; *Hindi* - Guma; *Marathi* – Deokhumba; *Sanskrit* - Dronpuspi; *Telugu* – Peddatumni.

Distribution - Punjab, Bengal, Assam, Himalaya, Rajputana Desert, Kathiawar, Gujarat, Tamilnadu.

Taxonomic description - Erect, hispid herbs, with ovate, crenate-serrate leaves. Flowers white, in large, globose terminal heads. Calyx-tube bent, mouth produced on the upper side, villous within. Nutlets obovoid-oblong.

Flowering & fruiting - September– December.

Habitat ecology - Common in weed wastelands and fields.

Regeneration - By seeds.

Field no. - DJ-165.

Parts used - Leaves.

Chemical constituents : Plant contains B- Sitosterol, Flavonoides, Glycosides, Glucose, Alkaloids, Baballenic acid, Aromatic oil.

Biological activity : Anthelmintic, antiseptic, antispasmodic, appetiser, carminative, cathartic, diaphoretic, expectorant, heating, indigestible, laxative, stimulant.

Folk uses

Food: Leaves are used as vegetable.

Uses in literature

Whole plant - Cold (Sharma *et al.,* 1979); Cough (Rajwar, 1983); Chronic fever, nervous disorders (Narain & Singh, 2006b); Malaria (Saxena & Vyas, 1981); Skin diseases (Deokule, 2006); Jaundice (Arinathan *et al.,* 2006; Dutta Choudhury & Shil, 2006).

Leaf - Headache (Maheshwari & Singh, J. P. 1984); Constipation (Srivastava & Varma, 1981); Fever, diarrhoea (Saxena & Vyas, 1981); Nasal bleeding (Sarma *et al.,* 2006); Scabies (Chatterjee & Bhatterjee, 2006); Cold (Arinathan *et al.,* 2006); Cough (Srivastava & Varma, 1981; Arinathan *et al.,* 2006); Tonsilitis (Rajendran & Rengamani, 2006); Stomach trouble (Das *et al.,* 2006); As vegetable (Jain & De, 1966).

Flower - Fever (Maheshwari & Singh, 1984).

Fruits - Fever (Maheshwari & Singh, 1984); Malaria (Deka *et al.,* 2006).

Limonia acidissima L. (Rutaceae)

Syn.: *Feronia elephantum* Corr.

Vernacular name : Kavit.

Regional and other names : *Bengali* – Katbel; *English* - Elephant apple; *Gujarati* - Kothun; *Hindi* - Kait, kavitha; *Kannada* – Bela; *Malayalam* - Vila, vilatti; *Marathi* – Kawatha; *Sanskrit* -Kapittha; *Tamil* - Velamarum; *Telugu* – Velega.

Distribution - Indigenous in S. India, cultivated in many parts of India.

Taxonomic description – Medium-sized trees, armed with spine. Leaves imparipinnate, with 3-9 leaflets. Flowers pentamerous, yellowish. Fruits globose, with woody rind.

Flowering & fruiting - February– May.

Habitat ecology - Rare, in wastelands and outstriks of forests.

Regeneration - By seeds & vegetative method.

Field no. - DJ-77.

Parts used : Fruits, Roots.

Chemical constituents : Fruits contain Riboflavine, Vitamin – C, Pectin, Iron, Phosphorus.

Leaves contain Volatile oil (Estragol).

Biological activity :

Fruit - Acrid, alexipharmic, aphrodisiac, refrigerant, sour, sweet.

Leaf - Astringent, carminative.

Folk uses

Food: Fruits are used for making chuteny. Fruit pulp is eaten by the children.

Medicine:

- Extract of root is taken in the cases of snake bite.
- Paste of root is applied on breast to cure tumor.
- The powder of unripe fruit is taken with water in the cases of obesity.
- Extract of fruit pulp is taken with water to cure menorrhoea.

Miscellaneous: Fruits are sold in the market.

Uses in literature

Root - Bodyache (Chaudhuri *et al.,* 1975).

Fruits - Diarrhoea (Bhalla *et al.,* 1982); Tonic (Deokule, 2006); Dysentery, dyspepsia (Bhalla *et al.,* 1982; Jadeja *et al.,* 2006); Edible (Maji & Sikdar, 1982; Arora, 1981; Singh & Singh, 1981; Shah *et al.,* 1981; Singh & Maheshwari, 1985).

Bark - Itching (Sexena & Vyas, 1983).

Linum usitatissimum L. (Linaceae)

Vernacular name : Alsi.

Regional and other names : *Bengali* – Tishi; *English* – Linseed, Flax plant; *Gujarati* – Alshi; *Hindi* – Alsi; *Kannada* – Alasibija; *Malayalam* – Akasi; *Marathi* – Javas; *Sanskrit* – Atasi; *Tamil* – Alivirai; *Telugu* – Avisi.

Distribution - Cultivated in the major part of India.

Taxonomic description - An erect annual. Leaves linear or lanceolate, 3-nerved at base. Flowers large, blue, in terminal, leafy, corymbose panicles. Fruits capsule. Seeds dark brown.

Flowering & fruiting - Februray-March.

Habitat ecology - Cultivated during the winter season.

Regeneration - By seeds.

Field no. - DJ-149.

Parts used - Seeds.

Chemical constituents: Seed - Proteins, Glucosides, Phytin, Lecithin, ß-carotenoids, Cholesterol, Campesterol, Stigmasterol, Sitosterol, 5-dehydro-avenasterol, Cyclosterol & 24-Methylenecycloartanol.

Seed oil - Proteins, Wax, Resin, Sugar, Phosphates, Linamarin.

Biological activity

Flowers - Cardiotonic.

Seeds - Acrid, bitter, diuretic, emollient, expectorant, laxative, sweet, thermogenic.

Folk uses

Medicine: Paste of seeds is applied on stomach to cure kidney stone problem.

Uses in literature

Seed – Wounds (Tarafder, 1984); Swelling (Deokule, 2006).

Root - To make black and silky hair (External application) (Jadeja *et al.*, 2006).

Luffa acutangula (L.) Roxb. (Cucurbitaceae)

Syn.: *Cucunis acutangula* L.

Vernacular name : Karvi taroi.

Regional and other names : *Bengali* – Titotorai; *English* - Bitter ribbed gourd; *Hindi* - Karvitori; *Kannada* – Kaduhire; *Malayalam* – Athanga; *Marathi* – Ranturai; *Sanskrit*- Kosataki; *Tamil* - Peypirkam; *Telugu* – Adavibira.

Distribution - Throughout India.

Taxonomic description - Extensive climbers, with 3-fid tendrils and palmetely 5 to 7- angled or lobed leaves. Flowers yellow; stamens 3. Fruits 10-angled.

Flowering & fruiting - September– November

Habitat ecology - Naturalized in the outstricks of forests and near habitations.

Regeneration - By seeds.

Field no. - DJ-161.

Parts used - Seeds.

Chemical constituents

Fruits contain bitter Alkaloids

Seed contains Oil & Albumin

Biological activity

Plant - Acrid, alexiteric, bitter, carminative, digestive, laxative, pungent.

Fruits - Cathartic, emetic.

Folk uses

Medicine: The extract of seeds is given with water in the case of the cattle do not like to eat anything.

Uses in literature

Fruits - Fever (Sharma & Malhotra, 1984); Diabetes (Deokule, 2006).

Madhuca indica Gmel. (Sapotaceae)

Syn.: *M. longifolia* (Koenig) MacBridev var. *latifolia* (Roxb.) Chev.; *Bassia latifolia* Roxb.

Vernacular name : Mauwdo, Mauwdi.

Regional and other names : *Bengali* – Mahua; *English* - Butter tree, mohwa tree; *Gujarati* - Mahuda; *Hindi* - Mahua; *Kannada* – Hippe; *Malayalam* – Poonamilupa; *Marathi* – Mahua; *Sanskrit* - Maduka; *Tamil* - Iiuppai; *Telugu* - Ippa.

Distribution - Indigenous chiefly in the moist forests in the western parts of India.

Taxonomic description - Large, deciduous trees, with elliptic-obovate leaves and white flowers in axillary fascicles. Berries ovoid.

Flowering & fruiting - April – June.

Habitat ecology - Found in the open plain area of the forest & along with road sides & rivers.

Regeneration - By seeds & coppices.

Field no. - DJ-35.

Parts used - Flowers, Leaves, Seeds, Stem-bark.

Chemical constituents : Flowers: Sugar, Cellulose.

Biological activity

Plant - Acrid, anthelmintic, aphrodisiac, bitter, cooling, flattening, sweet.

Flowers - Aphrodisiac, cooling, sweet.

Fruits - Aphrodisiac, cooling, indigestive, sweet.

Seed oil - Acrid, sweet.

Folk uses

Beverage: The tribals distil spirituous liquor from the flesh corollas (Mahuri).

Fodder: Leaves serve good fodder for cattle.

Medicine:

- Boiled flowers are tied on injured part of body to cure wounds.
- Poultice of boiled flowers is tied in the case of fracture.
- Extract of seed is taken in the cases of scorpion sting and snake bite.
- Decoction of stem bark is useful in diarrhoea.

Miscellaneous: The flesh corollas are sold in the market by the tribals.

Oil: The seed oil is used for making vagetables.

Totem & Taboo: This species is considered as a plant totem of Maida, Machar clan of Bhils and, therefore, preserved and whorshipped by them.

Uses in literature

Leaf - Burns (Negi *et al.,* 1985).

Seeds - Snake bite (Janaradhanan, 1963).

Seed oil - Pneumonia, chest pain (Tarafder & Rai Chaudhuri, 1981); Heal cracks (Narain & Singh, 2006).

Flower - Bone fracture, colic (Nautiyal, 1981); Scorpion bite (Maheshwasri & Singh, 1987); Rheumatism, swellings (Narain & Singh, 2006); cough (Nautiyal, 1981; Basak, 2006); Edible (Maheshwari & Singh, 1984; Maji & Sikdar, 1982; Goel *et al.,* 1984; Singh & Maheshwari, 1985; Nagar, 1985; Maheshwari *et al.,* 1981; Shah, 1984); As vegetable (Bedi, 1978; Jain, 1964; Negi *et al.,* 1985).

Bark - Diabetes, rheumatism (Bhalla *et al.,* 1982); Itching, ulcer, gum trouble (Nautiyal, 1981); Toothache (Goel *et al.,* 1984; Narain & Singh, 2006b); Snake bite (Bedi, 1978); Diabetes (Hembrom & Goel, 2005).

Fruit- Edible (Singh & Singh, 1981; Maheshwari & Singh, 1984; Jain, 1984; Singh & Maheshwari, 1985); As vegetable (Maji & Sikdar, 1982).

Mangifera indica L. (Anacardiaceae)

Vernacular name : Aam, Ambo.

Regional and other names : *Bengali* – Aam; *English* - Mango; *Gujarati* - Ambo; *Hindi* - Aam; *Kannada* - Mavinakayi, marv; *Malayalam* - Amram, cutam; *Marathi* – Amba; *Sanskrit* - Amra; *Tamil* - Ambiram, mau; *Telugu* - Mamidi, mavi.

Distribution - Throughout India.

Taxonomic description - Evergreen trees, with oblong-lanceolate leaves and polygamous flowers in panicles. Stamens and style solitary. Drupes yellow when ripe.

Flowering & fruiting - Febuary – June.

Habitat ecology - Commonly planted along roadsides & wastelands near habitations.

Regeneration - By seeds & grafts.

Field no. - DJ-85.

Parts used - Gum, Leaves, Fruits, Stem-bark, Wood.

Chemical constituents - Mangiferin, Amino acids, Gallotanin, Gallic acid, Ethylgallati, Isoquercetin, ß- Sitosterol, Riboflavin, Vitamin-C, Citronellal, Deterpene, Geraniol, Limonene, Mangiferol, Mangiferon, Nerol & Tannine, Vitamin- A, Ascorbic acid, Protein, Fat, Carbohydrate, Calcium, Phosphorous, Iron, Fructose, Sucrose, Glucose.

Biological activity

Bark - Acrid, astringent, cooling.

Leaf - Acrid, astringent.

Flower - Appetizer, astringent, cooling.

Unripe fruits - Acrid, astringent, sour.

Ripe fruits - aphrodisiac, appetiser, astringent, cooling, sweet.

Seed - Acrid, sweet.

Seed oil - Acrid, bitter, sweet.

Folk uses

Fodder: Leaves serve a good fodder for cattle.

Food: Unripe fruits are used for pickles and chutney. Ripe fruits are eaten by the tribals.

Medicine:

- The sap of unripe fruit is applied against scorpion sting.
- Extract of stem-bark is taken orally with water to cure diarrhoea.
- The mesocarp of boiled 2 unripe fruits is mixed in a glassful of water with one teaspoonful sugar and one teaspoonful salt. This juice is taken orally 2-3 times in a day to cure diarrhoea.
- The gum is filled in teeth cavity to cure toothache.

- The powder of dried small unripe fruits is taken with water to cure loose motion.
- Pickle of fruits is made into paste with 25 gm. Haldi (*Curcuma longa* L.) in 250 ml mustard oil and given to cattle in the case of Afra disease.
- Gargles of leaves decoction gives relief in mouth ulcer, toothache, pyorrhoea and gums trouble.
- The powder of seed is taken with water to cure leucorrhoea.

Miscellaneous: Unripe fruits (Keri) and seeds (Gutheli) are sold in the market by the tribals.

Timber: Bhils make musical instruments called Tambura, Dholak, Khanjri' & 'Dhol' from its wood. It is also used in making the gate of huts.

Totem & Taboo: This species is considered as a plant totem of Bhabhar, Aamliya clan of Bhils and, therefore, preserved and whorshipped by them.

Uses in literature

Bark - Abortifacient, antifertility (Billore & Audichya, 1978); Stomatitis (Rao & Jamir, 1982); Jaundice (Singh & Maheshwari, 1983); Dysentery (Deokule, 2006; Sarma *et al.,* 2006).

Fruit - Diarrhoea (Shah, 1984); Scorpion sting (Basak, 2006); Pickle (Nagendra Prasad & Abraham, 1984).

Gum - Toothache (Maheshwari & Singh, 1984); Scorpion sting (Pereira, 2006).

Seed- Gastric problems, indigestion (Udayan *et al.,* 2006); Diarrhoea (Kataria & Gauchan, 2006); Dysentery (Kumar *et al.,* 2006; Kataria & Gauchan, 2006); Stomachache (Arinathan *et al.,* 2006); Ringworm (Rajendran & Rengamani, 2006).

Leaves – Diabetes (Kumar *et al.,* 2006; Ghosh, 2006); Scorpion sting (Basak, 2006); Diarrhoea, leucorrhoea, ulcers (Das *et al.,* 2006).

Martynia annua L. (Martyniaceae)

Syn.: *M. diandra* Glox.

Vernacular name : Bicchu ankro.

Regional and other names- *Bengali* –Bagh noki; *English* –Devil's claw; *Hindi* -Shernui; *Malayalam* -Pulinakham; *Marathi* –Winchauri; *Tamil* -Nagathali; *Telugu* –Garudamukku.

Distribution - Native of mexico, becoming naturalized in India.

Taxonomic description - Glandular pubescent herbs, with fistular stem, ovate-cordate leaves & drooping, pinkish-white flowers. Fruits woody, beetle-shaped, hooked at the tip.

Flowering & fruiting - August-October.

Habitat ecology - Common along road-sides & wastelands.

Regeneration - By seeds.

Field no. - DJ-97.

Parts used - Root, Seeds.

Chemical constituents : Leaves- Chlorogenic acid.

Fruits/Seeds- Pale yellow semi drying Fatty oil, Fatty acids of fatty oil are- Palmitic, Stearic, Arachidic, Oleic, Linoleic acids.

Biological activity

Fruits- Alexeteric.

Folk uses

Food: Seeds are eaten by the tribals.

Medicine: Root extract is applied against scorpion sting.

Uses in literature

Seed- Eczema (Maheshwari *et al.,* 1986);

Leaf-Epilepsy (Sudhakar & Rolla, 1985; Shah *et al.,* 1983); Sore throat (Sudhakar & Rolla, 1985; Bedi, 1978).

Fruit- Rheumatism (Tribedi *et al.,* 1982); Itching (Saxena & Vyas, 1983).

Root- Tonsilitis (Joshi *et al.,* 1980).

Medicago sativa L. (Papilionaceae)

Vernacular name : Rizko.

Regional and other names- *Bengali* –Maina; *English* -Alfalfa; *Hindi* –Maina.

Distribution - A native of W. Asia. Extensively grown in many parts of India.

Taxonomic description - An erect, deep-green, annual or perennial herb; leaves 3-foliolate; leaflets linear-oblong; pod silky; seeds many, reniform, yellow to brown.

Flowering & fruiting - December-June.

Habitat ecology - Common as fodder plant, also as an escape.

Regeneration - By seeds.

Field no. - DJ-36.

Parts used : Leaves.

Chemical constituents : Amino acids, Minerals, Vitamins, Fatty acids, Enzymes, Toxic saponins, Alkaloids, Wax.

Folk uses

Food: Leaves are used as vegetable.

Medicine: Leaves are used as in the form of vegetable curries to cure night blindness.

Uses in literature

Leaf- Kidney diseases (Lal & Yadav, 1983).

Mentha spicata L. (Lamiaceae)

Syn.: *M. virdis* L.

Vernacular name : Pudino.

Regional and other names- *Bengali* –Pudina; *English* –Garden mint; *Hindi* -Podina; *Kannada* –Pudina; *Malayalam* -Putiyina; *Marathi* –Pudina; *Tamil* –Jech-chak; *Telugu* –Pudina.

Distribution - Northern temperate regions.

Taxonomic description - Aromatic, creeping or ascending herbs. Leaves ovate or ovate-elliptic. Flowers purplish.

Flowering & fruiting - Major part of the year.

Habitat ecology - Commonly grown in kitchen gardens.

Regeneration - By vegetative method.

Field no. - DJ-142.

Parts used : Leaves.

Chemical constituents : Leaves contain Diosmetin, Diosmin; Essential oil contains Carvacrol, L-carvone, Dihydrocarveol, Limeonene, Dipentene, Esters.

Biological activity : Stimulant, carminative, antispasmodic.

Folk uses

Food: The leaves are made into sauce (chutney).

Medicine:

- The leaves of pudina is ground with the leaves of marua (*Ocimum basilicum)* & the bulb of onion (*Allium cepa*) and a chutney is prepared. It is taken with food to cure stomach pain.
- The leaves are chewed in the cases of gastric disorders.

Uses in literature

Leaf- Edible (Lakshmanan *et al.,* 1986).

Merremia gangetica (L.) Cufod. (Convolvulaceae)

Syn.: *Convolvulus gangeticus* L.

Vernacular name : Undri, Undervel.

Regional and other names : *Hindi* – Musakari; *Malayalam* – Eliccevi; *Sanskrit* – Akhukarni; *Tamil* – Elikkatukkiari; *Telugu* – Elika jimudu.

Distribution - Throughout India.

Taxonomic description - A prostrate creeping herb with numerous filiform stems, rooting at the nodes. Leaves reniform or ovate-cordate. Flowers yellow, solitary or 2-3 together on a short peduncle. Capsule 2-celled.

Flowering & fruiting - Cold season.

Habitat ecology - Commonly in wastelands throughout the area.

Regeneration - By seeds & vegetative method.

Field no. - DJ-207.

Parts used - Leaves.

Chemical constituents*:* Seeds contain Caffic acid, Ferrulic acid, Sinapic acid, p-Coumaric acid.

Biological activity : *Plant* - Acrid, anthelmintic, bitter, carminative, digestive, diuretic, refrigerant.

Folk uses

Medicine: Paste of leaves is tied on forehead to cure headache during the complaint of Blood Pressure.

Uses in literature

Whole plant - Rheumatic pain (Sinha *et al.*, 2006).

Mimusops elengi L. (Sapotaceae)

Vernacular name : Maulsari.

Regional and other names : *Bengali* – Bakal; *English* - Bullet wood; *Gujarati* - Baphuli; *Hindi* - Mulsari; *Kannada* - Bakula, Pagademara; *Malayalam* - Batulam, Elanni; *Marathi* – Ovalli; *Sanskrit* - Bakulah; *Tamil* - Magulam, Magilam; *Telugu* - Vakulamu, Pogada.

Distribution - Throughout India.

Taxonomic description - Evergreen trees, with elliptic-oblong leaves. Flowers white with pinkish tinge, axillary, solitary or fascicled; corolla 8-lobed; stamens 8; berries elliptic-oblong.

Flowering & fruiting - April-June.

Habitat ecology - Found along the roadside for shade.

Regeneration - By Seeds & vegetative method.

Field no. - DJ-187.

Parts used - Fruits.

Chemical constituents

Bark contains Lupeol & Xylose.

Leaves contain Myrcetin, Quercetin.

Flower contains Glucosides, Quercitol, Ursolic acid, Glucose, Triterpene, Essential oil, Alcohol.

Seed contains Fatty oils, fatty acids – Palmitic acid, Stearic acid, Behenic acid, Oleic acid, Linoleic acid, Lauric acid, Myristic acid, Saponin.

Fruit contains Sugars.

Biological activity

Bark - Acrid, anthelmintic, astringent, cooling.

Flower - Acrid, anthelmintic, astringent, cooling.

Fruits - Acrid, anthelmintic, astringent, cooling.

Seed - Acrid, anthelmintic, astringent, cooling.

Folk uses

Food: Mature fruits are eaten by the tribals.

Uses in literature

Bark - Fever (Bhalla *et al.,* 1982; Deokule, 2006).

Flower - Headache (Bhalla *et al.,* 1982).

Fruit – Edible (Maji & Sikdar, 1982; Vartak, 1959; Ball, 1967).

Manilkara hexandra (Roxb.) Dub.(Sapotaceae)

Syn.: *Mimusops hexandra* Roxb.

Vernacular name : Rayna.

Regional and other names : *Bengali* – Khirkhejur; *English* - Rayan; *Gujarati* - Khirni; *Hindi* - Khirni; *Kannada* – Bakula; *Malayalam* – Pala; *Marathi* - Ranjan, Raini; *Sanskrit* - Kshirini; *Tamil* - Palla, Palai; *Telugu* - Pala, Manjipala.

Distribution - Upper Gangetic Plain, Central & S. India.

Taxonomic description - Small trees; bark rough, grey. Leaves oblong, elliptic or elliptic-ovate; flowers white, in fascicles of 2-6 and from the axils of follen leaves; fruits oblong-ellipsoid, reddish-yellow when ripe; seeds ovoid, reddish-brown, shining.

Flowering & fruiting - January – April.

Habitat ecology - Rare, in the deciduous forests; also planted along the roads.

Regeneration - By seeds.

Field no. - DJ-153.

Parts used - Fruits.

Biological activity - *Fruit* - Aphrodisiac, cooling, indigestion, sweet, tonic.

Folk uses

Food: Mature fruits are eaten by the tribals.

Miscellaneous: Fruits are sold in the market.

Uses in literature

Bark - Bodyache (Maheshwari *et al.,* 1986); Chest pain (Malhotra & Moorthy, 1973); Eye complaint (Saxena, 1986); Toothache (Mokat & Deokule, 2006).

Fruit - Toothache (Mokat & Deokule, 2006); Edible (Shah, 1984; Singh & Singh, 1981; Jain, 1964).

Seed - Stomatitis (Jain, 1965).

Momordica charantia L. (Cucurbitaceae)

Vernacular name : Karelo, Karela.

Regional and other names : *Bengali* – Baramasiya; *English* - Bitter gourd; *Gujarati* - Karelo; *Hindi* - Karela; *Kannada* – Karali; *Malayalam* – Kaypa; *Marathi* – Karli; *Sanskrit* - Karavelli; *Tamil* – Pakar; *Telugu* – Kakara.

Distribution - Cultivated throughout India.

Taxonomic description - Monoecious climbers, with 5 to 7-lobed leaves and yellow flowers. Bracts of male flowers born about or below the middle of peduncle. Fruits muricate-tuberculate.

Flowering & fruiting - April– October.

Habitat ecology - Cultivated for its fruits coocked as vegetable; often runs wild.

Regeneration - By seeds & vegetative method.

Field no. - DJ-171.

Parts used - Fruits and Leaves.

Chemical constituents : Plant contains Aromatic essential oil, Carotene, Glucoside, Saponine, Momoridicine alkaloid, Charantin (non nitrogenous substance).

Biological activity

Root - Acrid, astringent, bitter.

Leaf - Antipyretic, bitter, emetic, purgataive.

Fruits - Acrid, anthelmintic, antidiabetic, anti- inflammatory, appetiser, bitter, depurative, digestive, febrifuge, purgative, stimulant, stomachic, thermogenic.

Folk uses

Food: Fruits are used as vegetable.

Medicine:

- The paste of the leaves of *Ocimum basilicum* with *Momordica charantia* fruit and lime, is useful to cure ringworm.
- Extract of fruit is useful for diabetic patients.

Miscellaneous:

Fruits are sold in the market.

Uses in literature

Fruits - Rheumatism (Shah *et al.,* 1981); Oedema (Lal & Yadav, 1983); Eczema (Shah & Joshi, 1971); Diabetes (Shah *et al.,* 1981; Dwivedi *et al.,* 2005; Arinathan *et al.,* 2006; Ghosh, 2006); Acidity (Pereira, 2006); Malaria, rheumatism (Baruah *et al.,* 2006); Piles, leprosy (Basak, 2006); As vegetable (Jain & De, 1966; Shah *et al.,* 1981; Singh & Maheshwari, 1985; Shah, 1984).

Leaf - Malaria (Maheshwari & Singh, 1984; Baruah *et al.,* 2006; Deka *et al.,* 2006); Rheumatism (Baruah *et al.,* 2006); Piles, leprosy (Basak, 2006); Diabetes (Das *et al.,* 2006).

Momordica dioica Roxb. ex Willd. (Cucurbitaceae)

Vernacular name : Kikoda.

Regional and other names : *Bengali* - Bankarela; *English* - Spiny gourd; *Gujarati* – Kantolan; *Hindi* - Golkandra, kakra; *Malayalam* – Erimaposal; *Marathi* - Kartoli; *Sanskrit* –Karkotaki; *Tamil* – Paluppakkai; *Telugu* - Agakara, Puagakara.

Distribution - Throughout India.

Taxonomic description - Perennial, dioecious climbers, with tuberous roots. Leaves ovate, entire or 3 to 5-lobed. Flowers yellow. Fruits ovoid, beaked, densely echinate with soft spines.

Flowering & fruiting - June – August.

Habitat ecology - Common among the hedges of cultivated fields and outstricks of forests.

Regeneration - By seeds.

Field no. - DJ-195.

Parts used - Root, Fruits.

Chemical constituents : Plant contains Alkaloids, Mangenese.

Biological activity

Leaf - Anthelmintic, aphrodisiac.

Fruits - Alexiteric, bitter, laxative, pungent, stomachic.

Folk uses

Detergent: The tribals use the root as a detergent & substitute for washing the cloths.

Food: Fruits are used as vegetable.

Miscellaneous: Fruits are sold in the market.

Uses in literature

Root - Fever (Saxena & Vyas, 1981); Piles (Puri, 1983); Jaundice, bleeding piles (Deokule, 2006).

Whole plant - Injury (Sebasitian, 1984).

Fruit – Edible (Jain & De, 1966; Singh & Maheshwari, 1985; Shah, 1984; Maji & Sikdar, 1982; Arora, 1981; Singh & Singh, 1981; Maheshwari *et al.*, 1986; Vartak, 1959).

Moringa oleifera Lam. (Moringaceae)

Syn.: *M. pterygosperma* Gaertn.

Vernacular name : Surajno.

Regional and other names : *Bengali* –Sajina; *English* - Horse–radish tree; *Gujarati* - Midhosaragavo; *Hindi* - Sahijan; *Kannada* – Nuggi; *Malayalam* – Murinna; *Marathi* – Munagachajhada; *Sanskrit* - Sobhanjanah; *Tamil* - Murnakai; *Telugu* – Mulaga.

Distribution - Indigenous in the sub-Himalayan tract from the Chenab to the Sarda, also in the Oudh forests. Cultivated throughout India.

Taxonomic description - Small trees, with 3-pinnate leaves and creamy-yellow or white flowers without red streaks on petals. Fruits slender, 3-gonous, 9-ribbed.

Flowering & fruiting - Almost throughout the year.

Habitat ecology - Found along roadsides & also as wild.

Regeneration - By seeds & vegetative method.

Field no. - DJ-98.

Parts used - Fruits, Stem bark, Gum, Wood.

Chemical constituents

Root bark contains Alkaloids, Moringine, Moringinine, Spirochin, Ptrigospermine.

Leaf & fruits contain Vitamin- A, Nicotinic acid, Ascorbic acid, Vitamin- B, Fatty acid.

Flower & fruits contain Amino acids, Glucose & Sucrose, Citric acid, Malic acid, Succinic acid, Fumaric acid, Oxalic acid.

Biological activity

Root - Acrid, anodyne, anthelmintic, anti- inflammatory, bitter, carminative, constpating, diuretic, dysentery, expectorant, stimulating, thermogenic.

Bark - Abortifacient, acrid, antifungal, bitter, thermogenic.

Leaf - Anodyne, anthelmintic, anti- inflammatory.

Seed - Acrid, anodyne, anti- inflammatory, antipyretic, bitter, purgative.

Folk uses

Food: Fruits are used as vegetable.

Medicine:

- Paste of stem bark is applied on forehead to cure headache.
- The crushed gum is put into water for whole night, next day its extract is taken orally with water to cure diarrhoea.
- The extract of root is given with water by the tribals to their cattle in the case of cattle donot like to eat anything.
- Extract of inner bark of the stem is taken with water to cure leucorrhoea.

Miscellaneous: Fruits are sold in the market.

Timber: The wood is used for making a musical instrument 'Tambura.'

Uses in literature

Root - Rheumatism, asthma, liver complaint (Rajwar, 1983); Paralysis, epilepsies (Karnick *et al.*, 1981); Scorpion bite (Maheshwari *et al.*, 1986); Snake bite, toothache (Purkayastha & Subhan, 2006); astringent (Shah, 2005).

Gum - Ear complaint (Karnich *et al.*, 1981).

Bark - Stomach trouble, fever (Nagendra Prasad & Abraham, 1984); Eczema (Sexena & Vyas, 1983); Typhoid (Tarafder, 1986); Boils, toothache (Kumar *et al.*, 2006); Diabetes (Ghosh, 2006); Poisonous bites (Rajendran & Rengamani, 2006); Rheumatism (Das *et al.*, 2006).

Leaf - Scurvy, wound (Shah, 1982); Stye (Kumar & Chauhan, 2005); Night blindness (Deokule, 2006); Diarrhoea, dysentery (Kataria & Gauchan, 2006); Cough (Arinathan *et al.*, 2006); Cold, diseases of liver, spleen (Basak, 2006).

Flowers- Eye ailments (Pereira, 2006); Cold (Arinathan *et al.*, 2006); As vegetable (Maheshwari & Singh, 1984; Shah, 1984; Maji & Sikdar, 1982).

Pods- Eye ailments (Pereira, 2006); Cold, diseases of liver, spleen (Basak, 2006).

Seed- Indigestion, gout (Jadeja *et al.*, 2006).

Seed oil – Gout, rheumatism (Kumar *et al.*, 2006).

Fruit- Pickle (Arora, 1981); As vegetable (Maheshwari & Singh, 1984; Shah, 1984; Maji & Sikdar, 1982; Das *et al.*, 1983; Singh & Singh, 1981; Jain, 1981).

Morus alba L. (Moraceae)

Vernacular name : Sentut, Sentus.

Regional and other names : *Bengali* – Tut; *English* – Mulberry; *Gujarati* - Shetur; *Hindi* - Sehtut; *Kannada* – Hipnerle; *Malayalam* – Pattunulppulucceti; *Marathi* –Tut; *Sanskrit* –Tutam; *Tamil* - Kampilippuccicceti; *Telugu* – Pippalipanducettu.

Distribution - Wild in the sub-Himalayan tract, cultivated elsewhere.

Taxonomic description - Deciduous tree. Leaves ovate to lanceolate, acute or acuminate, dentate serrate or lobed. Flowers monoecious. Male spikes elongated catkins. Female spikes short, ovoid. Fruits white or dark purple to red when ripe.

Flowering & fruiting - March-May.

Habitat ecology - Found in garden & along roadside.

Regeneration - By vegetative method.

Field no. - DJ-172.

Parts used - Inflorescence.

Chemical constituents - Fruits contain Carotene (Vitamin – A), Thiamine, Nicotanic acid, Riboflavin, Ascorbic acid, Calcium, Phosphorus, Iron.

Biological activity

Root - Anthelmintic, astringent, purgative.

Leaf - Antibacterial, diaphoretic, emollient.

Fruits - Aphrodisiac, cooling, diuretic, anthelmintic, laxative, sour, sweet.

Folk uses

Food: Inflorescence eaten by the tribals.

Miscellaneous: Inflorescence sold in the market.

Uses in literature

Root - Appetiser (Bennet, 1985).

Latex - Ulcer (Kumar *et al.,* 1987).

Fruit- Edible (Shah, 1984).

Musa paradisiaca L. (Musaceae)

Syn.: *M. sapientum* L.

Vernacular name : Kero, Kell.

Regional and other names : *Bengali* – Kala; *English* – Banana; *Hindi* – Kela; *Kannada* – Balehannu; *Malayalam* – Kadalivala; *Marathi* – Kele; *Sanskrit* – Kadali; *Tamil* – Kadali; *Telugu* – Ariti.

Taxonomic description - Tree like, stoloniferous herbs. Leaves spirally arranged, oblong. Flowers yellow, lower female, upper male in drooping spathaceous spikes. Fruits 3-gonous, oblong.

Flowering & fruiting - Throughout the year.

Habitat ecology - Cultivated for its edible sweet fruits.

Regeneration - By sucker.

Field no. - DJ-73.

Parts used - Fruits, Root.

Chemical constituents : Serotonin & Norepinephrine, Dopamine and an unidentified Catecholamine.

Biological activity

Root - Anthelmintic, depurative.

Fruits- Anthlemintic, aphrodisiac, astringent, cooling, demulcent, emollient, sweet.

Folk uses

Food: Mature fruits are eaten by the tribals.

Medicine:

- Extract of root is taken with water during menstruation period to prevent conception.
- 2 drops of latex of *Calotropis gigantea* (L.) R. Br. are dropped in ripe fruit of Banana *(Musa paradisiaca)* and eaten alternate day early in the morning to cure jaundice.
- Unripe fruit is eaten in loose motion.
- The extract of spathe is given to cattle in the case of prolapse of uterus.

Uses in literature

Leaf - Asthma, whooping cough (Jain, 1984); Boils (Rai, 1985); Gout (Jadeja *et al.,* 2006).

Flowers - Antifertility (Sharma, 1981); Asthma, whooping cough (Virjee *et al.,* 1984); Edible (Nagendra Prasad & Abraham, 1984).

Root - Diarrhoea (Megoneitso & Audichya, 1978); Antifertility (Billore & Audichya, 1978); Joint pain (Janmeda *et al.,* 2006).

Fruits - Diarrhoea, intestinal worms (Janmeda *et al.,* 2006); Hepatitis (Dutta *et al.,* 2006); Edible (Nagendra Prasad & Abraham, 1984); Diabetes (Ghosh, 2006).

Juice from central trunk - Kidney stone (Rajendra & Rengaswami, 2006).

Stem - Otalgia, haemoptysis (Deokule, 2006).

Mucuna pruriens (L.) DC. (Papilionaceae)

Syn.: *Dolichos pruriens* L.

Vernacular name : Kevuch, Kemuch, Sarsoi, Sarsai.

Regional and other names : *Bengali* – Akolshi; *English* - Cowhage; *Gujarati* - Kancha; *Hindi* - Kaunc; *Kannada* – Nasuganni; *Malayalam* – Corivalli; *Marathi* – Karacha; *Sanskrit* - Kapikacchuh; *Tamil* - Punaippidukkan; *Telugu* – Pilliadugu.

Distribution - Punjab plains, from the base of the Himalaya to SriLanka, Myanmar. Often cultivated.

Taxonomic description - Twining herbs, with 3-foliolate leaves and purple flowers in racemes. Pods densely bristly, slightly curved at both ends.

Flowering & fruiting - August– October.

Habitat ecology - Common along hedges & shrubs and in open areas of the forests.

Regeneration - By seeds & vegetative method.

Field no. - DJ-188.

Parts used - Seeds, Fruits.

Chemical constituents : Seed contains Carbohydrates, Steroids, Phenols, Glycosides, Tannin, Aluminium, Iron, Zinc, Calcium, Magnessium, Potassium.

Biological activity

Root - Anthelmintic, aphrodisiac, bitter, diuretic, emollient, febrifuge, purgative, stimulant, sweet, thermogenic.

Leaf - Anthelmintic, aphrodisiac.

Seed - Anthelmintic, aphrodisiac.

Sour - Anthelmintic, aphrodisiac, astringent, laxative.

Folk uses

Medicine:

- Extract of seed is applied in the cases of scorpion sting.
- 1-2 roasted seeds are chewed by the children to cure pneumonia fever.
- Extract of seeds is taken with water to cure dog bite.

- Bhils scrape the hairs from the fruits and give orally once to their cattle suffering form intestinal worms.

Uses in literature

Root - Cough, (Jain & De, 1966); Dysentery (Chaudhuri *et al.,* 1975); Worm infection, urinary problem (Pereira, 2006); Bleeding piles (Deokule, 2006).

Stem - Eye diseases (Pal, 1984).

Leaf - Blister, boils, ulcers (Goel *et al.,* 1984).

Seed - Urine complaint (Saxena & Vyas, 1986); Snake bite (Rajwar, 1983); Scorpion bite (Bhatt & Sabnis, 1987); Skin diseases (Deokule, 2006); As vegetable (Shah *et al.,* 1981; Bedi, 1978; Ball, 1967).

Fruit- As vegetable (Maji & Sikdar, 1982; Arora, 1981).

Mukia madraspatana (L.) Roem. (Cucurbitaceae)

Syn.: *Cucumis madraspatanus* L.

Vernacular name : Charpoti.

Regional and other names : *Hindi* - Bilari; *Malayalam* – Mukkalpiram; *Sanskrit* - Trikosaki; *Tamil* - Musumusukkai; *Telugu* – Pottibudamu.

Distribution - Throughout India.

Taxonomic description - Annual climbers with wiry branches clothed with white hairs; Leaves orbicular, 3 to 5-lobed. Fruit globose, the size of pea, red when ripe; seeds 1-4, ellipsoids, turgid, white and scabrous.

Flowering & fruiting - July – February

Habitat ecology - Common in wastelands, scrub forests and outstricks of deciduous forests.

Regeneration - By seeds.

Field no. - DJ-167.

Parts used - Fruits.

Chemical constituents: Root contain Columbin.

Biological activity : Anodyne, bitter, carminative, expectorant, refrigerant, sweet.

Folk uses

Food: Mature fruits are eaten by the tribals.

Uses in literature

Seed - Diarrhoea, dibetes (Singh & Pandey, 1980).

Root - Toothache (Singh & Pandey, 1980).

Fruit- Edible (Sen *et al.,* 1985; Maji & Sikdar, 1982).

Neptunia oleracea Lour. (Mimosaceae)

Vernacular name : Loyadi.

Regional and other names : *Bengali* – Paninajak; *English* – Sensitive water plant; *Hindi* – Lajjalu; *Kannada* – Nidrayam; *Malayalam* – Nirtottavati; *Sanskrit* – Alambusa; *Tamil* – Suntaikkirai; *Telugu* – Nirutalavapu.

Distribution - Throughout India.

Taxonomic description - An annual herb. Leaves bipinnate; leaflets 8-15 pairs, small, sensitive, narrow-oblong, obtuse, glabrous; flowers minute, yellow, in axillary oblong heads; pods brown.

Flowering & fruiting - August-October.

Habitat ecology - Common in grasslands throughout the area.

Regeneration - By seeds & vegetative method.

Field no. - DJ-83.

Parts used - Root.

Biological activity - *Plant* - Anodyne, anthelmintic, antidiarrhoeal, astringent, diuretic, refrigerant, sweet.

Folk uses

Medicine: Extract of root is taken with water to cure dysentery.

Nerium indicum Mill. (Apocynaceae)

Syn.: *N. odorum* Soland.; *N. oleander* L.

Vernacular name : Kaner.

Regional and other names : *Bengali* – Karabi; *English* - Indian oleander; *Gujarati* - Kanera; *Hindi* - Kaner; *Kannada* – Kanagilu;

Malayalam – Arali; *Marathi* – Kaneri; *Sanskrit* - Karavira; *Tamil* - Arali; *Telugu* – Karaviramu.

Distribution - Throughout India.

Taxonomic description - Evergreen shrubs, with linear-elliptic leaves and white or red flowers in cymes.

Flowering & fruiting - Throughout the year

Habitat ecology - Cultivated for ornamental purposes.

Regeneration - By cuttings.

Field no. - DJ-74.

Parts used - Flowers, Roots.

Chemical constituents: Root bark contains Neriodorin, Nerioderin, Karabin, Glucoside (Odorin).

Leaf contains Oleandrin, Neriodin, Ursolic acid, Adynerin.

Flower contains Parafin, Benzyl alcohol, Compesterol, Stigmasterol, Fatty acids.

Biological activity :

Root/leaf - Acrid, anthelmintic, aphrodisiac, astringent, bitter, diuretic, febrifuge, stomachic, thermogenic.

Folk uses

Medicine: Extract of flowers/root is used to promote breast milk of mother.

Uses in literature

Root - Abortifacient (Tarafder, 1983); Veterinary diseases (Sebastian & Bhandari, 1984).

Flower - Eye diseases (Lal. & Yadav, 1983).

Latex - Eye diseases (Chaudhuri & Pal, 1976); Blisters, boils (Sebastian & Bhandari, 1984); Ringworm (Kumar & Chauhah, 2005).

Leaf- Skin diseases, eye troubles (Das *et al.,* 2006).

Nicotiana tabacum L. (Solanaceae)

Vernacular name : Tambakhu, Tamaku.

Regional and other names : *Bengali* – Tamaku; *English* – Tobacco; *Gujarati* – Tabacu; *Hindi* – Tamaku; *Kannada* – Tambaku; *Malayalam* – Pokala; *Marathi* – Tambaku; *Sanskrit* – Tamakhuh; *Tamil* – Pugaiyilai; *Telugu* – Pogaku.

Distribution - Native of South America; cultivated throughout the India.

Taxonomic description - Glandular-pubescent herbs, with pinkish flowers and ovate capsules.

Flowering & fruiting - November-March.

Habitat ecology - This plant is not naturally growing in the area, but the plant products/organs are available in the market.

Field no. - DJ-181.

Parts used - Leaves.

Chemical constituents : Leaves contain Nicotine alkaloid.

Biological activity : *Leaves* - Abortifacient, acrid, anodyne, anthelmintic, anti-inflmmatory, bitter, carminative, emetic, laxative, narcotic, sedative, styptic, thermogenic.

Folk uses

Medicine:

- Leaf extract is taken with water in the case of snake bite.
- The paste of leaves is applied on wound.

Miscelleneous: The tobacco smoking habit is rampant throughout the tribal population in the form of bidi, chilam & hookaha.

Uses in literature

Leaf - Eczema, itching (Purohit *et al.,* 1985); Snake bite (Kharkongor & Joseph, 1981); Toothache (Megoneisto & Rao, 1983; Malhotra & Moorthy, 1973); Wounds (Jain, 1963, 1965); Gout (Jadeja *et al.,* 2006).

Seed- Veterinary diseases (Sebastian & Bhandari, 1984).

Nyctanthes arbor-tristis L. (Oleaceae)

Vernacular name : Hiyadi, Tamath, Simhali.

Regional and other names : *Bengali* – Sheuli; *English* - Night jasmine; *Gujarati* - Jayaparvatai; *Hindi* - Harsingar; *Kannada* – Parijata; *Malayalam* – Parijatam; *Marathi* – Parijatak; *Sanskrit* -Sephali; *Tamil* - Pavilamalligai; *Telugu* – Pagadamalli.

Distribution - Cultivated in many parts of India.

Taxonomic description - Large shrubs, with 4-angled branches and ovate or ovate-lanceolate leaves. Flowers white to orange, fragrant. Capsule orbicular, compressed.

Flowering & fruiting - August – November

Habitat ecology - Frequently found on rocky hills.

Regeneration - By seeds, root suckers & cuttings.

Field no. - DJ-105.

Parts used - Leaves, Stem, Root.

Chemical constituents : Leaf & stem contains Beta- Sitosterol, Nyctanthic acid, Iridoid, Glycosides- Arbortristosides A, B, C, D & E, Glucomannan, Glycerides of Lignocetric, Linocetric, Myristic acid, Oleic acid, Palmitic acid, Stearic acid (In Seeds), Astragalin, Nictoflorin, Nyctanthoside, Arborsides, Ascorbic acid, Carotene, Glucose, Fructose, Manitol.

Biological activity

Leaf - Acrid, anodyne, anthelmintic, antibacterial, anti-inflammatory, bitter, depurative, digestive, diuretic, expectorant, febrifuge, laxative, thermogenic.

Flower - Astringent, bitter, carminative stomachic.

Folk uses

Fodder: Leaves serve a good fooder for cattle.

Medicine:

- Leaf juice is applied on ringworm.
- Extract of root is used to cure malaria fever.
- Decoction of leaves is useful in sciatica.
- Paste of leaves is tied on anus to cure piles.
- Extract of seeds is taken with water to cure leprosy.

Timber: The wood is used for thatching the roofs and for making baskets (Topla), supra & broom (Jharu).

Uses in literature

Flower - Skin diseases (Agrawal, 1981).

Bark - Bone fracture (Jain, 1987); Internal Injury (Jain, 1965).

Leaf - Fever (Singh & Maheshwari, 1983; Purkayastha & Subhan, 2006; Das *et al.,* 2006); Malaria (Singh, & Maheshwari, 1983; Narain & Singh, 2006b);Sciatica (Bhalla *et al.,* 1982; Hembrom & Goel, 2005); Rheumatism (Shah, 1982); cough (Chaudhri *et al.,* 1985); Snake bite (Murthy *et al.,*1986); Anthelmintic (Negi *et al.,* 1985); Hiccups (Tarafder, 1983); Joint diseases (Dwivedi *et al.,* 2005).

Leaf juice- Fever (Deokule, 2006); in Curries (Chaudhuri *et al.,* 1985).

Seed- Edible (Shrivastava & Varma, 1981).

Twig- Basketry (Singh & Maheshwari, 1985).

Ocimum basilicum L. (Lamiaceae)

Vernacular name Maruo.

Regional and other names : *Bengali* – Babuitulsi; *English* - Sweet basil; *Gujarati* – Damaro; *Hindi* - Babuitulsi; *Kannada* – Kamakasturi; *Malayalam* – Ramattulasi; *Marathi* – Marva; *Sanskrit* – Barbari; *Tamil* - Tirunittru; *Telugu* – Bhutulasi.

Distribution - Throughout greater part of India.

Taxonomic description - An erect, branching herb. Stem and branches green or purplish. Leaves ovate-lanceolate, entire or toothed. Flowers whitish-pink or purplish, in whorls on simple or branched racemes. Corolla glabrous or variously pubescent. Nutlets black, pitted.

Flowering & fruiting - Throughout the year.

Habitat ecology - Common in waste places.

Regeneration - By seeds & vegetative method.

Field no. - DJ-37.

Parts used - Leaves.

Chemical constituents : Plant contains Ocimene, Methyl chavicol, Ceneol, Linoalool, Pinene- Camphor terpin hydrate, Aromatic oil.

Biological activity: Acrid, anodyne, anthelmintic, antibacterial, antidiarrhoeal, antiemetic, anti-inflammatory, antipyretic, antispasmodic, appetiser, aromatic, bitter, carminative, depurative, digestive, diuretic, expectorant, insecticidal, pectoral, stimulating, stomachic, thermogenic.

Folk uses

Medicine:

- The paste of leaves with bitter gourd (*Momordica charantia*) fruit and lime is useful to cure ringworm.
- A drop of extract of leaves is dropped in ear to check pus formation.

Miscellaneous: Commonly planted along the boundaries of their houses to keep away the snakes.

Uses in literature

Leaf - Cough (Deokule, 2006; Baruah *et al.,* 2006; Gupta, 2006); Skin diseases (Baruah *et al.,* 2006); Headache (Arinathan *et al.,* 2006); Cold, Fever (Gupta, 2006); Gout (Jadeja *et al.,* 2006).

Seed - Diarrhoea (Das *et al.,* 1983; Kataria & Gauchan, 2006); Dysentery (Malhotra & Moorthy, 1973; Kataria & Gauchan, 2006).

Flower - Cholera, dysentery, stomachache (Rao & Jamir, 1982); Cough (Das *et al.,* 1983); Headache (Srivastava & Varma, 1981); Wounds (Pal & Srivastava, 1976).

Ocimum tenuiflorum L. (Lamiaceae)

Syn.: *O. sanctum* L.

Vernacular name : Tulsa, Tulsi.

Regional and other names : *Bengali* – Kalatusi; *English* - Holy basil; *Gujarati* - Talasi; *Hindi* - Tulsi; *Kannada* – Karitulasi; *Malayalam* – Karuttatrttavu; *Marathi* – Tulasa; *Sanskrit* - Tulasi; *Tamil* - Tulaci; *Telugu* – Tulasi.

Distribution - Throughout India.

Taxonomic description - A much branched, woody herb. Stems and branches clothed with soft spreading purplish hairs. Leaves opposite-decussate, ovate, elliptic-oblong, entire or crenate-serrate, hairy, minutely gland-dotted. Flowers purplish-pink, whorld, in racemes, often forming panicles. Nutlets broad ellipsoid, smooth.

Flowering & fruiting- Most part of the year.

Habitat ecology - Cultivated & running wild throughout the area.

Regeneration - By seeds & vegetative method.

Field no. - DJ-38.

Parts used - Leaves.

Chemical constituents : Plant contains Essential oil Eugenol, Eugenol methyl ether, Carvacrol, Caryophyllene, Ursolic acid, Apigenin, Luteolin, Ascorbic acid, Carotene, Alkaloid, Glycosides, Saponins & Tannins.

Biological activity : Acrid, aromatic, bitter, diaphoretic, digestive, diuretic, expectorant, febrifuge, stomachic, vermifuge.

Folk uses

Medicine:

- The extract of leaves with cloves and jaggery is used in cough.
- 2 drops of leaf extract are dropped in eyes to cure conjunctivitis.
- The leaves are chewed in stomach pain.

Uses in literature

Root - Fever (Nagendraprasad & Abraham, 1984).

Seed - Laxative (Dar *et al.,* 1984).

Leaf - Wound (Sharma & Malhotra, 1984); Malaria (Bhalla *et al.,* 1982); Fever (Shah, 1984; Purdyastha & Subhan, 2005); Bronchistis (Sudhakar & Rolla, 1985); Ringworm, leprosy, blood purifier (purohit *et al.,* 1985); Cold (Singh & Pandey, 1980); Cough (Singh & Pandey, 1980); Abortifacient (Tarafder, 1983); Stomach pain (Purkayastha & Subhan, 2005); As tea (Singh & Pandey, 1980).

Opuntia elatior Mill. (Cactaceae)

Vernacular name : Hasla-Thuar.

Distribution - Throughout India.

Taxonomic description - Spreading bush, with flat, succulent, jointed, obovate branches having several straight spines on each cushion.

Flowering & fruiting - January-June.

Habitat ecology - Occasionally found in wastelands; often planted for fencing.

Regeneration - By cuttings.

Field no. - DJ-189.

Parts used : Latex/Juice.

Chemical constituents : Flower contains Glycoside of Isorhamnetin and Quercertain and Glucoside Narcissi. Isolation of Isoquercitrin, Beta-sitosterol, Opintiol. a Polysaccharide contains Galactose and Arabinose from pods.

Biological activity

Fruit- Refrigerant.

Latex- Purgative.

Folk uses

Biofencing: Commonly planted along the boundaries of fields for fencing purposes.

Food: Mature pear-shaped fruits are eaten raw by the tribal people.

Medicine:

- Flower is roasted in the burning dried cow dung. After the flower is roasted, it is eaten for 7 days to cure Asthma.
- 11 seeds of *Piper nigrum* L. are penetrated in stem of opuntia and are pulled back after 5 days. The powder of these seeds is used as "sunghni" / "Naswar" for the treatment of mentally weak patient.
- Two tea spoonful latex is taken with water and also applied in cases of dog bite.
- 5 drops of juice obtained from the plant is mixed with flour of genhu (*Triticum aestivum* L.) and some pills are prepared. One pill taken in the case of dog bite.

Uses in literature

Stem –Boils (Shekhawat & Sunil Anand, 1984); Burns, muscular swelling, wounds (Shah *et al.,* 1983).

Leaf- Veterinary medicine (Sebastian & Bhandari, 1984).

Fruit- Edible (Shah, 1984; Ramachandran & Niar, 1981).

Oroxylum indicum (L.) Vent. (Bignoniaceae)

Syn.: *Bignonia indica* L.

Vernacular name : Faro, Fara.

Regional and other names : *Bengali* – Sanpatti; *English* - Indian trumpet tree; *Gujarati* - Podval; *Hindi* - Arlu; *Kannada* – Tattuna; *Malayalam* – Palakappayyani; *Marathi* – Taitu; *Sanskrit* - Syonakah; *Tamil* - Palaiyudaycci; *Telugu* – Pampana.

Distribution - Throughout India.

Taxonomic description - Trees, with 2 to 3-pinnate leaves. Flowers in racemes; corolla fleshy; stamens 5. Capsules elongate, compressed.

Flowering & fruiting - June – July.

Habitat ecology – Rare, in the forests; often planted.

Regeneration - By seeds & vegetative method.

Field no. - DJ-208.

Parts used - Root.

Chemical constituents

Root bark contain Oroxylin- A bitter Glycoside, Acrid substance, Pectin Chrysie, Baicalein oil.

Seed contain Tetuin, Flavin, Glucoside, Fatty acid & Amino acids.

Bark contain Sodium salicilate.

Biological activity

Root - Acrid, anodyne, anthelimintic, anti-inflammatory, aphrodisiac, appetiser, astringent, bitter, carminative, diaphoretic, digestive, diuratic, expectorant, febrifuge, refrigerant, sweet.

Leaf - Anodyne, stomachic.

Fruits - Acrid, anthelmintic, stomachic, sweet.

Folk uses

Medicine: The extract of root is taken in the cases of jaundice, typhoid and small pox.

Uses in literature

Seed - Piles (Saxena, 1986).

Leaf - Headache & stomachache (Baruah *et al.,* 1987).

Root - Snake bite & stomachache (Maheshwari *et al.,* 1986).

Root bark - Diarrhoea, dysentery (Nagendra Prasad & Abraham, 1984); Induce conception (Saxena *et al.,* 1981); Miscarriage (Tarafder, 1984); Veterinary medicine (Shah *et al.,* 1983).

Stem bark - Body chest pain (Jain, 1965); Scabies (Jain *et al.,* 1973); Jaundice (Bhatt & Sabins, 1987; Mokat & Deokule, 2006); Indigestion (Chetri, 2005); Diarrhoea & dysentery (Purkayastha & Subhan, 2005); Veterinary medicine (Bedi, 1978).

Fruit- Edible (Lal & Yadav, 1983).

Flower- As vegetable (Bhujel *et al.,* 1984).

Orthosiphon pallidus Royle ex Benth. (Lamiaceae)

Vernacular name - Damdo.

Distribution - Throughout India.

Taxonomic description - A suffrutescent herb; leaves ovate, glabrous, serrate or crenate, gland-dotted; flowers white or pale lilac in 6-flowered distant whorls, arranged in terminal racemes; calyx enlarging to twice as long in fruit; nutlets subglobose, smooth, yellowish-brown.

Flowering & fruiting - July-October.

Habitat ecology - Common on gravelly soil in wastelands, throughout the area.

Regeneration - By seeds.

Field no. - DJ-191.

Parts used - Seeds.

Folk uses

Medicine:

- The smoke obtained by burning of dried seeds is given through nasal tract to cure typhoid.
- 2 leaves of *Piper betel* L. are ground with one clove of *Syzygium aromaticum* and seeds of *Orthosiphon pallidus* and a paste is prepared. This paste is applied as ointment on whole body to cure itching.
- Extract of root is taken with water to cure menorrhoea.

Uses in literature

Whole plant- Fever, urinary complaint (Pandey *et al.,* 1981).

Oryza sativa L. (Poaceae)

Vernacular name : Sawal, Chawal.

Regional and other names- *English* -Paddy; *Hindi* -Chawal; *Kannada* –Akki; *Malayalam* -Ari; *Marathi* –Tandula; *Tamil* -Arishi; *Telugu* –Urlu.

Distribution - Much cultivated in wet fields throughout India.

Taxonomic description - Tall, annual grass; panicles lax, at length drooping; spikelets persistent, pale-brown at maturity.

Flowering & fruiting - September-October.

Habitat ecology - Cultivated at very small scale in the area.

Regeneration - By seeds.

Field no. - DJ-190.

Parts used : Seeds.

Chemical constituents : Grain- Protein, Carbohydrate, Fibre, Minerals, Essential amino acids, Free amino acids, Vitamins.

Biological activity : *Grain*- Acrid, sweet, tonic, aphrodisiac, diuretic.

Folk uses

Fodder: The husk (Chawal ko Paral) is a good source of fodder for cattle.

Food:

- The boiled rice are used as food.
- Khichri (a food preparation of rice and pulse) and Kheer (a preparation of rice, milk & sugar) are other food prepared from the rice.

Medicine: 1 kg. rice are put into "Chika" of cow-milk for whole night. Next day the rice grains are dried in shady conditions. After drying them, a powder is prepared from it. One teaspoonful powder is taken with cow milk in the case of leucorrhoea.

Uses in literature

Flower- Antifertility (Sharma, 1981).

Grain – Edible (Maji & Sikdar, 1982).

Pandanus odoratissimus L. f. (Pandanaceae)

Syn.: *P. fascicularis* Lam.

Vernacular name : Kavdo.

Regional and other names : *Bengali* – Keori; *English* - Screw pine; *Gujarati* - Kewoda; *Hindi* - Kevda; *Kannada* – Kedage; *Malayalam* – Kaita; *Marathi* – Keora; *Sanskrit* - Ketaki; *Tamil* - Talal; *Telugu* – Ketaki.

Distribution - Seacoast of the Indian peninsula on both sides, Andamans.

Taxonomic description - Much-branched, small trees, with aerial roots. Leaves drooping, spirally arranged, ensiform, with recurved marginal spines. Flowers white, sweet scented. Fruits woody, yellowish-brown.

Flowering & fruiting - Rainy season.

Habitat ecology - Rare, along the streams in the forests.

Regeneration - By seeds & vegetative method.

Field no. - DJ-39.

Parts used - Root.

Chemical constituents : Plant contains Physcion, Circilineol, Circilineol, n- tricontanol, ß- Stiosterol, Stigmasterol, Compesterol, Daucosterol, Palmitic acid, Stearic acid.

Biological activity

Root - Acrid, anodyne, antiseptic, aphrodisiac, bitter, carminative, depurative, emollient, febrifuge, stomachic, sweet, thermogenic.

Leaf - Acrid, aphrodisiac, bitter, depurative.

Flower - Acrid, anodyne, aphrodisiac, bitter.

Folk uses

Medicine:

- Powder of roots is useful for the prevention of miscarriages.
- Extract of root is taken with water to cure typhoid.

Uses in literature

Root - Snake bite (Apparanantham & Chelladurai, 1986); Miscarriage (Tarafder, 1984).

Leaf - Scabies & syphilis (Nagendraprasad & Abraham, 1984); leprosy & skin diseases (Rao, 1981).

Oil of bracts- Rheumatic pain & headache (Deokule, 2006).

Fruit- Edible (Bhargava, 1983; Singh & Singh, 1981).

Papaver somniferum L. (Papaveraceae)

Vernacular name : Ammal, Afim.

Regional and other names : *Bengali* – Pasto; *English* - Opium poppy; *Gujarati* - Aphina; *Hindi* - Aphim; *Kannada* – Gasagase; *Malayalam* – Avin; *Marathi* – Aphu; *Sanskrit* – Ahiphena, Aphuka; *Tamil* - Kasakasa; *Telugu* – Nallamandu.

Distribution - Cultivated in warmer parts of India.

Taxonomic description - Annual herbs, with oblong leaves and large, showy, white, purple or scarlet flowers.

Flowering & fruiting - January-April

Habitat ecology - Cultivated for opium in Mandsaur district of Western Madhya Pradesh.

Regeneration - By seeds & vegetative method.

Field no. - DJ-40.

Parts used - Leaves.

Chemical constituents : Root & fruits contain Reframidine, Thebain, Codeine, Cryptopine, Narcotine.

Latex contain Papavarine, Morphine, Oxalic acid.

Seed contains Trigly ceride ester, Alkaloid- Somniferine.

Biological activity

Petals - Bitter, expectorant, sedative.

Fruits - Opium obtained from fruits is anodyne, antispasmodic, aphrodisiac, astringent, bitter, constipating, narcotic, sedative, sweet.

Seed - Aphrodisiac, constipating, sweet.

Folk uses

Food: Leaves are used as vegetable.

Medicine: Leaves are tied on eyes to cure swelling.

Uses in literature

Abortifacient, body pain, cholera, dysentery (Rao & Jamir, 1982).

Seed- Diarrhoea, dysentery (Jadeja *et al.,* 2006); Edible (Sharma *et al.,* 1979).

Paspalum scrobiculatum L. (Poaceae)

Syn.: *P. commersonii* Lam.

Vernacular name : Kodra.

Regional and other names- *Bengali* –Kodoadhan; *English* –Punctured paspalum; *Gujarati* -Harik ; *Hindi* -Koda; *Kannada* –Haraka; *Malayalam* -Varaku; *Marathi* –Kadra; *Sanskrit* –Hudala; *Tamil* –Allu.

Distribution - Throughout India.

Taxonomic description - Erect or ascending, perennial grass, often rooting at lower nodes. Leaves tapering to a fine point. Racemes digitate, with 2-ranked spikelets. Caryopsis subglobose.

Flowering & fruiting - September-February.

Habitat ecology - Commonly found near marshy habitats.

Regeneration - By Seeds & vegetative method.

Field no. - DJ-209.

Parts used : Whole plant.

Chemical constituents : Grain- Amino acids: Arginine, Histidine, Isoleucine, Leucine, Lysine, Methionine, Phenylalanine, Threonine, Tryptophan, Valine, n-Lysine, Hentriacontanol, Hentriacontanone, Sitosterol, Campesterol.

Biological activity: *Grain-* Astringent, alexeteric, bitter, cooling, constipating, diuretic, sedative, styptic, sweet, tonic.

Folk uses

Medicine: Boiled plant is tied in sprain.

Uses in literature

Grain- Cold, cough (Maheshwari & Singh, 1987); Edible (Ramachandran & Nair, 1981; Chaudhuri & Pal, 1978).

Phoenix sylvestris (L.) Roxb. (Arecaceae)

Syn.: *Elate sylvestris* L.

Vernacular name : Khajoor.

Regional and other names : *Bengali* – Khajur; *English* – Date sugar palm; *Gujarati* – Kajuri; *Hindi* – Khajuri; *Kannada* – Ichalu; *Malayalam* – Itta; *Marathi* – Shindi; *Sanskrit* – Kharjuri; *Tamil* – Icham; *Telugu* – Itha.

Distribution - Throughout India.

Taxonomic description - Usually unbranched palms, with straight or crooked trunk & large pinnate leaves; leaflets conduplicate, spine-tipped. Spathes woody, separating into boat-shaped valves. Drupes ellipsoid-oblong, orange-yellow.

Flowering & fruiting - Februray-August.

Habitat ecology - Abundant in low-lying areas.

Regeneration - By seeds.

Field no. - DJ-104.

Parts used - Leaves, Fruits, Stem.

Chemical constituents : Thiamine, Riboflavin, Nicotinic acid, Vitamin-C, calcium, Phosphorus, Iron, Sodium, Potassium, Copper, Cobalt, Nickel, Magnessium, Molybdenum.

Biological activity : *Fruit* - Aphrodisiac, cardiotonic, constipating, cooling, restorative, sweet.

Folk uses

Fodder: Leaves serve a good fodder for cattle.

Food:

- Mature fruits are eaten by the tribals and the tribals also distil liquor called 'Tari' from the sap obtained from stem for their own use.
- The jaggery (Gur) is also prepared from the juice obtained from the tree.

Medicine:

- Extract of roots is mixed with mishri (Crystalline sugar) and taken with water to cure menorrhoea.
- The young and fresh leaflet is chewed in the cases of burning in urethra during micturition.

- The juice obtained from the tree (tari) is considered a cooling beverage. It is also useful in the case of undesirable discharge of semen with urine.

Miscellaneous: Fruits are sold in the market. The leaves are used for making brooms (Jharu), Tokri (bask-ets), mat (Chatai), supra, fans (Pankha) etc. & for thatching.

Timber: The stem is used in their huts like a beam and rafter.

Uses in literature

Leaf - Diarrhoea, dysentery (Rai, 1985); Gout (Jadeja *et al.,* 2006).

Stem - Gonorrhoea (Tarafder, 1984).

Root - Toothache (Janardhanan, 1963); Menstrual complaint (Hemadri & Rao, 1983).

Fruit- Edible (Shah, 1984; Maji & Sikdar, 1982; Jain & De, 1966; Bhujel *et al.,* 1984; Singh & Singh, 1981; Goel *et al.,* 1984; Vartak, 1959; Nagar, 1985; Datar & Vartak, 1975; Singh, 1945).

Phyllanthus emblica L. (Euphorbiacae)

Syn.: *Emblica officinalis* Gaertn.

Vernacular name : Amlo.

Regional and other names : *Bengali* – Amalki; *English* - Emblic myrobalan; *Gujarati* - Ambri; *Hindi* - Amvala; *Kannada* – Nellka; *Malayalam* – Nellikka; *Marathi* – Anvala; *Sanskrit* - Amalaki; *Tamil* - Nelli; *Telugu* – Amalakamu.

Distribution - Throughout India.

Taxonomic description - Large, deciduous trees, with distichous, linear-oblong leaves. Flowers greenish-yellow, in fascicles on leafless branches. Fruits globose, fleshy; seeds bony.

Flowering & fruiting - February – April.

Habitat ecology - Not very common; planted along the boundaries of fields.

Regeneration - By seeds & vegetative method.

Field no. - DJ-84.

Parts used - Fruits, Stem-bark, Seeds.

Chemical constituents : Fruits contain Protein, Fats, Fibres, Carbohydrates, Vitamin- C, Nicotinic acid, Tannins.

After drying fruits contain Gallic acid, Ellagic acid, Flavin & Glucose.

Seed contains Linoleic acid, Linolenic acid, Oleic acid.

Biological activity

Root bark - Astringent.

Fruits - Acrid, alterant, anodyne, antipyretic, aphrodisiac, astringent, bitter, carminative, cooling, digestive, diuratic, laxative, sour, stomachic, sweet.

Folk uses

Firewood: Twigs are used for fire.

Food: Mature fruits are eaten by the tribals.

Medicine: Extract of stem bark is taken orally with water to cure diarrhoea.

Miscellaneous: Fruits & seeds are sold in the market.

Uses in literature

Seed - Asthma, bronchistis (Sharma *et al.,* 1979); Madness (Jain & De, 1966); Menstrual trouble (Das *et al.,* 2006).

Leaf - Diabetes, dysentery (Negi *et al.,* 1985); Scorpion bite, wounds (Maheswari & Singh, 1987); Urine problems (Baruah *et al.,* 2006); Veterinary diseases (Tarafder, 1983).

Fruits - Fever & indigestion (Bhalla *et al.,* 1982); Headache, liver complaint (Jain & De, 1966); Stomach complaint (Virjee *et al.,* 1984); Eye complaint (Megoneitso & Rao, 1983); Diabetes (Tiwari *et al.,* 1973; Ghosh, 2006; Das *et al.,* 2006); Bronchitis (Joshi *et al.,* 1980); Dysentery (Purkayastha & Subhan, 2006; Rajendran & Rengamani, 2006); Inflammation of the eyes (Upadhyay & Jaswant Singh, 2005); Diuretic, tonic (Deokule, 2006); Flatulence (Jadeja *et al.,* 2006); Urine problems (Baruah *et al.,* 2006); Malaria (Deka *et al.,* 2006); Anaemia (Kadavul & Jisha, 2006); Hepatitis (Dutta Choudhury & Shil, 2006); Scurvy (Basak, 2006); Constipation (Jain & De, 1966; Das *et al.,* 2006); Edible (Maji & Sikdar, 1982; Jain & De, 1966; Arora, 1981; Bhujel *et al.,* 1984; Singh & Singh, 1981; Vartak, 1981; Nagar, 1985; Maheswari *et al.,* 1986; Datar & Vartak, 1975).

Phyllanthus fraternus Webster (Euphorbiaceae)

Vernacular name : Bekhri.

Regional and other names : *Bengali* - Bhui amla; *Gujarati* – Bhonyaanmali; *Hindi* - Jarmla, bhonyabali; *Kannada* - Nela nelli; *Malayalam* – Kirganellia; *Marathi* – Bhuiavali; *Sanskrit* –Bhudhatri; *Tamil* – Kilanelli; *Telugu* - Nela usirika.

Distribution - Throughout India.

Taxonomic description - Erect, annual herbs; stem subterete or obscurely angled; leaves oblong; stipules lanceolate; male flowers 1-3 together; disc lobes 6, glandular; female flowers solitary; perianth segments oblong-spathulate; seeds brown.

Flowering & fruiting - July- November.

Habitat ecology - Commonly distributed throughout the area.

Regeneration - By seeds & vegetative method.

Field no. - DJ-178.

Parts used - Whole plant.

Chemical constituents : Leaves contain Bitter, Phyllanthin & Hypophyllanthin, Alkaloid, Lignas, Vitamin- C, Quercetin, Astrogaln, Querscitrin, Rutin.

Root contains Glucoflavon.

Seed contains Oil which contains Linoleic & Linolenic acid.

Biological activity

Acrid, alexipharmic, bitter, cooling, diuretic, sour, sweet.

Folk uses

Medicine:

- Extract of whole plant is taken with water to cure viral fever.
- Extract of the plant is given by the tribals to their oxen to heal up the wound caused due to ploughing.
- The chutney of the plant is eaten with wheat bread to cure liver disorders and jaundice.

Uses in literature

Root - Abortifacient (Tarafder, 1983); Jaundice (Mishra & Sahu, 1984).

Leaf - Allergy (Bhalla *et al.,* 1982); Gonorrhoea (Mishra & Sahu, 1984); Dysentery (Banerjee, 1977); Jaundice (Sharma *et al.,* 1977).

Stem - Dysentery (Mishra & Sahu, 1984).

Whole plant - Jaundice (Nagendra Prasad & Abraham, 1984); Dropsy & gonorrhoea (Mishra & Sahu, 1984); Ulcer, dysentery (Rajwar, 1983); Boils (Maheshwari *et al.,* 1980).

Physalis minima L. (Solanaceae)

Vernacular name : Popcho.

Regional and other names : *Bengali* – Bantepariya; *English* - Country gooseberry; *Gujarati* – Parpoti; *Hindi* - Cirpoti; *Kannada* – Guddehannu; *Malayalam* – Nottanotiyan; *Marathi* – Chroboti; *Sanskrit* – Cirapotha; *Tamil* – Tottakkali; *Telugu* – Kuppante.

Distribution - Throughout India.

Taxonomic description - Erect herbs, with sinuate-dentate leaves. Flowers yellow, axillary, solitary; calyx 5-angled, enclosing yellow berries.

Flowering & fruiting - August- January.

Habitat ecology - Common in moist and shady habitats throughout the area.

Regeneration - By seeds & vegetative method.

Field no. - DJ-177.

Parts used - Fruits.

Chemical constituents : Plant contains Steroidal physalin A, B & C. Epaxyphysalin- B, Plamitic acid, Stearic acid, Oleic acid, Linoleic acid, Hexadeconoid acid, Epoxy & Hydroxy fatty acid.

Biological activity : Appetiser, bitter, cooling, diuretic, expectorant, luxative, sweet.

Folk uses

Food: Mature fruits are eaten by the tribals.

Medicine:

- Extract of leaves and unripe fruit is taken with water to cure fever.
- Paste of leaves is tied on swollen body part.

Uses in literature

Whole plant - Earache (Sudhakar & Rolla, 1985).

Leaf - Fever (Maheshwari *et al.,* 1986).

Root - Gastric diseases (Krishna & Singh, 1987).

Fruits - Diarrhoea & toothache (Chhetri, 2005); Edible (Singh & Singh, 1981; Maheshwari *et al.,* 1986; Ramachandran & Nair, 1981; Singh & Maheshwari, 1985).

Piper betle L. (Piperaceae)

Vernacular name : Nagervelo, Paan.

Regional and other names : *Bengali* – Pan; *English* – Betel; *Gujarati* –Nagervel ; *Hindi* – Pan; *Kannada* – Villayadele; *Malayalam* – Vettilakkoti; *Marathi* – Pan; *Sanskrit* – Tambulavalli, Tambulah; *Tamil* – Vettilai; *Telugu* – Tamalapaku.

Distribution - Cultivated in the hotter and damper parts of India.

Taxonomic description - This species is not found wild in the area. The leaves are available in the market.

Flowering & fruiting - September-April.

Field no. - DJ-129.

Parts used - Leaves.

Chemical constituents : Leaves contain Nicotinic acid, ascorbic acid, carotin, essential amino acids.

Biological activity: *Plant* - Acrid, anthelmintic, aphrodisiac, astringent, bitter, carminative, expectorant, febrifuge, laxative, stomachic, sweet.

Folk uses

Medicine:

- 2-3 petioles are chewed at night to cure cough.
- Boiled leaves are tied on stomach to cure stomach pain.
- 3 leaves are ground with kattha (*Acacia catechu*) and a paste is prepared. This paste is applied on the body of the child who is suffered from rickets.

Miscellaneous: The leaves are chewed with lime, kattha & betel-nuts (supari) etc. as mouth-freshner.

Uses in literature

Leaves - Antiseptic (Rao & Jamir, 1982); Eye diseases (Kharkongor & Joseph, 1981); Stomach disorders (Nagendra Prasad & Abraham, 1984); Wounds (Tarafder, 1984); Cuts, injuries (Baruah *et al.,* 2006); Gout (Jadeja *et al.,* 2006); Dysmenorrhoea (Sarma *et al.,* 2006); Inflammation of eyes (Deokule, 2006); Masticatory (Nagendra Prasad & Abraham, 1984).

Root - Abortifacient (Tarafder, 1983); Anitfertility (Majumdar *et al.,* 1978; Tarafder, 1983).

Piper longum L. (Piperaceae)

Vernacular name : Pipramur.

Regional and other names : *Bengali* – Piplamor; *English* – Indian long pepper; *Gujarati* –Pipli ; *Hindi* – Pipli; *Kannada* – Hipli; *Malayalam* – Tippali; *Marathi* – Pimpli; *Sanskrit* – Pippali; *Tamil* – Pippli; *Telugu* – Pippallu.

Distribution - Hotter provinces of India, SriLanka, Malay Pennisula.

Taxonomic description - This species is not found wild in the area. The leaves are available in the market.

Field no. - DJ-192.

Parts used - Root.

Chemical constituents : Roots - Piperlongumine, Piperlongu-minine, Sesamin, Piperine, Methyl-3, 4, 5-trimethoxycinnamate.

Biological activity

Root - Anthlemintic, bitter, digestive, diuretic, expectorant, purgative, stomachic, thermogenic.

Dried spikes - Acrid, aphrodisiac, carminative, digestive, emollient, expectorant, febrifuge, laxative, stomachic.

Folk uses

Medicine:

- Powder of root is taken with tea to cure general fever and pneumonia.
- Extract of root is useful for maintaining body heat during pregnancy.

Uses in literature

Root - Child birth (Tarafder, 1984); Cold, cogh (Jain & De, 1966); Indigestion (Hajra & Baishya, 1981); Snake bite (Mukherjee & Sutapa Deb Roy, 1987).

Fruits - Cold (Maheshwari *et al.*, 1980, 1981); Paralysis, rheumatism (Mukherjee & Sutapa Deb Roy, 1987); Fever (Chhetri, 2005); Cough (Chhetri, 2005; Maheshwari *et al.*, 1980, 1981).

Piper nigrum L. (Piperaceae)

Vernacular name : Kari-Mirach.

Regional and other names : *Bengali* – Golmoirch; *English* – Black pepper; *Gujarati* – Kalamari; *Hindi* – Kalimirc; *Kannada* – Miri; *Malayalam* – Kurumulaku; *Marathi* – Kalimire; *Sanskrit* – Maricam; *Tamil* – Marisam; *Telugu* – Miriyalu.

Distribution - Throughout India.

Taxonomic description - This species is neither cultivated nor grows wild in the area, but its seeds are locally available in the market.

Field no. - DJ-41.

Parts used - Seeds.

Chemical constituents: Fruits contain α-cis-bergamotene & β-cis-berg from essential oil, Pipercide.

Biological activity: *Fruits* - Acrid, alexeteric, alterant, anthelmintic, aphrodisiac, bitter, carminative, digestive, diuretic, rubefacient, stimulant.

Folk uses

Medicine: Powder of 5 seeds of *Piper nigrum* is boiled with tea. This tea is useful in the case of hoarseness of the voice.

Uses in literature

Fruits/Seed - Antifertility (Tiwari *et al.*, 1982); Cold, fever, sore throat, stomachic (Nagendra Prasad & Abraham, 1984); Carminative, stim-ulant, stomachic (Mukherjee & Sutapa Deb Roy, 1987); Constipition, flatulence, gout (Jadeja *et al.*, 2006); Malaria (Deka *et al.*, 2006); Skin diseases (Mokat & Deokule, 2006); Cough (Nagendra Prasad & Abraham, 1984; Sarma *et al.*, 2006; Gupta, 2006); To make smooth and silky hair (External application) (Jadeja *et al.*, 2006); Cold, fever (Gupta, 2006).

Pithecellobium dulce (Roxb.) Benth. (Mimosaceae)

Syn.: *Mimosa dulcis* Roxb.

Vernacular name : Sarkari imli, Angreji-amli.

Regional and other names- *Bengali* –Dekhani babul; *English* – Mamla tamarind; *Hindi* –Vilayati babul; *Telugu* –Simachinta.

Distribution - Throughout India.

Taxonomic description - Medium-sized trees, armed with spines. Pinnae & leaflets one pair each. Flower-heads dirty white; stamens 1-adelphous. Pods circinate or twisted.

Flowering & fruiting - March-August.

Habitat ecology - Found along roadsides.

Regeneration - By seeds & by cuttings.

Field no. - DJ-103.

Parts used : Leaves, Fruits.

Folk uses

Fodder: Leaves are good fodder for cattle.

Food: The aril of seeds is eaten by the tribals.

Medicine: Decoction of stem bark is useful in diarrhoea.

Miscellaneous: Fruits are sold in the market.

Uses in literature

Leaf- Leprosy (Rajwar, 1983); Promotes hair growth (Nagendra Prasad & Abraham, 1984; Rajwar, 1983).

Fruit- Edible (Nagendra Prasad & Abraham, 1984; Maji & Sikdar, 1982; Singh & Singh, 1981; Ramachandran & Nair, 1981; Shah *et al.*, 1981).

Plectranthus amboinicus (Lour.) Spreng. (Lamiaceae)

Syn.: *Coleus amboinicus* Lour.

Vernacular name : Ajman.

Regional and other names : *Bengali* – Paterchur; *English* - Indian borage; *Hindi* - Patharcur; *Kannada* – Karpurahalli; *Malayalam* –

Kannikkurkka; *Marathi* – Pathurchur; *Sanskrit* – Karpuravalli, Ashmabheda; *Tamil* - Karpuravalli; *Telugu* – Sugandhavalkam.

Distribution - Cultivated throughout India.

Taxonomic description - A small, perennial, hairy herb, about 1-3 ft. high with thick and fleshy aromatic leaves; flowers light purple.

Habitat ecology - Cultivated in gardens, but not observed as wild in the area.

Regeneration - By cuttings.

Field no. - DJ-122.

Parts used - Leaves.

Chemical constituents

Leaves contain Diterpene, Methylene – quinone, Couleon, Cyclobuta tucin.

Root contains Naphtho pyrone (Rocetin) and Dialdehyde, Avolatile oil, Carvacrol is also found.

Biological activity - *Leaf* - Acrid, anodyne, anthelmintic, appetiser, aromatic, bitter, carminative, constipating, digestive, diuretic, expectorant, stomachic, thrmogenic.

Folk uses

Medicine:

- 1-2 leaves chewed daily in the cases of vomiting, gastric comlaints and stomach pain.
- Extract of leaves is given to the children in the case of cough, cold and pneumonia.
- Extract of leaves is given to promote breast milk of mother.

Plumbago zeylanica L. (Plumbaginaceae)

Vernacular name : Sitawal, Chitawal.

Regional and other names : *Bengali* – Chita; *English* - Ceylon leadtwort; *Gujarati* – Chitaro; *Hindi* – Chita; *Kannada* – Chitramula; *Malayalam* – Tumbakoduveli; *Marathi* – Chitramula; *Sanskrit* – Chitrak; *Tamil* – Cithiramulam; *Telugu* - Tella chitramulamu.

Distribution - Throughout India.

Taxonomic description - Erect or straggling undershrubs, with ovate-lanceolate leaves and white flowers in panicled racemes. Seed oblong, smooth.

Flowering & fruiting - July- March.

Habitat ecology - Found in rocky places and along the roadside hedges.

Regeneration - By seeds & vegetative method.

Field no. - DJ-42.

Parts used - Root.

Chemical constituents

Plant contains Volatile oil, Chitranone, Alpha & Beta- Amyrin, Lupeol, Taraxasterol, Fructose, Glucose, Ivertase, Protease, Chloro-plumbagin, Droserone, Elliptinone, Zeylanone, Zeylinone, Meritone, Catechol, Tannin, Amino acids, Plumbagic acid.

Root contains Plumgagin alkaloids.

Biological activity : *Root bark* - Alterative, anthelmintic, astringent, bitter, carminative, dry, hot, stomachic.

Folk uses

Medicine:

- Extract of root is taken with tea to cure headache, cough, cold, fever, skin itching & pneumonia.
- Extract of root is taken with curd to cure stomachic during mensus.
- Extract of root is boiled with jaggery and a syrup (Kadha) is prepared. It is taken at night (Sleeping time) in the case of fever.

Uses in literature

Root - Skin diseases (Janardhanan, 1963); Hydrocele (Hembrom & Goel, 2005); Piles & paralysis (Bhalla *et al.,* 1982); Rheumatism, dyspepsia (Shah, 1984); Abortifacient (Tarafder, 1983); Wounds & abdominal pain (Joshi *et al.,* 1980); Antifertility (Goel *et al.,* 1984; Purkayastha & Subhan, 2006); Diarrhoea (Sharma *et al.,* 1979; Kataria & Gauchan, 2006); Dysentery (Sebastian & Bhandari, 1984; Kataria & Gauchan, 2006); Eczema (Shah *et al.,* 1983); Muscular pain (Jain, 1965); Fever (Gupta, 1981); Headache (Venugopal & Reghunathan, 1982); Leucoderma (Bhatt & Sabnis, 1987); Skin diseaes (Deokule, 2006); Veterinary diseases (Sebastian & Bhandari, 1984).

Root bark- Constipation (Kadavul & Jisha, 2006).

Fruit- Gout (Jadeja *et al.*, 2006).

Pongamia pinnata (L.) Pierre (Papilionaceae)

Syn.: *Cytisus pinnatus* L.

Vernacular name : Karanj.

Regional and other names : *Bengali* – Karanja; *English* - Indian beech; *Gujarati* – Kanaji; *Hindi* - Karanj; *Kannada* – Hongemara; *Malayalam* – Ponnu; *Marathi* – Karanja; *Sanskrit* - Karanjah; *Tamil* - Ponka; *Telugu* – Kanugacettu.

Distribution - Throughout India.

Taxonomic description - Trees, with imparipinnate leaves. Flowers pinkish-white. Pods obliquely oblong, woody, indehiscent, glabrous.

Flowering & fruiting - February – May.

Habitat ecology - Found along the roadsides & wastelands.

Regeneration - By seeds & vegetative method.

Field no. - DJ-102.

Parts used - Seed, Seed-oil, Leaves.

Chemical constituents

Seed contains Bitter fatty acid oil, Pongamol (Hongay oil) and Protein.

Plant contains Alkaloids, Tannins, Sugar, Resin, Fatty oil (Karanjin).

Biological activity

Root bark - Acrid, alexeteric, anthelmintic, bitter.

Leaf - Anthelmintic, digestive, laxative.

Seed - Anthelmintic, acrid, bitter, depurative.

Seed oil - Anthelmintic, depurative, styptic.

Folk uses

Medicine:

- Powder of seeds is taken with cow milk to cure

- Bodyache.
- Seed oil is used as massage oil; it is useful in skin itching.
- The seed oil is very effective massage oil for arthritis.
- Leaf juice is applied on ringworm.
- Extract of leaves is applied as ointment in the case of urticaria.
- The seed oil is applied as ointement in the case of cervical pain.

Miscellaneous: Seed are sold in the market.

Poison: The leaves are crushed and dropped in water as a fish poison by the Bhils.

Uses in literature

Leaf - Diabetes (Bhalla *et al.,* 1982); Whooping cough (Ganeshan *et al.,* 2006); Gout (Jadeja *et al.,* 2006).

Fruit - Cough (Kumar & Chauhan, 2005).

Root - Fistula, skin diaseses, ulcer (Bhalla *et al.,* 1982); Sore (Jain & Sahu, 1993).

Seed - Whooping cough, bronchitis (Bhalla *et al.,* 1982); Skin diseases (Shah, 1984; Pereira, 2006); Joint pain (Saxena, & Vyas, 1983); Ear complaint (Shah *et al.,* 1983); Rheumatism (Saxena, & Vyas, 1986); Chest pain (Roy & Chaturvedi, 1986); Burns (Jain, 1965); Edible (Bhargava, 1981, 1983).

Seed oil - Eczema, itch (Jain, 1965); Cold (Jain, & De, 1966); Scabies (Janardhanan, 1963); Fever (Tarafder, 1986); Rheumatims (Kumar *et al.,* 2006); Fish poison (Vishnu Mittre, 1981).

Portulaca oleracea L. (Portulacaceae)

Vernacular name : Lunkio, Lunsio.

Regional and other names : *Bengali* – Kulfi; *English* - Common purslane; *Gujarati* – Loni; *Hindi* - Badi loni; *Kannada* – Budagora; *Malayalam* – Koluppa; *Marathi* – Bhuigholi; *Sanskrit* - Brhat loni; *Tamil* – Karikkirai; *Telugu* – Peddapavilakura.

Distribution - All over India.

Taxonomic description - Diffuse herbs, with succulent, obovate-spathulate leaves and yellow flowers. Capsules obovoid.

Flowering & fruiting - Throughout the year.

Habitat ecology - Found in sandy riverbeds, cultivated fields & in moist habitats.

Regeneration - By seeds & vegetative method.

Field no. - DJ-43.

Parts used - Leaves.

Chemical constituents - Plant contains Crude fibres, Carbohydrates, Calcium, Magnesium, Phosphorus, Iron, Potassium, Copper, Sulphur & Chlorine.

Biological activity

Stem & leaves - Alexeteric, alterant, antibacterial, bitter, diuretic, salty, sour, stomchic, thermogenic.

Seed - Antidysenteric, cooling, diuretic.

Folk uses

Food: Leaves are used as vegetable.

Medicine: Boiled leaves are tied on injured body part.

Uses in literature

Whole plant - Burning sensation (Das *et al.,* 1983); Headache (Shah *et al.,* 1981); Blood purifier (Rao & Neogi, 1980); As vegetable (Shah, 1984; Maji & Sikdar, 1982; Das *et al.,* 1983; Vartak, 1959; Chaudhuri *et al.,* 1985; Singh, 1945).

Leaf - Fever & headache (Shash *et al.,* 1983); Stomach trouble (Purkayastha & Subhan, 2005); Burns, inflammation (Subhangi & Patil, D.A. 2006).

Seed- Dysentery (Jadeja *et al.,* 2006).

Psidium guajava L. (Myrtaceae)

Vernacular name Jamphal.

Regional and other names : *Bengali* – Goaachhi; *English* - Common guava; *Gujarati* - Jamrud; *Hindi* - Amrud; *Kannada* – Balehannu; *Malayalam* - Pera, koyya; *Marathi* – Jamba; *Sanskrit* - Perukah; *Tamil* - Koyya; *Telugu* – Jampandu.

Distribution - Throughout India.

Taxonomic description - An erect, woody shrub or small tree, with reddish-brown wood and scaly, brownish bark.Leaves oblong or oblong-elliptic. Flowers on axillary peduncles of 1-3-flowered. Fruits berries.

Flowering & fruiting - July-September.

Habitat ecology - Grown commonly in gardens for edible fruits.

Regeneration - By cuttings.

Field no. - DJ-45.

Parts used - Fruits, Leaves.

Chemical constituents

All parts contain Leucocyanidin, Querecetin, β- sitosterol.

Flower & leaves contain Ellagic acid, β- Caryophyllene, Guaijaverin, Umonene, Pinene, Niacin, Vitamin- C & B, Cadinene, β-Cophaene, Curcumene, Farnesene, Citric, Cinnamic, Malic, Oxalic, Quinic, Succinic, Tartaric acids, Riboflavin, Thiamine, Carotene, Tannins, Sucrose, Glucose, Fructose, Rhamnose, Starch & Amino acids.

Biological activity

Root - Anthelmintic, astringent, constipating.

Leaves - Anodyne, antispasmodic, astringent, febrifuge.

Flower - Cooling, laxative

Fruits - Aphrodisiac, astringent, cooling, laxative, sour, sweet.

Folk uses

Food: Fruits are used as vegetable. Ripe fruits are eaten by the tribals.

Medicine:

- Ripe berry fruit is used after roasting to cure cough and cold.
- Gargles of leaves decoction relieves mouth ulcer, toothache, pyrrhoea and gums trouble.

Miscellaneous: Fruits are sold in the market.

Uses in literature

Leaf - Blisters in mouth, cold, cough (Maheshwari & Singh, 1984); Dysentery (Singh & Maheshwari, 1983; Krishna & Singh, 1987; Maheshwari *et al.,* 1986; Basak, 2006); Gonorrhoea (Tarafder, 1984); Fever & headache (Rao & Jamir, 1982); Menstrual diseases (Bhattacharjee *et al.,* 1980); Cold (Udayan *et al.,* 2006 b); cholera

(Upadhyay & Jaswant Singh, 2005); Fever (Deokule, 2006; Udayan *et al.*, 2006 b); Diarrhoea (Upadhyay & Jaswant Singh, 2005; Sarma *et al.*, 2006); Toothache (Aditya *et al.*, 2006).

Fruits - Jaundice (Tiwari *et al.*, 1979); Constipation (Arinathan *et al.*, 2006); Dysentery (Gupta, 2006); Edible (Shah, 1984).

Bark - Stomachache (Rao & Jamir, 1982); Diarrhoea, leucorrhoea (Das *et al.*, 2006).

Quiscualis indica L. (Combretaceae)

Vernacular name : Tameswer.

Regional and other names : *Bengali* – Brahmolata; *English* – Rangoon Creeper; *Gujarati* – Barmasini bel; *Hindi* – Rangunachi veli; *Marathi* –Lal chameli; *Tamil* – Irangun-malli; *Telugu* – Rangonimalli.

Distribution - A native of Java & Malaya, grown ornamentally elsewhere.

Taxonomic description - A large, subscandent shrub. Leaves elliptic-oblong. Flowers, deep pink to red, fragrant, in terminal drooping corymbose spikes.

Flowering & fruiting - Summer & Monsson.

Habitat ecology - Planted in gardens for ornamental purposes.

Regeneration - By cuttings.

Field no. - DJ-46.

Parts used - Leaves.

Chemical constituents : Leaves contain Trigonelline, L-proline, L-asparagine, Quisqualic acid.

Seed oil - Linolenic, Oleic, Palmitic, Stearic, Arachidic acids, Sterol, Fatty acids.

Biological activity

Seeds - Anthelmintic, antidiarrhoeal, febrifuge.

Leaves - Carminative.

Folk uses

Medicine:

- Boiled leaves are tied on throat to cure tonsilitis.
- A drop of leaf extract is dropped in eyes to cure conjunctivitis.

- Leaves are heated a bit with edible oil and tied on boils.

Uses in literature

Leaves- Anthelmintic (Deokule, 2006).

Raphanus sativus L. (Brassicaceae)

Vernacular name : Muli, Muro, Mulo, Mula.

Regional and other names : *Bengali* – Mula; *English* – Radish; *Gujarati* –Mura ; *Hindi* – Muli; *Kannada* – Mulangi; *Malayalam* – Mullanki; *Marathi* – Mula; *Sanskrit* – Mulika; *Tamil* – Mullanki; *Telugu* – Mullangi.

Distribution - Throughout India.

Taxonomic description - Annual herbs, with fusiform roots & white or lilac flowers.

Flowering & fruiting - Winter season.

Habitat ecology - Cultivated throughout the area for edible roots & fruits.

Regeneration - By seeds & vegetative method.

Field no. - DJ-128.

Parts used - Leaves, Fruits and Roots.

Chemical constituents

Seeds- Glycoside, Enzyme, Methyl mercaptan, Sulforaphene, Essential oil.

Biological activity

Roots- Acrid, anodyne, anti-inflammatory, appetizing, bitter, depurative, digestive, laxative, stomachic, thermogenic.

Leaves- Antibacterial, diuretic, laxative.

Seeds - Carminative, diuretic, expectorant, laxative.

Folk uses

Food: Leaves, fruits and roots are used as vegetable. Fusiform roots are also used as salad.

Medicine: Extract of 5- 10 leaves is taken with sugar daily 3 times in the case of jaundice.

Uses in literature

Leaf - Acidity (Shah, 1984); As vegetable (Shekhawat & Sunil Anand, 1984; Shah, 1984).

Seed - Ringworm (Purohit *et al.,* 1985); Skin diseases (Saxena & Vyas, 1986).

Root - Acidity (Kumar & Chauhan, 2005); Jaundice, kidney stone (Ballabh & Chaurasia, 2006); As vegetable (Shekhawat & Sunil Anand, 1984; Shah, 1984; Lakshamanan *et al.,* 1986).

Ricinus communis L.(Euphorbiaceae)

Vernacular name : Arandi.

Regional and other names : *Bengali* – Rehri; *English* - Castor; *Hindi* - Erandi; *Kannada* – Haralu; *Malayalam* – Avanakku; *Marathi* – Erand; Sanskrit- Svetaeranda; *Tamil* - Avanakku; *Telugu* – Amudamuchepta.

Distribution - Throughout India.

Taxonomic description - Large shrubs or small trees, with fistular stem and large, palmetly 5 to 10-lobed leaves. Flowers in pyramidal panicles. Capsules globose, spiny, of 3- cocci; seeds caruncled.

Flowering & fruiting - Throughout the year.

Habitat ecology - Found as an escape in wastelands near villages.

Regeneration - By Seeds & vegetative method.

Field no. - DJ-47.

Parts used - Seed oil, Leaves.

Chemical constituents : Root contains Resin, Tannin, Starch.

Seeds contain 50% Oil, Protein, Ricinine, Ricin, Lipase enzyme, Fixed oil.

Fixed oil contains Glycerides, Ricinolesis, Isoricinoleic acid, Stearic acid, Crystals of Calcium oxalate, Nitrate.

Biological activity

Alterative, anti-inflammatory, aphrodisiac, carminative, febrifuge, purgative, stomachic.

Folk uses

Biofencing: Commonly planted along the boundaries of fields for fencing purposes.

Medicine:

- Seed oil is used for teething trouble in children.
- Seed oil is taken with warm water at night in constipation.
- Leaves are heated a bit with edible oil and tied on wounds and boils.
- Leaves are heated a bit with edible oil and tied on forehead to cure migrain.

Miscellaneous: Seeds are sold in the market.

Oil: Seed oil is also used as fuel in lamps.

Uses in literature

Leaves - Dyspepsia, gonorrhoea (Bhalla *et al.*, 1982); Bronchitis (Lal & Yadav, 1983); Piles (Lal & Yadav, 1983; Mishra & Billore, 1983; Shah & Joshi, 1971); Skin dieases (Rao, 1981; Nautiyal, 1981); Cuts, ear complaint (Shah & Joshi, 1971); Pimples, cuts, wounds, boils (Purkyastha & Subhan, 2005); Headache (Deokule, 2006; Das *et al.*, 2006); Muscular pain (Sarma *et al.*, 2006); Bodyache (Das *et al.*, 2006).

Fruits - Laxative (Rajwar, 1983); Edible (Maji & Sikdar, 1982).

Seeds - Labour (Tarafder, 1984); Intestinal warm (Jadeja *et al.*, 2006); Gout (Jadeja *et al.*, 2006); Edible (Lakshamanan *et al.*, 1986).

Seed oil – Cough, cold, asthma, cardiac pain (Kumar *et al.*, 2006).

Saccharum officinarum L. (Poaceae)

Vernacular name : Santto, Hantta, Ganno.

Regional and other names : *Bengali* –Ganna; *English* –Sugar-cane; *Gujarati* –Sheradi; *Hindi* – Ganna, Ikh; *Kannada* –Basarimara; *Malayalam*–Karimba; *Marathi*–Aos; *Sanskrit*–Asipatra; *Tamil*– Karimpu; *Telugu* – Charki.

Distribution - Throughout India.

Taxonomic description - Tall, perennial, rhizomatous plants, with juicy culms and leaves wide and laminate right to the base. Spikelets awnless; glumes glabrous.

Flowering & fruiting - January -onwards.

Habitat ecology - Cultivated for culm juice used for making jaggery.

Regeneration - By stem-cuttings.

Field no. - DJ-194.

Parts used : Culm-juice.

Chemical constituents

Stem- Flavonoids, Sucrose, Glucose, Fructose, Fibres, Nitrogenous substances, Fats, Waxes, Gums, Pectins, Free acids, Combined acids, Saccharetin.

Cane juice- Glycans as Saccharans A,B,C,D,E & F, Neocarlinoside, Neoisoschaftoside, Orientin, Ferulic acid, Caffeic acid, Sugars, Carbohydrates, Hemicellulose, Pentosans, Starch, Pectins, Proteins, Peptose, Carboxylic acids, Inorganic salts, Silica, Amino acids, Vitamins as Thiamine, Riboflavin, Niacin, Biotin, Gums, Phenols, Calcium oxalate, Vitamin-D.

Biological activity

Stem- Apetizre, aphrodisiac, cooling, diuretic, emollient, expectorant, laxative, stomachic, sweet.

Root- Cooling, demulcent, diuretic, stimulent.

Folk uses

Fodder: The leaves & chopped stem (ganna ki varundi) are used as fodder for cattle.

Medicine: Sugarcane juice is taken orally to cure jaundice.

Uses in literature

Cane juice- Gall bladder (Rao & Jamir, 1982); Constipation (Maheshwari & Singh, 1984); Jaundice (Deokule, 2006; Dutta *et al.*, 2006); Leprosy, intestinal trouble (Kumar & Yadav, 2006).

Santalum album L. (Santalaceae)

Vernacular name: Chandan, Sandan.

Regional and other names : *Bengali* – Chandan; *English* - Sandal wood; *Hindi* - Safed chandan; *Kannada* – Srigandha; *Malayalam* – Chandanam; *Marathi* – Chandan; *Sanskrit* - Chandanam; *Tamil* - Sandanam, ulocidam; *Telugu* – Srigandhamu.

Distribution - W. peninsula, cultivated elsewhere.

Taxonomic description - Small, evergreen trees, with drooping branches. Leaves ovate-lanceolate. Flowers reddish-brown, in axillary and terminal cymes. Fruits globose, dark purple.

Flowering & fruiting - September – December.

Habitat ecology - Rare, in mixed deciduous forests.

Regeneration - By Seeds.

Field no. - DJ-143.

Parts used - Wood oil.

Chemical constituents

Heart wood contains Resin, Tanic acid & Volatile oil which is thick, yellow & aromatic. It contains œ & ß – Santalol, Aldehyde, Ketone, Santenone, Santalone.

Biological activity:

Wood - Alexeteric, antipyretic, aphrodisiac, bitter, cooling.

Folk uses

Medicine: The wood oil is applied in the case of muscular pain.

Uses in literature

Seed - Abortifacient (Bhalla *et al., 1982*).

Wood - Fever, skin diseases (Bhalla *et al.,* 1982); Headache (Nagendra Prasad & Abraham, 1984); Boils (Lal & Yadav, 1983).

Wood oil- Gonorrhoea, cough, urinary troubles (Deokule, 2006).

Stem bark- Acidity (Jadeja *et al.,* 2006).

Stem- Diabetes (Ghosh, 2006); Gout (Jadeja *et al.,* 2006).

Senna occidentalis (L.) Link (Caesalpiniaceae)

Syn.: *Cassia occidentalis* L.

Vernacular name : Sanp wari rukhdi.

Regional and other names : *Bengali* - Kalkashunda; *English* - Negro coffee; *Gujarati* - Kasodari; *Hindi* - Kasaumdi; *Kannada* - Doddaagace; *Malayalam* - Ponnariviram; *Marathi* - Kasoda; *Sanskrit* - Kasamardah; *Tamil* - Nattam takarai; *Telugu* – Kasinda.

Distribution - Throughout India.

Taxonomic description - Diffuse undershrubs, with 3-6 pairs of leaflets and a gland on the petiole. Flowers yellow, in corymbose panicles. Pods linear-oblong.

Flowering & fruiting - October – March.

Habitat ecology - Found in wastelands.

Regeneration - Seeds.

Field no. - DJ-106.

Parts used - By Seeds.

Chemical constituents

Plant contains Emodin, Tannic acid, Fatty oils, Mucilage, Aminoacids, Chrysophanol, Sitosterol, Xanthon, Cassiodin, Nitrogen bases, Betain, Choline, Stachydrine and Trigonelline.

Seed contains Galactomanon.

Biological activity

Root - Acrid, anti – inflammatory, bitter, digestive, diuretic, stomachic.

Leaf - Acrid, aphrodisiac, bitter, depurative, expectorant.

Seed - Acrid, bitter, depurative, diuretic, expectorant, febrifuge, purgative, stomachic.

Folk uses

Magico-religious beliefs: Commonly planted along the boundaries of their houses to keep away snakes.

Medicine: Paste of seeds is applied against snake bite.

Uses in literat'ire

Leaf - Bone fracture (Apparanantham & Chelladurai, 1986); Eczema, ringworm, throat infection (Shah *et al.,* 1983); Fever (Tarafder, 1986); Skin diseases (Malhotra & Moorthy, 1973; Das *et al.,* 2006); Wound (Singh *et al.,* 1987); Stomach trouble (Ganeshan *et al.,* 2006); Rheumatism (Kumar *et al.,* 2006).

Root - Gastric complaint (Goel *et al.,* 1984); Bodyache, urinary problems (Narain & Singh, 2006b); Skin diseases (Jain *et al.,* 1973); whooping cough (Singh & Pandey, 1980).

Seed - Skin diseases (Maheshwari & Singh, 1987; Das *et al.,* 2006); Whooping cough (Singh & Paney, 1980).

Senna tora (L.) Roxb.(Caesalpiniaceae)

Syn.: *Cassia tora* L.

Vernacular name : Phuario.

Regional and other names : *Bengali* – Chakunda; *English* - Foetid cassia, wild senna; *Gujarati* - Kawario; *Hindi* - Cakunda, Punwar; *Kannada* - Taragasi, Gandutogache; *Malayalam* -Takara, Chakramand-arakam; *Marathi* - Takala, tarota; *Sanskrit* - Cakramardah; *Tamil* - Tagarai; *Telugu* – Tantiyamu.

Distribution - Throughout India.

Taxonomic description - Suffruticose herbs, with 3 pairs of leaflets and glands between them. Flowers yellow or reddish tinged, geminate; all 7 anthers rounded at apex. Pods linear, turgid.

Flowering & fruiting - September- November

Habitat ecology - Extensively found throughout the area as a weed.

Regeneration - By seeds.

Field no. - DJ-72.

Parts used - Leaves, Seeds, Root, Flowers.

Chemical constituents : Seed contains Glucosides, (Emodin) Proteins, Steroids, Amino acids, Reducing sugars, Triterpenes, Tannins, Oleic acid, Linotic acid, Palmitic acid, Lignocenic acid, Rubrofusarin, Fixed oils, Sodium, Potassium & Iron.

Biological activity : *Leaves & Seeds* - Acrid, anthelmintic, antiperiodic, depurative, expectorant, laxative, ophthalmic.

Folk uses

Beverage: The powder of roasted seeds is used as a substitute for tea.

Food: Leaves are used as vegetable.

Medicine:

- Leaves are used as vegetable curries to cure skin itching, ascariasis & arthritis.
- Paste of 4 leaves is eaten with jaggery for 3 days thrice a day to cure typhoid.
- Paste of 6 flowers is eaten with jaggery for 3 days thrice a day to cure jaundice.
- The powder of seeds is used as coffee powder, and a teasp-oonful powder is taken with cow milk to cure skin itching.

- Paste of flowers is tied on boils/cuts as poultice.
- Root extract is useful in snake-bite.

Miscellaneous: Seeds are sold in the market by the tribals.

Uses in literature

Leaf - Stomach trouble; indigestion (Tarafder, 1983); Nightblindness (Singh & Maheshwari, 1983); Bone fracture (Jain *et al.,* 1973); Boils (Sexena & Vyas, 1981); Jaundice (Hemadri & Rao, 1984); Skin diseases (Shah *et al.,* 1981; Purkayastha & Subhan, 2005; Kumar *et al.,* 2006; Yesodharan & Sujana, 2006; Das *et al.,* 2006); Wound (Jain, 1963; Yesodharan & Sujana, 2006); Dysentery (Ganeshan *et al.,* 2006); Worms (Kadavul & Jisha, 2006); Cuts (Yesodharan & Sujana, 2006); As vegetable (Shah, 1984; Jain & De, 1966; Vartak, 1981; Shah *et al.,* 1981; Ramachandran & Nair, 1981; Jain, 1963, 1964; Mishra & Billore, 1983; Goel *et al.,* 1984; Singh & Maheshwari, 1985; Vartak, 1959).

Seeds - Eczema, ringworm (Shah *et al.,* 1983); Cold (Shah & Joshi, 1971); Itching (Sexena & Vyas, 1981); Scorpion bite (Nagar, 1985); Stomach trouble (Rajwar, 1983); Snake bite (Kiruba *et al.,* 2006); Skin diseases (Kumar *et al.,* 2006; Das *et al.,* 2006); Worms (Kadavul & Jisha, 2006); Gout (Jadeja *et al.,* 2006); Substitute for coffee (Shah, 1984).

Root - Ringworm (Bhalla *et al.,* 1982); Irregular menstruation (Sarma *et al.,* 2006); Skin diseases (Janardhanan, 1963); Fever (Tarafder, 1987); Scorpion bite (Maheshwari *et al.,* 1986); Snake bite (Kumar *et al.,* 2006).

Fruit- Anthelmintic (Deokule, 2006); As vegetable (Singh & Singh, 1981; Goel *et al.,* 1984).

Flower- As vegetable (Chaudhuri *et al.,* 1985).

Sesbania bispinosa (Jacq.) Wight (Papilionaceae)

Syn.: *Aeschynomene bispinosa* Jacq.

Vernacular name - Dhandhan.

Distribution - Cultivated in many parts of India.

Taxonomic description - Herbs or undershrubs, with prickly branches and leaf-rachis. Flowers yellow. Pods linear, subterte, beaked.

Flowering & fruiting - August- October.

Habitat ecology - Common weeds in crop-lands and other swampy places.

Regeneration - By seeds.

Field no. - DJ-183.

Parts used - Seeds.

Folk uses

Medicine:

- The powder of seeds is taken with water in the case of arthritis.

Uses in literature

- Seed- Edible (Gunjatkar & Vartak, 1982; Vartak, 1981).

Sida cordifolia L. (Malvaceae)

Vernacular name : Balbij.

Regional and other names : *Bengali* - Swetberela, kureta; *English* - Country mallow; *Gujarati* - Baladana; *Hindi* - Bariara, kharenta; *Kannada* - Hettuthi, hettugigada; *Malayalam* - Kurunthotti, velliurum; *Marathi* – Chikana; *Sanskrit* - Bala; *Tamil* - Karumthotti, milatutti; *Telugu* - Chirwbenda, tellantisa.

Distribution - Tropical and sub-tropical regions of both hemispheres.

Taxonomic description - Undershrubs, with ovate-cordate leaves and axillary yellow flowers. Mericarps 10, 2-awned.

Flowering & fruiting - August– November.

Habitat ecology - Along cultivated field.

Regeneration - By seeds.

Field no. - DJ-133.

Parts used : Seeds.

Chemical constituents

Root contains Beta-sitosterol, Alkaloids- Visicinonoe, Vasicine, Vasicinol, Choline & Betain.

Seeds contain Ephedrine, Fatty oils, Steroids, Resin, Mucin, Sterculic acid, Malvalic acid, Linoleic acid, Potassium nitrate.

Biological activity

Plant - Aphrodisiac, astringent, bitter, emollicent, sweet, tonic.

Fruits - Acrid, aphrodisiac, astringent, cooling, digestive, sweet.

Folk uses

Medicine: The seeds are cooked (~50 gm) in pure cow milk (250 gm) to prepare 'Kheer"(Sugar may be added) and taken on 7 days in the cases of sexual debility in male.

Uses in literature

Root - Boils (Jain *et al.,* 1973); Wound (Das *et al.,* 1983; Kumar *et al.,* 2006); Stomatitis (Shah & Gopal, 1986); Leucorrhoea (Upadhyay & Jaswant Singh, 2005); Post delivery trouble (Sarma *et al.,* 2006); Colic pain (Chatterjee & Bhatterjee, 2006).

Whole plant – Dysentery (Singh & Maheshwari, 1983); Rheumatism (Malhotra & Moorthy, 1973); Fever, leucorrhoea, nervous disorders (Kumar *et al.,* 2006); Gonorrhoea (Singh & Maheshwari, 1983; Kumar *et al.,* 2006); Spermatorrhoea (Prakash & Singh, 2006.).

Seed - Gonorrhoea, leucorrhoea (Singh, 1945).

Leaf - Gonorrhoea, sex weakness (Tarafder & Chaudhari, 1981); Wounds & cuts (Kiruba, 2006).

Sida rhombifolia L. (Malvaceae)

Vernacular name : Balbij.

Regional and other names : *Bengali* - Lal–berela; *Gujarati* – Baladana; *Hindi* - Bariara, lal berela; *Kannada* - Kallangadale, bolamgadale; *Malayalam* - Anakurunthotti, vatturum; *Marathi* – Saha-devi; *Sanskrit* –Mahabala; *Tamil* - Sirramutti, chitra mutti; *Telugu* - Athibala, chettu.

Distribution: Tropical and sub-tropical regions of both hemispheres.

Taxonomic description : Much-branched herbs, with obovate-rounded or rhomboid leaves and axillary, solitary, yellow flowers. Mericarps 1-awned or awnless.

Flowering & fruiting : October – December.

Habitat ecology - Along cultivated fields.

Regeneration - By seeds.

Field no. - DJ-137.

Parts used - Seeds.

Chemical constituents : Plant contains Ephedrine, Cellulose, Lignin, Quanozolines, ß- Phenethylamines, Tryptomine, Sterculic acid, Malvalic acid, Linoleic acid.

Biological activity

Root - Aphrodisiac, sweet, tonic.

Leaf - Aphrodisiac, sweet, tonic.

Folk uses

Medicine:

- The seeds are cooked (~50 gm) in pure cow milk (250 gm) to prepare 'Kheer"(Sugar may be added) and taken on 7 days in the cases of sexual debility in male.
- The seeds are very effective to check hair falling, for which the tribals apply the paste of seeds on the head for an hour or so before taking bath.

Uses in literature

Whole plant - Gonorrhoea (Tarafder, 1984); Rheumatism, urinary diseases (Das *et al.,* 2006); Detergent (Nagendra Prasad & Abraham, 1984).

Leaf - Dysentery (Sebastian & Bhandari, 1984); Swelling (Malhotra & Moorthy, 1973); Wound (Upadhye *et al.,* 1986); Spermatorrhoea (Prakash & Singh, 2006.).

Root - Rheumatism (Nagendra Prasad & Abraham, 1984); Boils (Tarafder, 1984); Asthma & fever (Singh & Pandey, 1980); Leucorrhoea (Upadhyay & Jaswant Singh, 2005).

Solanum nigrum L. (Solanaceae)

Vernacular name : Charpoti.

Regional and other names : *Bengali* - Kakmachi, Gurukamai; *English* - Black night shade; *Gujarati* - Piludi; *Hindi* - Gurkamai, Markol; *Kannada* – Ganika; *Malayalam* – Mulagthakkali; *Marathi* – Kangani; *Sanskrit* -Kakamaci; *Tamil*- Munathakkali; *Telugu* - Kachi, kachchipundu.

Distribution - Throughout India.

Taxonomic description - Unarmed, glabrous or pubescent herbs, with ovate-oblong leaves, white flowers and globose, yellow, red or purple fruits.

Flowering & fruiting - June – February.

Habitat ecology - Common in shady habitats.

Regeneration - By seeds.

Field no. - DJ-144.

Parts used : Fruits.

Chemical constituents : Plant contains Alkaloids, Reducing sugars, Glycosides, Saponins, Steroids, Leutein, Lycopene, Vitamin – C, Glucose, Fructose, Caffeic – solasodine, Tomatidenel, Solamargine, Solasomine, Trigogenine, Potassium, Iron, Sulphur, Calcium, Phosphorus.

Biological activity : *Fruits* - Alterative, aphrodisiac, appetiser, bitter, diuretic, laxative, pungent.

Folk uses

Food: Mature fruits are eaten by the tribals.

Uses in literature

Seed - Gouitre (Dar *et al.,* 1984); Liver trouble (Das *et al.,* 2006).

Whole plant - Fever (Shekhawat & Anand, 1984); Jaundice (Billore, & Audichya, 1978; Dutta Choudhury & Shil, 2006); Piles, swelling (Deokule, 2006); Diarrhoea (Kataria & Gauchan, 2006); Dysentery (Deokule, 2006; Kataria & Gauchan, 2006).

Fruits - Diarrhoea, fever (Bhalla *et al.,* 1982); Cough & throat trouble (Megoneitso & Rao, 1983); Swelling, (Shah *et al.,* 1983); Liver trouble (Purkayastha & Subhan, 2005; Baruah *et al.,* 2006); Indigestion (Baruah *et al.,* 2006); Stomach pain (Arinathan *et al.,* 2006); Edible (Das *et al.,* 1983; Vartak, 1981; Singh & Singh, 1981; Abraham, 1981; Shah & Joshi, 1971; Maheshwari *et al.,* 1981, 1986); As vegetable (Chaudhuri *et al.,* 1985).

Leaf - Swelling (Maheshwari & Singh, 1984; Purkayastha & Subhan, 2005); Boils, skin diseases, earache (Sebastian & Bhandari, 1984); Liver complaint (Lal & Yadav, 1983; Baruah *et al.,* 2006); Rheumatism (Malhotra & Morrthy, 1973); Dysentery (Maheshwari & Singh, 1987); Stomach trouble (Abraham, 1981); Piles (Singh,1988); Eye complaint (Shah & Joshi, 1971); Indigestion (Baruah *et al.,* 2006); Constipation (Das *et al.,* 2006); As vegetable (Shekhawat & Sunil Anand, 1984; Arora, 1981).

Solanum virginianum L. (Solanaceae)

Syn.: *S. xanthocarpum* Schrad. & Wendl

Vernacular name : Bhuri-ringdi.

Regional and other names : *Bengali* – Kantakari; *English* - Yellow berried, night shade; *Gujarati* - Bhoyaringani; *Hindi* - Kateli, katal; *Kannada* – Kandan; *Malayalam* – Karychunda; *Marathi* - Kateringanii, bhuiringani; *Sanskrit* -Kantakari; *Tamil* - Kandankattiri; *Telugu* - Pinnamulaka, nelamulakaa.

Distribution - Throughout India.

Taxonomic description - Prostrate, prickly herbs, with sinuately lobed to subpinnatifid leaves. Flowers violet-purple or white. Fruits globose, yellow when ripe.

Flowering & fruiting - Feburary – June.

Habitat ecology - Occasional in moist and shady places.

Regeneration - By seeds.

Field no. - DJ-75.

Parts used - Fruits, Seeds, Root.

Chemical constituents

Plant contains Alkaloids (Solasonine), Diosgenin. Solacinine.

Seed contains Essential oil, Solanocarpine.

Flower contains Flavonol, Glycoside, Queretin, Apigenin, Sitosterol.

Leaves contain Coumarine, Scopolin, Esculin, Esculetin.

Biological activity

Root - Anthelmintic, appetiser, bitter, laxative, pungent, stomachic.

Fruits - Anaphrodisiac, anthelmintic, appetiser, bitter, digestive.

Folk uses

Medicine:

- The smoke obtained by burning of dried seeds is given through ear tract to cure toothache.
- Extract of root is useful in scorpion sting.
- Extract of root is taken with curd to cure dysentery.

- Ash of the root is mixed with coconut oil and applied in the case of mouth ulcers.

Uses in literature

Seed - Blisters, boils (Maheshwari & Singh, 1987); Toothache (Jain & De, 1966).

Leaf - Eye complaint (Saxena & Vyas, 1983).

Flower - Blister, boils (Maheshwari & Singh, 1984).

Whole plant - Asthama, cough, dropsy, gonorrhoea, fever (Bhalla *et al.,* 1982); Toothache (Jain & De, 1966); Swelling (Virjee *et al.,* 1984).

Root - Asthama, fever (Shah *et al.,* 1983); Bronchitis (Mahaswari & Singh, 1984); Cold, cough, scorpion bite (Maheshwari *et al.,* 1986); Toothache (Gupta, 1981); Cough (Upadhyay & Jaswant Singh, 2005).

Fruits - Asthma (Nagendra Prasad & Abraham, 1984); Piles (Shekhawat & Anand, 1984); Paralysis (Sudhakar & Rolla, 1985); Gum trouble, toothache (Shah *et al.,* 1983); Earache (Jain, 1963); Cough (Jain & De, 1966); As vegetable (Jain, 1964).

Sonchus brachyotus DC. (Asteraceae)

Vernacular name : Jungli rukhdi.

Regional and other names- *Bengali* –Banpalang; *English* - Hogweed; *Hindi* -Sadhi; *Telugu* –Nallatapata.

Distribution - Throughout India.

Taxonomic description – Erect,perennial herbs; basal leaves in a rosette, upper ones remote, entire or lyrate-pinnatilobed, semiamplexicaulat base; heads pale yellow, in corymbose panicles; achenes linear-oblong; pappus hairs white.

Flowering & fruiting - March-November.

Habitat ecology – Occasional, in moist & shady places.

Regeneration - By seeds.

Field no. - DJ-48.

Parts used : Leaves.

Chemical constituents : Aerial parts of the plants contain Fatty acids, Srearic & Palmic acids, Ceryl alcohol, Choline, Invert sugar, Tartaric acids, Cobalt, Copper, Iron, Zinc.

Latex- Rubber, Mannitol, a bitter substance Alpha- & Beta-Lactucerols.

Biological activity

Bitter, diuretic.

Folk uses

Medicine: Extract of leaves is used to cure bodyache.

Sorghum bicolor (L.) Moench (Poaceae)

Syn.: *Holeus bicolor* L.

Vernacular name : Juwar.

Regional and other names- *Bengali* –Juar; *English* –Indian Millet; *Hindi* -Juar; *Kannada* –Jola; *Malayalam* -Cholam; *Marathi* –Jwari; *Tamil* -Cholam; *Telugu* –Jonnalu.

Distribution - Cultivated throughout India.

Taxonomic description - A tall annual grass. Leaves broad-linear, with a prominent white midrib. Spikelets large, broad, hairy; pedicelled usually neuter, pedicels short.

Flowering & fruiting - September-October.

Habitat ecology - Cultivated throughout the area.

Regeneration - By seeds.

Field no. - DJ-210.

Parts used : Seeds.

Chemical constituents

Root- Hexacosanol, Beta-sitosterol, Glucose.

Shoot- Cyanogenetic glucoside.

Leaves- HCN.

Biological activity *: Seeds/Grains-* Diuretic, demulcent, aphrodisiac, cooling, constipating.

Folk uses

Medicine: The flour is fed to the animals to cure dysentery.

Uses in literature

Grain- Baldness (Lal & Yadav, 1983).

Syzygium aromaticum (L) Merr. & Perry (Myrtaceae)

Syn.: *Caryophyllus aromaticum* L..

Vernacular name : Laung.

Regional and other names : *Bengali* – Laung; *English* - Clove; *Gujarati* – Lavang; *Hindi* - Laung; *Kannada* – Laung; *Malayalam* – Krambu; *Marathi* – Laung; *Sanskrit* - Lavang; *Tamil* - Krambu; *Telugu* – Lavangalu.

Taxonomic description - This species is not found wild in the area. The Cloves are available in the market.

Field no. - DJ-49.

Parts used - Cloves.

Chemical constituents : Dried flower bud contains Beta-caryophyllene, Eugenol, Acetate, Methyl Salicylate, N-Amylcarbinol, Benzyl alcohol, Furfural, Furfury alcohol, Vanillin, Eugenitine, Eugenon, Galactose, Glucose, Fructose, Rhamnose, Sucrose, Gallotanic acid.

Biological activity : *Dried flower bud* - Appitizer, aromatic, carminative, stimulant, stomachic.

Folk uses

Medicine:

- A syrup madeup of roasted cloves powder with jaggery is used in the post delivery period.
- 5-10 cloves of *Syzygium aromaticum* and 5 flowers of *Calotropis gigantea* are burnt with cotton fibre to prepare ash. About 10 gm. ash is taken with water to cure jaundice.

Uses in literature

Cloves - Anti- fertility (Tarafder, 1983).

Syzygium cumini (L.) Skeels (Myrtaceae)

Syn.: *Myrtus cumini* L.

Vernacular name : Jambu.

Regional and other names : *Bengali* – Katajam; *English* – Jambu fruits, black plum; *Gujarati* - Jambu; *Hindi* - Jamun; *Kannada* – Nerali; *Malayalam* - Naval, nagapazham; *Marathi* – Jambhool; *Sanskrit* - Jambula; *Tamil* - Navel peru, negapazham; *Telugu* - Jamburu, neredupandu.

Distribution - Throughout India.

Taxonomic description - Trees, with elliptic-lanceolate leaves and pale-green or white flowers in axillary paniculate cymes. Fruits globose or oblong, purple when ripe.

Flowering & fruiting - March – May.

Habitat ecology - Found along the riversides.

Regeneration - By seeds and by cuttings.

Field no. - DJ-86.

Parts used - Leaves, Fruits, Stem-bark.

Chemical constituents : Bark contains Friedelin, Kaempferol, Tannins, Quercetin, β- sitosterol, Betullinic acid.

Fruits contain Anthocyanin, Eugin.

Seed contains Ellagic acid, IAA, Oxalic acid, Citric acid, Glycollic acid, Glucose,Fructose, Gallic acid.

Leaves contain Glycine, Alanin, Leucin, Tyrosin.

Biological activity

Bark - Acrid, anthelmintic, digestive, sweet.

Fruits - Acrid, astringent, cooling, sweet.

Seed - Astringent, sweet.

Folk uses

Food: Mature fruits are eaten by the tribals.

Medicine:

- Extract of stem-bark is taken orally with water to cure diarrhoea and ascariasis.
- Gargles of leaves decoction relieves mouth ulcers.

Miscellaneous: Fruits are sold in the market.

Poison: The stem-bark is crushed and dropped in water as a fish poison by the Bhils.

Timber: The timber is used for making "Khat"(Cot).

Uses in literature

Seed - Diabetes (Shah *et al.,* 1981; Deokule, 2006; Rajendran & Rengamani, 2006; Das *et al.,* 2006); Pimples (Roy & Chaturvedi, 1986).

Leaf - Dysentery (Lal & Yadav, 1983); Diabetes (Purkyastha & Subhan, 2005; Das *et al.,* 2006); Wound, itching (Sharma *et al.,* 2006); As vegetable (Nagar, 1985).

Bark - Dysentery (Joshi, 1982; Deokule, 2006; Kataria & Gauchan, 2006; Sharma *et al.,* 2006); Piles (Singh, 1988); Diabetes (Purkyastha & Subhan, 2005); Diarrhoea (Deokule, 2006; Kataria & Gauchan, 2006).

Fruits - Stomatitis (Mahshwari & Singh, 1984); Edible (Nagar, 1985; Mahshwari & Singh, 1984; Maji & Sikdar, 1982; Lakshamanana *et al.,* 1986; Arora, 1981; Vartak, 1959, 1981; Bhujel *et al.,* 1984; Singh & Singh, 1981; Shah *et al.,* 1981; Ramamachandran & Nair, 1981; Jain, 1963, 1964; Mahshwari & Singh, 1985; Abraham, 1981; Mahshwari *et al.,* 1986; Datar & Vartak, 1975).

Root- Dysentery (Jadeja *et al.,* 2006).

Tamarindus indica L. (Caesalpiniaceae)

Vernacular name : Amli, Imli.

Regional and other names : *Bengali* – Tentul; *English* - Tamarind; *Gujarati* - Ambli; *Hindi* - Imli; *Kannada* - Hunise chigun, huli, amli; *Malayalam* – Amlam; *Marathi* – Chinch; *Sanskrit* - Amlika; *Tamil* - Puli, amilan; *Telugu* - Chinta chettu.

Distribution - Throughout India.

Taxonomic description - Evergreen trees, with 6-12 pairs of leaflets and yellow flowers with purple streaks. Pods subtorulose, falcate.

Flowering & fruiting - April – June.

Habitat ecology - Found along roadsides & wastelands.

Regeneration - By seeds.

Field no. - DJ-101.

Parts used - Fruits, Seeds.

Chemical constituents : Polysaccharides, Balsamine, Catechin, Nasturtium, Tamarin, Phosphatidic acid and its Choline, Ethanollamine, Serine, Inositol, Alkaloid, Citric acid, Tartaric acid, Potassium bitartrate.

Biological activity

Bark - Astringent.

Leaf - Bitter, laxative, sweet.

Flower - Acrid, appetiser, sour, sweet.

Fruits - Anthelmintic, appetiser, sour, sweet.

Folk uses

Food: Mature fruits are eaten by the tribals.

Medicine:

- Paste of seeds is applied against scorpion sting.
- 1 teaspoonful powder of seeds is taken with cow milk to cure appendicitis.

Miscellaneous: Fruits are sold in the market.

Uses in literature

Fruits - Sun stroke (Shekhawat & Anand, 1984); Carminative (Deokule, 2006); Swelling (Rajendran & Rengamani, 2006); Edible (Maji & Sikdar, 1982; Bhandari, 1974; Malhotra & Moorthy, 1973; Singh & Singh, 1981; Shah *et al.,* 1981; Jain, 1964; Goel *et al.,* 1984; Nagar, 1985; Mahshwari *et al.,* 1981, 1986; Datar & Vartak, 1975).

Leaf - Fever, urinary complaint (Bhalla *et al.,* 1982); Stomatitis (Tarafder, 1983); Swelling (Shah *et al.* 1983); Eye complaint (Maheshwari *et al.,* 1986); Toothache (Shah & Gopal, 1986); Gum problem (Aditya *et al.,* 2006); Vomiting (Hembrom & Goel, 2005); Rheumatism (Das *et al.,* 2006).

Bark - Stomatitis (Tarafder, 1983).

Seed - Scorpion sting (Ganeshan *et al.,* 2006; Das *et al.,* 2006); Snake bite (Das *et al.,* 2006).

Tectona grandis L.f. (Verbenaceae)

Vernacular name Sagad, Sagdo, Sagwan.

Regional and other names : *Bengali* – Segun; *English* - Teak; *Gujarati* - Saga; *Hindi* - Sagun; *Kannada* - Sagwani, jadi; *Malayalam* – Thekku; *Marathi* – Sag; *Sanskrit* – Anila; *Tamil* - Tekku; *Telugu* – Teku.

Distribution - Konkan, W. Ghats of Bombay and Madras Presidencies, Central India.

Taxonomic description - Deciduous trees, with broadly elliptic large leaves. Flowers white or pinkish; stamens all equal. Fruits enclosed in bladder-like inflated calyx.

Flowering & fruiting - September – December.

Habitat ecology - Chief component of deciduous forests.

Regeneration - By seeds.

Field no. - DJ-100.

Parts used - Tender twigs, Stem-bark, Seeds, Wood.

Chemical constituents : Tactograndon, Betulin, Lupeol, Betulinic acid, Betulinaldehyde, Carbohydrates, Fatty acids, Lapachanone, Deoxylapachol, Tectoquinine.

Biological activity

Flower - Acrid, bitter, dry.

Bark - Acrid, sweet.

Wood - Acrid, cooling, laxative.

Folk uses

Fodder: The tender twigs & stem bark are used as fodder for animals.

Medicine:

- Extract of bark is taken with water to cure dysentery & vomiting.
- 1 gm. ash of stem bark is taken with water on empty stomach to cure menorrhoea.
- Extract of seeds is taken with water to cure small pox.
- Extract of stem bark is taken with water to cure injury / wounds. This extract is also applied on wounds as ointment.

Tannin: Bhil tribals use the stem bark for tanning the leather which gives black to brown colour.

Timber: The wood is used in making of agricultural implements, huts, carts and household articles. The wood is also used in the formation of some musical instruments called 'Khanjri'; 'Shahnai' & 'Dhol'.

Uses in literature

Stem bark - Stomatitis (Joshi *et al.,* 1980); Dysentery (Kataria & Gauchan, 2006); Diarrhoea (Joshi *et al.,* 1980; Kataria & Gauchan, 2006).

Wood - Eye diseases (Sharma *et al.,* 1979); Swelling (Janardhanan, 1963; Basak, 2006); Musical instrument (Goel *et al.,* 1984).

Fruits - Hair tonic (Sudhakar & Rolla, 1985).

Seed - Urinary problems (Rajwar, 1983).

Terminalia cuneata Roth (Combretaceae)

Syn.: *T. arjuna* (Roxb. ex DC.) Wt. & Arn.

Vernacular name : Kavdo.

Regional and other names : *Bengali* – Arjun; *English* - Arjuna; *Gujarati* - Arunsadada; *Hindi* - Arjun; *Kannada* – Maddi; *Malayalam* – Vellilavu; *Marathi* – Arjuna; *Sanskrit* – Arjuna, Kukubha; *Tamil* - Vellaimarudamarum, vellamatta; *Telugu* - Tellamadoi, yerramaddi

Distribution - Throughout the greater part of India.

Taxonomic description - Trees, with smooth bark and leaves cordate at base. Flowers white, in axillary spikes or terminal panicles. Fruits 5-winged, wings curved upwards.

Flowering & fruiting - May – December.

Habitat ecology - Commonly found near the banks of rivers & stream.

Regeneration - By seeds.

Field no. - DJ-87.

Parts used - Stem bark, Root.

Chemical constituents : Bark contains Bita- sitosterol, Arjunic acid, Friedlene, Glucoside, (Arjunetic), Tannins, Sugars, Sodium, Magnessium, Aluminium, Calcium carbonate.

Biological activity : *Bark* - Acrid, alexiteric, antidysenteric, cooling, heating, styptic, sweet, tonic.

Folk uses

Medicine:

- Extract of stem bark is taken with curd to cure leucorrhoea.
- Extract of stem bark is taken with curd to cure diarrhoea.
- Extract of root is taken with water in the case of typhoid.

Poison: The stem bark is crushed and dropped in water as a fish poison by the Bhils.

Tannin: Bhil tribals use the stem bark for tanning the leather which gives black to brown colour.

Uses in literature

Bark - Dysentery (Mahshwari *et al.,* 1986); Heart diseases (Saxena & Vyas, 1983); Fish poison (Joshi, 1986).

Leaf - Wound (Sharma & Malhotra, 1984).

Terminalia bellirica (Gaertn.) Roxb.(Combretaceae)

Syn.: *Myrobalanus bellirica* Gaertn.

Vernacular name : Bedo, Beda.

Regional and other names : *Bengali* – Bahera; *English* - Beleric myrobalan; *Gujarati* - Bahedo; *Hindi* - Bahera; *Malayalam* – Thanni; *Marathi* – Beheda; *Sanskrit* – Aksha, Bibhitaka; *Tamil* - Tanari; *Telugu* – Tandra

Distribution - Throughout the forests of India.

Taxonomic description - Deciduous trees, with broadly obovate leaves clustered at the end of branches. Flowers yellow or creamy white, in spikes. Fruits ellipsoid, 5-ridged.

Flowering & fruiting - February-September.

Habitat ecology - Comman in wastelands.

Regeneration - By seeds.

Field no. - DJ-145.

Parts used - Fruits, Leaves.

Chemical constituents : Fruits contain β- Sitosterol, Gallic acid, Ellagic acid, Chebulagic acid, Galloyl, Glucose, Many free sugars, Manitol, Galactose, Rhamnose, Glucoside.

Biological activity : *Fruits* - Acrid, anthelmintic, bitter, digestive, laxative, pungent.

Folk uses

Food: Fruits are eaten by the children.

Fodder: The leaves are used as fodder for cattle.

Miscellaneous: Fruits are sold in the market by the tribals.

Tannin: Bhil tribals use the stem bark for tanning the leather which gives black to brown colour.

Poison: The fruit is crushed and dropped in water as a fish poison by the Bhils.

Uses in literature

Seed - Gastric trouble (Singh & Pandey, 1980); Edible (Vartak, 1981; Bhujel *et al.,* 1984; Ramamachandran & Nair, 1981; Datar & Vartak, 1975).

Bark - Cold (Jain & De, 1966).

Seed oil - Swelling, skin diseases (Kumar *et al.,* 2006).

Fruits - Diarrhoea, dropsy, fever (Bhalla *et al.,* 1982); Stomachic (Tribedi *et al.,* 1982); Liver complaint (Janardhanan, 1963); Asthma, bronchitis (Shah *et al.,* 1983); Diabetes (Tiwari *et al.,* 1979); Leprosy (Pal & Jain, 1989; Basak, 2006); Cough (Sharma & Malhotra, 1984); Constipation (Roy & Chaturvedi, 1986); Tonic, headache, diarrhoea (Nath *et al.,* 2006); Swelling, skin diseases (Kumar *et al.,* 2006); Malaria (Deka *et al.,* 2006); Piles (Bhalla *et al.,* 1982; Basak, 2006); Edible (Singh & Maheshwari, 1985; Vartak, 1959).

Tinospora cordifolia (Willd.) Miers. ex Hk. f. & Th. (Menispermaceae)

Syn.: *Menispermum cordifolium* Willd.

Vernacular name : Limbvel, Kari amarvel.

Regional and other names : *Bengali* – Golancha; *English* - Heart leaved moonseed; *Gujarati*-Gado; *Hindi* - Amrita; *Kannada* – Amrutaballi; *Malayalam* – Amrytu; *Marathi* – Gulvel; *Sanskrit* – Amrita, Guduchi; *Tamil* - Shindil, Kodi, Amudam; *Telugu* – Tippateege.

Distribution - Throughout tropical India.

Taxonomic description - Succulent, glabrous, deciduous climbers, with corky and lenticelled bark. Leaves ovate-cordate. Flowers yellow, in axillary racemes. Fruits globose, red when ripe.

Flowering & fruiting - December – June.

Habitat ecology - As a climber on tree & shrub.

Regeneration - By seeds & stem cutting.

Field no. - DJ-50.

Parts used - Stem, Leaves.

Chemical constituents : Starch, Glycoside (Giloin), Tinosporic acid, Vasa alcohol, Essential oil, Vasa acid, Proteins, Calcium, Phosphorus, Berberine, Gilosterol.

Biological activity : *Stem* - Bitter, diuretic, stimulant, stomachic.

Folk uses

Medicine:

- Juice of stem is used to cure leucorrhoea, snake bite, leprosy, asthma and dysentery.
- Juice of stem is taken with water every day early in the morning for 7 days to cure leucorrhoea, menorrhoea, typhoid, eczema, ringworm, skin itching and malaria fever.
- Extract of stem is taken with water to cure arthritis.
- Juice of stem is useful in malaria fever and psoriasis.
- Juice of stem is useful in dog bite.
- Extract of stem is taken with water to cure arthritis.
- Juice of stem is taken with water in the cases of burning in urethra during micturition.
- Juice of stem is taken with water daily 3 times for a week to check undesirable discharge of semen with urine.
- The juice of leaves of *Tinospora cordifolia* is shaked with the juice of leaves of Bhang (*Cannabis sativa*), water and cow urine. After well shaking (101 times),a glassful of juice is prepared. This juice is given in the case of miscarriages.
- The stem juice is used for a month to cure stomach ulcer.
- Extract of stem is given with water to cattle in the case of dog bite.

Uses in literature

Root - Diarrhoea, dysentery (Bhatt & Sabins, 1987); Diabetes (Ghosh, 2006).

Leaf - Fever, headache, piles (Lakshamanan & Sankaranarayanan, 1988); Joint pain (Ganeshan *et al.,* 2006); Diabetes (Kadavul & Jisha, 2006); Gastric trouble (Rajendran & Rengamani, 2006).

Whole plant - Bone fracture (Molla & Roy, 1985); Piles (Singh, 1988); Fever (Vedprakash & Mehortra, 1988); Jaundice (Jadeja *et al.,* 2006).

Stem - Skin diseases, jaundice (Sharma *et al.,* 1979); Stomachic, asthma (Shah & Joshi, 1971); Abdominal gripes (Sarma *et al.,* 2006); Diphtheria (Roy & Chaturvedi, 1987); Cough (Malhotra & Moorthy, 1973); Joint diseases (Dwivedi *et al.,* 2005); Fever (Kumar & Chauhan. 2005); Jaundice (Deokule, 2006); Dysentery (Kataria & Gauchan, 2006); Malaria (Deka *et al.,* 2006); Diarrhoea (Kataria & Gauchan, 2006; Gupta, 2006); Ulcers, swelling (Das *et al.,* 2006); Veterinary medicine (Shah & Joshi, 1971; Sebastian & Bhandari, 1984).

Trachyspermum ammi (L.) Sprague (Apiaceae)

Syn.: *Carum copticum* Benth.; *Sison ammi* L.

Vernacular name : Ajwan, Ajwain, Ajmo.

Regional and other names- *Bengali* –Javan; *Gujarati* -Ajma ; *Hindi* -Ajwain; *Kannada* –Oma; *Malayalam* -Ayanudakan; *Marathi* –Onva; *Sanskrit* -Ajamodika; *Tamil* -Omam; *Telugu* –Vamu.

Distribution - Throughout India.

Taxonomic description - Annual herbs, with whitish or pinkish flowers and muricated fruits.

Flowering & fruiting - November-March.

Habitat ecology - Cultivated for the fruits used as condiment.

Regeneration - By seeds.

Field no. - DJ-51.

Parts used : Seeds.

Chemical constituents : Fruits contain an Aromatic volatile oil, Stearoptene, Cumene, Terpene as Thymene; Phenolic glucosides, Thymol, p-Cymene, r-Terpene, Stearoptin, Cymon, Terpene.

Biological activity

Fruits- Antidiarrhoeal, appetiser, anthelmintic, digestive, stomachic, stimulant, tonic.

Leaves- Anthelmintic, appetiser, digestive, heart tonic, stimulant, stomachic.

Folk uses

Medicine:

- Powder of seeds is mixed with jaggery, this mixture is warmed a bit with ghee and small pills are prepared. One pill is taken everyday after pregnancy for strength and vigour.
- Powder of roasted seeds is mixed with powder of turmeric (*Curcuma longa*) and salt. This mixture is taken with water to cure vomiting.
- Power of seeds is mixed with jagerry and some small pills are prepared. 1-2 pills taken daily with water to cure dysmenorrhoea.

Uses in literature

Fruit- Abortifacient, pneumonia (Singh & Pandey, 1980).

Seed- Boil, cuts, colic (Saxena & Vyas, 1986).

Trapa natans L. var. ***bispinosa*** (Roxb.) Makino. (Trapaceae)

Syn.: *Trapa bispinosa* Roxb.

Vernacular name : Singhora, Singarha, Singoda, Hingoda.

Regional and other names : *Bengali* – Pani-phal; *English* – Water chestnut; *Hindi* – Singaadha; *Kannada* – Singhara; *Malayalam* – Karimpolam; *Marathi* – Singarha; *Sanskrit* – Srngataka; *Tamil* – Singhara; *Telugu* – Kubyakama.

Distribution - Throughout the India.

Taxonomic description - This species is neither cultivated nor grows wild in the area but its fruits are locally available in the market.

Flowering & fruiting - August-December.

Habitat ecology - This species is neither cultivated nor grows wild in the area but its fruits are locally available in the market.

Field no. - DJ-134.

Parts used : Fruits.

Chemical constituents : Fruit contains - Citric acid, Tannin, Amylose, Amylopectin, Carbohydrates, β-amylase, Phosphorylase Enzymes, Protein, Fat, Nicotinic acid, Riboflavin, Thiamine, Vitamine A & C, Mangenese.

Biological activity : Anti-inflammatory, astringent, refrigerant, stomachic, tonic.

Folk uses

Food: The fresh fruits are eaten by the tribals.

Medicine: The small pieces of dried seeds are chewed with small pieces of dried endosperm of *Cocos nucifera* L. daily 3 times to cure lucorrhoea.

Miscellaneous: The fruits are sold in the market.

Uses in literature

Seed - Gonorrhoea (Singh & Pandey, 1998); Bronchitis, cough, dropsy (Jain & Tarafder, 1970).

Fruit- Edible (Shah, 1984; Maji & Sikdar, 1982).

Trianthema portulacastrum L. (Aizoaceae)

Vernacular name : Dhankni.

Regional and other names- *Bengali* –Sabuni; *English* –Horse purslane; *Hindi* -Sabuni; *Marathi* –Pundharighentuli; *Sanskrit* -Chiratika; *Tamil* -Shavalai; *Telugu* –Ambatimadu.

Distribution - Throughout India.

Taxonomic description - Succulent, prostrate herbs, with spathulate leaves. Flowers white or pink, solitary; stamens 10. capsules mitrifrom.

Flowering & fruiting - August-December.

Habitat ecology - Common in wastelands.

Regeneration - By seeds.

Field no. - DJ-158.

Parts used : Leaves.

Chemical constituents : Protein, Fat, Carbohydrate, Crude fibre, Calcium, Phosphorus, Iron, Ascorbic acid, Carbonate, Oxalates.

Biological activity : Bitter, hot, alexeteric, analgesic, stomachic, laxative, alterative.

Folk uses

Food: Leaves are used as vegetable.

Uses in literature

Root- Jaundice (Shekhawat & Sunil Anand, 1984).

Leaves- Liver complaint (Sudhakar & Rolla, 1985); As vegetable (Maheshwari & Singh, 1984; Shekhawat & Sunil Anand, 1984; Maji & Sikdar, 1982; Singh & Singh, 1981; Ramachandran & Nair, 1981).

Tribulus terrestris L. (Zygophyllaceae)

Vernacular name : Bekhrio, Gekhru.

Regional and other names : *Bengali* – Gakhura; *English* - Land caltrops; *Gujarati* - Betagokhry; *Hindi* - Gokhru; *Kannada* – Negalu; *Malayalam* – Nerunji; *Marathi* – Lahangokhru; *Sanskrit* – Goksura, Bahukantaka; *Tamil* - Nerunji; *Telugu* – Chinnipalleru.

Distribution - Throughout India.

Taxonomic description – Procumbent, annual herbs, with pinnate leaves and yellow flowers. Mericarps dorsally tuberculate, spines 4.

Flowering & fruiting - March – November.

Habitat ecology - Abundant in open wastelands in sandy and gravelly soils.

Regeneration - By seeds.

Field no. - DJ-136.

Parts used - Whole plant.

Chemical constituents : Fruits contains Alkaloid, Fixed oil, Essential oil, Resin, Steroids, Saponins, Vitamin- C, Protein, Nitrates.

Biological activity

Root - Alterative, aphrodisiac, appetiser, cooling, fattening, sweet, tonic.

Fruits - Alterative, aphrodisiac, appetiser, cooling, fattening, sweet, tonic.

Seed - Aphrodisiac, cooling, diuretic, fattening.

Folk uses

Medicine:

- Extract of plant is taken with water by the women to cure leucorrhoea & menorrhoea.
- Leaves are eaten in the cases of fever.
- Extract of whole plant is useful in typhoid.

Uses in literature

Fruits - Leucorrhoea (Saxena & Vyas, 1983); Cough (Malhotra & Moorthy, 1973); Asthma (Singh & Pandey, 1980); Urinary complaint (Shah *et al.,* 1981; Sharma *et al.,* 2006); Liver complaint (Gopal & Shah, 1985); Kidney diseases (Deokule, 2006); Edible (Bhandari, 1974; Bhargava, 1959; King, 1969).

Whole plant - Impotency (Shekhawat & Anand, 1984); As vegetable (Singh, 1945).

Leaf - Jaundice (Ganeshan *et al.,* 2006); Gonorrhoea, rheumatism (Shah, 2005); Gout (Jadeja *et al.,* 2006); As vegetable (Shah *et al.,* 1981; Singh & Singh, 1981).

Tridax procumbens L. (Asteraceae)

Vernacular name : Kalio.

Regional and other names- *Bengali* –Tridaksha; *English* –Coat buttons; *Kannada* –Savanthi; *Tamil* -Vel; *Telugu* –Balapaku.

Distribution - Throughout India.

Taxonomic description - Straggling herbs, with ovate-lanceolate leaves & yellow heads terminating long peduncles. Achenes silky.

Flowering & fruiting - Most part of the year.

Habitat ecology - Common among bushes in wastelands and on the hill rocks throughout the area.

Regeneration - By seeds.

Field no. - DJ-112.

Parts used : Leaves, Root.

Chemical constituents

Leaves contain Fumaric acid.

Flowers contain Glucoluteolin, Luteolin, Isoquercetin, Quercetin.

Biological activity : Hypotensive

Folk uses

Medicine:

- Paste of leaves is applied on cuts and scratches.
- Juice of leaves is applied against scorpion sting.
- Paste of the leaves is mixed with the powder of *Foeniculum vulgare* and crystalline sugar. This mixture is taken with water early in the morning for the period of 7 days to cure kidney stone problem.
- Root is tied on ear to cure migraine.
- Paste of leaves is fried in pure ghee and this paste is tied in the case of corn.
- Extract of leaves is taken with mishri (crystalline sugar) and water in the case of inflammation in stomach.
- A drop of leaf extract is dropped in the ear of opposite direction in case of toothache.

Uses in literature

Leaf- Cuts & wounds (Nagendra Prasad & Abraham, 1984; Sudhakar & Rolla, 1985; Ramachandran & Nair, 1981; Maheshwari *et al.,* 1986; Upadhye *et al.,* 1986; Goel & Mudgal, 1988); Diarrhoea & dysentery (Sebastian & Bhandari, 1984); Eczema (Bhatt & Sabnis, 1987); Eye diseases (Goel *et al.,* 1984); Leprosy (Goel & Mudgal, 1988); Scorpion bite (Jain & De, 1966); Toothache (Sudhakar & Rolla, 1985; Dhasmana, 1986); Veterinary medicine (Sebastian, 1984).

Root- Skin diseases (Jain & De, 1966).

Whole plant – Ulcers (Pandey *et al.,* 1981).

Trigonella foenum-graecum L. (Papilionaceae)

Vernacular name : Methi.

Regional and other names : *Bengali* – Methi; *English* – Fenugreek; *Gujarati* –Methi ; *Hindi* – Methee; *Kannada* – Menthe; *Malayalam* – Venthiam; *Marathi* – Methi; *Sanskrit* – Methika; *Tamil* – Vendayam; *Telugu* – Mentulu.

Distribution - Cultivated throughout India.

Taxonomic description - Herbs, with 3-foliolate leaves, pale-yellow flowers & straight, turgid, beaked pods.

Flowering & fruiting - November-March.

Habitat ecology - Cultivated during winter season for the seeds & leaves used as vegetable.

Regeneration - By seeds.

Field no. - DJ-160.

Parts used - Seeds, Leaves.

Chemical constituents : Seed contains Quercetin, Luteolin, Trigonelline, Choline, Nicotinic acid, Fixed oil, Essential oil, Mucilage, Bitter extractive, Colouring matter, 2-steroidal Saponins.

Biological activity : *Seeds* - Anti-inflammatory, aphrodisiac, aromatic, astringent, bitter, carminative, diuretic, emollient, stomachic, thermogenic, tonic.

Folk uses

Fodder: The husk (methi ko suklo) is a good fodder for cattles.

Food: Leaves and seeds are used as vegetable.

Medicine: The powder of roasted seeds is taken with water to cure arthritis.

Uses in literature

Seeds - Carminative, gas trouble (Shah *et al.,* 1984); Piles (Saxena & Vyas, 1986); Dyspepsia, dysentery (Jadeja *et al.,* 2006); Diarrhoea, dysentery (Kataria & Gauchan, 2006); Diabetes (Ghosh, 2006); Rheumatism (Shah *et al.,* 1984; Saxena & Vyas, 1986; Kumar *et al.,* 2006); Swelling (Deokule, 2006).

Shoot- As vegetable (Shah, 1984).

Triticum aestivum L. (Poaceae)

Vernacular name : Gehu, Gaun, Genhu.

Regional and other names- *Bengali* –Gom; *English* –Bread wheat; *Hindi* -Genhu; *Kannada* –Godhi; *Malayalam* -Gendum; *Marathi* –Gahum; *Tamil* -Godumai; *Telugu* –Goodhumalu.

Distribution - Largely cultivated throughout India.

Taxonomic description - Medium-sized annuals. Inflorescence a terminal distichous spike; spikelets solitary, awned. Grains elliptic-oblong.

Flowering & fruiting - October-April.

Habitat ecology - Widely cultivated for its grains and straw in the area.

Regeneration - By seeds.

Field no. - DJ-135.

Parts used : Seeds.

Chemical constituents : Grain contain Protein, Crude fibres, Sugars, Fat, Minerals, Vitamins, Enzymes.

Biological activity : *Grain-* Sweet, refrigerant, emollient, laxative, rejuvenating, appetiser, vulnerary, demulcent, aphrodisiac.

Folk uses

Fodder: The husk (genhu ko suklo) is a good fodder for cattle.

Medicine:

- A preparation made with wheat flour and jaggery-locally called 'Gud-rab' is given to the ladies after child-birth to clear the uterus.
- Latex of the leaves of Vadli (*Ficus benghalensis*) is mixed with wheat flour and small pills are prepared. One pill taken every day early in the morning for 7 days with cow milk to check undesirable discharge of semen with urine.
- 5 drops of juice obtained from the plant of Haslathuwar (*Opuntia elatior*) is mixed with wheat and some pills are prepared. One pill taken in the case of dog bite.
- The pulp of leaves of Gwarpatha (*Aloe vera*) is mixed with wheat flour and some small balls are prepared. These balls are roasted in dried cow dung. One ball is taken early in the morning with pure ghee in winter season for a period of 7 days for strength and vigour.

Triumfetta rhomboidea Jacq. (Tiliaceae)

Vernacular name : Lapti, Chikni jhad.

Regional and other names- *Bengali* –Ban-okra; *English* –Spiny cockle burr; *Hindi* -Chikti; *Kannada* –Kadubinde; *Marathi* –Jhinjudi; *Tamil* - Ottupullu; *Telugu* –Tutturubenda.

Distribution - Throughout India.

Taxonomic description - Stout herbs, with ovate-rounded, 3-lobed & entire leaves. Flowers yellow, with 5-10 stamens. Capsules pubescent, spines glabrous.

Flowering & fruiting - July-November.

Habitat ecology - Common in hilly tracts in wastelands.

Regeneration - By seeds.

Field no. - DJ-159.

Parts used : Root.

Chemical constituents : 4-Hydroxyisoxazole and Triumboidin.

Biological activity : *Root-* Bitter, acrid, tonic, styptic, galactagogue, aphrodisiac, cooling.

Folk uses

Medicine:

- Extract of root is taken with water to cure leucorrhoea.
- Extract of root is taken with water in the cases of prolepsy of uterus and hydrocoel.
- Extract of root is taken with water to check undesirable discharge of semen with urine.
- Extract of root or stem is taken with water in the cases of snake bite.
- Extract of root is taken with curd in the case of loose motion.
- The tender twigs are used as tooth-brush.
- Extract of root is given with water in the case of bone fracture.

Uses in literature

Leaf- Diarrhoea (Sebastian & Bhandari, 1984); Stomachache (Kumar *et al.,* 1987); Impotency (Shah *et al.,* 1983); As vegetable (Vartak, 1981).

Typha elephantina Roxb. (Typhaceae)

Vernacular name : Her, Aer, Her ko bur, Her ko Bhutto.

Regional and other names- *Bengali* –Hogla; *English* –Elephant grass; *Hindi* -Mothitrina; *Kannada* –Apu; *Marathi* –Rambdna; *Tamil* -Chamba; *Telugu* –Kandra.

Distribution - Throughout India.

Taxonomic description - Erect, rhizomatous herbs, with spongy linear leaves flat above and convex beneath. Spikes cylindrical, male flowers above, female below and up to 3 cm long gap in between.

Flowering & fruiting - August-June.

Habitat ecology - Common in marshy and aquatic habitats throughout the area.

Regeneration - By seeds.

Field no. - DJ-131.

Parts used: Inflorescence.

Biological activity *: Whole plant-* Cooling, aphrodisiac.

Folk uses

Medicine: The ripe fruit is applied on a fresh cut/ injury; bleeding stops forthwith and the injury heal up quickly. It also heals wounds.

Timber: The plants are used as rope for bundling the wheat crop.

Uses in literature

Fruit- Stop bleeding (Shah, 1984).

Whole plant- Broom, mats (Maheshwari *et al.,* 1990).

Vernonia anthelmintica (L.) Willd. (Asteraceae)

Syn.: *Conyza anthelmintica* L.

Vernacular name : Ghora-jiru.

Regional and other names : *Bengali* – Somraj; *English* - Purple fleabane; *Gujarati* - Kalijiri; *Hindi* - Somraj; *Kannada* - Kadu – jirage; *Malayalam* – Kalajirakam; *Marathi* – Kalajira; *Sanskrit* – Agnibija, Aranyajirkam; *Tamil* - Kattuchiragam; *Telugu* – Nelavavili.

Distribution - Throughout India.

Taxonomic description - Erect, aromatic, punctate or glandular hairy annuals. Leaves elliptic-obovate or ovate, coarsely serrate. Heads purple. Achenes terete. 10 to 12-ribbed, glandular hairy.

Flowering & fruiting - August– November.

Habitat ecology - A weed occasionally found in forest areas.

Regeneration - By seeds.

Field no. - DJ-151.

Parts used - Seeds.

Chemical constituents - Seeds contains β- Sitosterol, Vernolic acid, Flavinoids, Glycosides, Trivernolin, Vernodolol, Terpineols, Sabenene, Isolimonene, Caryophyllene, Selinene.

Biological activity - *Seeds* - Acrid, anthelmintic, astringent, Sharp taste.

Folk uses

Medicine:

- Extract of seeds is given by the tribals to their cattle in gastric trouble.
- Extract of seeds is taken with water in gastric trouble in human-being.

Uses in literature

Seed - Leucoderma & skin diseases (Singh & Pandey, 1980); Acidity (Chaudhuri & Pal, 1976); Asthma (Bedi, 1978); Cough (Rajwar, 1983).

Fruits - Paralysis & Swelling (Shah, 1984).

Whole plant - Constipation (Goel & Mudgal, 1988).

Vigna mungo (L.) Hepper. (Papilionaceae)

Syn.: *Phaseolus mungo* L.

Vernacular name : Urad.

Regional and other names- *Bengali* –Mashkalai; *English* –Black gram; *Hindi* -Urid; *Kannada* –Uddinabee; *Malayalam* -Uzhunnuparippu; *Marathi* –Uddachi; *Tamil* -Ulunnu; *Telugu* –Minapapappu.

Distribution - Commonly cultivated throughout India.

Taxonomic description - Trailing, hispid herbs, with 3-foliolate leaves & yellow flowers. Pods linear, turgid.

Flowering & fruiting - August-October.

Habitat ecology - Cultivated as a pulse crop throughout the area.

Regeneration - By seeds.

Field no. - DJ-130.

Parts used : Seeds.

Chemical constituents

The plant contains albuminoids, Starch, Oil.

Seed contains Saponin I, II & III. It contains Genistein, alpha-Hydroxygenistein, Dalbergioidin.

Biological activity

Root- Narcotic.

Seed- Sweet, tonic, laxative, aphrodisiac, appetiser, diuretic.

Folk uses

Fodder: The husk (urd ko suklo) is used as fodder for cattle.

Food: Seeds are used as vegetable.

Medicine: The paste of seeds is tied on anus to cure piles.

Vitex negundo L. (Verbenaceae)

Vernacular name : Negad

Regional and other names : *Bengali* - Nisinda, samalu; *English* - Chaste tree, Indian private; *Gujarati* - Nagaol; *Hindi* - Shamalu, sambhalu; *Kannada* - Zakkigida, nekki; *Malayalam* - Indrani, vellanocchi; *Marathi* – Nirgundi; *Sanskrit* -Nirgudi; *Tamil* - Vellai nocohi, nirkkundi; *Telugu* - Nallavavilli, vaavilli.

Distribution - Throughout India.

Taxonomic description - Evergreen shrubs, with 3 to 5-foliolate leaves. Flowers violet, in peduncled cymes. Fruits ovoid, 4-celled.

Flowering & fruiting - September – April.

Habitat ecology - Commonly found along roadsides, banks and in hedges.

Regeneration - By seeds & vegetative method.

Field no. - DJ-99.

Parts used - Leaves, Stem.

Chemical constituents : Volatile oil, Alkaloids (Nergundin & Hydrocotylon), Vitamin- C, Carotene, Flavon, Tannic acid, Hydroxy Benzoic acid, Resin, Mallic acid.

Biological activity : Acrid, anthelmintic, astringent, bitter, cephalic, heating, Pungent, stomachic.

Folk uses

Medicine:

- Boiled leaves are tied on forehead to cure headache and migrain.
- Decoction of leaves is useful in gout.

Timber: The stem-strips are used for making "Topla" (basket).

Uses in literature

Root - Piles (Singh, 1988); Rheumatism (Gupta, 1981); Diarrhoea, dysentery (Kataria & Gauchan, 2006).

Seed - Gout (Singh & Pandey, 1980).

Flower - Pneumonia (Virjee *et al.,* 1984); Diarrhoea (Malhotra & Moorthy, 1973); Dysentery (Sharma *et al.,* 1979); Eczema, syphilis, skin diseases, leprosy (Purohit *et al.,* 1985).

Stem - Bone fracture (Gupta, 1981); Fever (Tribedi *et al.,* 1985); Sciatica (Hembrom & Goel, 2005); Basketry (Singh & Maheshwari, 1985); Veterinary diseases (Tribedi *et al.,* 1985).

Leaf - Boils, blister (Maheshwari & Singh, 1984); Paralysis (Nagendra Prasad & Abraham, 1984); Cold & cough (Jain & De. 1966); Colic (Rajwar, 1983); Bodyache (Das *et al.,* 1983); Gum trouble (Tarafder, 1978); Ulcer (Bhalla *et al.,* 1982); Wound (Tarafder, 1984); Swelling (Rao, & Jamir, 1982; Deokule, 2006); Fever (Shah *et al.,* 1983); Headache (Shah, 2005); Sinusitis (Chhetri, 2005; Rajendran & Rengamani, 2006); Jaundice (Purkayastha & Subhan, 2005; Jadeja *et al.,* 2006); Joint diseases (Dwivedi *et al.,* 2005); Bodyache, muscular pain (Pereira, 2006; Deokule, 2006); Rheumatism (Tarafder, 1978; Mokat & Deokule, 2006; Das *et al.,* 2006); Malaria (Deka *et al.,* 2006); Gout (Jadeja *et al.,* 2006); Migrain (Sharma *et al.,* 2006).

Whole plant – Asthma, lung diseases, tonsilitis, rheumatism, urinary troubles (Kumar *et al.,* 2006).

Withania somnifera (L.) Dunal (Solanaceae)

Syn.: *Physalis somnifera* L.

Vernacular name : Ashing.

Regional and other names : *Bengali* – Ashvaganda; *Gujarati* - Asan; *Hindi* - Ashvagandha; *Kannada* – Viremaddlinagadde; *Malayalam* – Pevetti; *Marathi* – Askondhatilli; *Sanskrit* - Ashvagandha; *Tamil* - Amukkura; *Telugu* - Putivendram, pannerugadda.

Distribution - In the drier part of India.

Taxonomic description - Hoary-tomentose undershrubs, with elliptic-lanceolate or obovate leaves, bisexual flowers and berries enclosed in the persistent calyx.

Flowering & fruiting - August- December.

Habitat ecology - Found in wasteland localities.

Regeneration - By seeds.

Field no. - DJ-132.

Parts used - Root.

Chemical constituents : Root contains Alkaloids (Somniferon), Withaminon, Wasamin, Sugars, Glycosides, Amino acids, Essential oils, Withaniol, Hentriacontane, Phytosterol & Oils.

Leaves contain Withanoloids alkaloids, Glycosoids, Glucose, Amina acids.

Stem contains Proteins, Amino acids, Tannin, Flavonoids, Calcium, Iron.

Biological activity

Tuber - Acrid, alterative, anthelmintic, aphrodisiac, bitter, tonic.

Folk uses

Medicine:

- Powder of dried roots is taken with cow milk to cure arthritis and epilepsy.

Uses in literature

Seed - Chest complaint (Bhalla *et al.,* 1982).

Leaf - Boils, eye complaint (Bhalla *et al.,* 1982); Skin diseases (Lal & Yadav, 1983).

Root - Cough, dropsy, dyspepsia, rheumatism, (Bhalla *et al.,* 1982); Asthma, bronchitis (Shah *et al.,* 1983; Sharma *et al.,* 2006); Wound (Tarafder, 1984); Epilepsy (Tiwari *et al.,* 1979); General debility, spermatorrhoea (Kumar & Chauhan, 2005); Tonic (Deokule, 2006); Sexual vigour (Mokat & Deokule, 2006; Chatterjee & Bhatterjee, 2006); Diabetes (Ghosh, 2006).

Xanthium strumarium L.(Asteraceae)

Vernacular name : Adosisi.

Regional and other names : *Bengali* – Chotadhatura; *English* - Burweed; *Gujarati* - Gadriyun; *Hindi* - Banokra; *Malayalam* – Maruluumatti; *Marathi* – Dutuudi; *Sanskrit* - Arishta; *Tamil* - Maruloomadtam; *Telugu* - Marula, matangi.

Distribution : Throughout India.

Taxonomic description : Foetid-smelling, unarmed herbs, about 1m high. Leaves ovate-suborbicular. Heads greenish yellow, in terminal and axillary racemes.

Flowering & fruiting : January – February.

Habitat ecology : Common weed around cultivated field, wasteland & forest edges.

Regeneration : By seeds.

Field no. : DJ-76.

Parts used : Fruit.

Chemical constituents : Seed contains Xanthostrumarin acid, Oxalic acid.

Fruits contain Strumaroside, Stigmasterol, Xanathin, Xanthinin, Xanthumin, Xanthanol.

Biological activity

Acrid, alexiteric, alterative, anthelmintic, antipyretic, appetiser, cooling, digestive, fattening, laxative, tonic.

Folk uses

Magico-religious beliefs: The fruit is tied on 'Sikha' with the help of thread by the tribals to protect them from evil spirits.

Medicine: A fruit is tied on ear with the help of thread to cure migraine.

Uses in literature

Seed - Wound (Pal & Jain, 1989); Headache (Maheshwari *et al.*, 1986).

Fruit - Eye diseases (Maheshwari & Singh, 1987); Small pox (Deokule, 2006); Rheumatism (Das *et al.*, 2006).

Leaf - Malaria (Pal & Jain, 1989); Ringworm (Maheshwari & Singh, 1987).

Flower - Toothache (Sudhakar & Rolla, 1985).

Root - Ulcer (Malhotra & Moorthy, 1973); Snake bite (Kumar & Chauhan, 2005).

Zea mays L. (Poaceae)

Vernacular name - Makki.

Regional and other names : *Bengali* – Bhutta; *English* – Corn; *Hindi* – Makka; *Kannada* – Musukojola; *Malayalam* – Colam; *Sanskrit* – Makayah; *Tamil* – Makkaccolam; *Telugu* –Mokkajonnalu.

Distribution - Cultivated throughout India.

Taxonomic description - Monoecious annuals. Male spikelets usually in terminal racemes, white. Female inflorescence consists many spikelets seated on a spongy axis.

Flowering & fruiting - November-February.

Habitat ecology - Cultivated during rainy season for its grains.

Regeneration - By seeds.

Field no. : DJ-193.

Parts used : Root.

Chemical constituents - Grain contains Starch, Fructose, Fucose, Sucrose, Raffinose, Vitamin B, C, E, K; Carotene, Maysin, Sorbitol, Zeanin.

Biological activity - *Grain* - Appetiser, astringent, diuretic, nutritious, sweet.

Folk uses

Food: Makka is the main food of the bhil tribe.

Medicine: Extract of one year's old root is given with water to cure dysentery.

Uses in literature

Grain - Stomachcahe (Lal & Yadav, 1983); Piles (Kumar & Yadav, 2006); Edible, drink (Dam & Hajra, 1981).

Inflorescence - Kidney trouble (Virjee *et al.,* 1984); Whooping cough (Ved Prakash & Mehrotra, 1988).

Styles - Diseases of kidney and bladder (Kumar & Yadav, 2006).

Zingiber officinale Rosc. (Zingiberaceae)

Vernacular name - Adrak.

Regional and other names - *Bengali* – Ada; *English* - Ginge;r *Hindi* - Ada; *Kannada* – Shunti; *Malayalam* – Inchi; *Marathi* – Ale; *Sanskrit* - Ardraka; *Tamil* - Allam; *Telugu* – Allamu.

Distribution - Widely Cultivated eleswhere.

Taxonomic description - A small herb with horizontal, fleshy rhizome; leaves distichous, narrow, oblong-lanceolate. Flowers on long spike and are of purple colour.

Flowering & fruiting - November-March.

Habitat ecology - Commonly cultivated for its rhizome.

Regeneration - By rhizome.

Field no. - DJ-53.

Parts used - Rhizome.

Chemical constituents : Rhizome contains Essential oil, Fibres, Proteins, Starch, Carbohydrates, Resin, Glutamin, Thrionin, Free amino acids, Zingiberol, Zingiberin, Glutamic acid, Aspartic acid.

Biological activity : *Rhizome* - Analgesic, aphrodisiac, appitizer, aromatic, blood purifier, carminative, digestive, expectorant, febrifuge stimulant, thermogenic.

Folk uses

Medicine:

- Powder of dried rhizome is mixed with jaggery and safed musli (*Chlorophytum tuberosum*). This mixture is warmed a bit with ghee and small balls (Laddoo) are prepared. One ball (Laddoo) is taken everyday after pregnancy for strength and vigour.
- Powder of dried rhizome is taken with tea to cure cold.

- Hing (Gum resin obtained from roots of *Ferula asafoetida* L.) is filled in the dried rhizome of Adrak through hole. This rhizome is roasted in the burning cow dung. After it is roasted, it is made into a powder, 1 gm. of powder with honey is taken in early morning daily to cure asthma.

Uses in literature

Rhizome - Cough, insect sting, snake bite (Kharkongor & Joshep, 1981); Cough, cold (Pereira, 2006; Deokule, 2006); Abortifacient (Tarefder, 1953); Headache (Deokule, 2006); Gastric trouble (Gupta, 2006); Constipation, dyspepsia (Jadeja *et al.,* 2006); Malaria (Deka *et al.,* 2006); Gout (Jadeja *et al.,* 2006)

Ziziphus mauritiana L. (Rhamnaceae)

Vernacular name - Bor.

Regional and other names - *Bengali* - Kul, borui; *English* - Indian plum; *Gujarati* - Bor; *Hindi* - Ber; *Kannada* – Yelachi; *Malayalam* – Elantha; *Marathi* – Bor; *Sanskrit* - Badari; *Tamil* - Elantharandor; *Telugu* - Regu, pandu, pazham.

Distribution - Throughout India.

Taxonomic description – Evergreen, small trees or shrubs, with droopling branches armed with prickles. Leaves ovate-elliptic to suborbicular, tomentose beneath. Flowers greenish- yellow, in axillary cllusters. Drupes fleshy.

Flowering & fruiting - March – May

Habitat ecology - Throughout the area.

Regeneration - By seeds, coppice & by stem cutting.

Field no. - DJ-55.

Parts used - Fruits, Leaves.

Chemical constituents - Fruits contains Sugars & Mucilage.

Biological activity

Root - Bitter, cooling.

Leaf - Antipyretic, bitter, cooling.

Fruits - Aphrodisiac, cooling, indigestible, laxative, tonic.

Seed - Aphrodisiac, acrid, sweet.

Folk uses

Food : Fruits are edible and also used as vegetable.

Medicine:

- Powder of roots is useful in diarrhoea and dysentery.
- The paste of leaves is applied in the cases of scorpion sting.
- Extract of root is taken with water to cure fever.

Miscellaneous: Fruits are sold in the market.

Uses in literature

Fruits - Blood purifier (Bhalla *et al.,* 1982); Stomachache (Kharokongor & Joseph, 1981; Baruah *et al.,* 2006); Hepatitis (Dutta Choudhury & Shil, 2006); Edible (Maheshwari & Singh, 1984; Shekhawat & Sunil Anand, 1984; Shah, 1984; Maji & Sikdar, 1982; Jain & De, 1966; Jain, 1964; Singh & Singh, 1981; Ramachandran & Nair, 1981; Goel *et al.,* 1984; Vartak, 1959; Bhargava, 1959; Nagar, 1985; Datar & Vartak, 1975).

Bark - Diarrhoea (Joshi *et al.,* 1980; Kataria & Gauchan, 2006); Piles (Deokule, 2006); Dysentery (Kataria & Gauchan, 2006).

Leaf - Rheumatism (Dar *et al.,* 1984); Whooping cough (Virjee *et al.,* 1984);Conjuctivitis (Bhalla *et al.,* 1982); Scorpion sting (Shah *et al.,* 1981); Dizziness (Sarma *et al.,* 2006); Hepatitis (Dutta Choudhury & Shil, 2006).

Root - Stomachic (Tarafder, 1983); Ulcer, wound (Janardhanan, 1963); Dysentery (Basak, 2006).

Z. nummularia (Burm. f.) Wt. & Arn. (Rhamnaceae)

Syn.: *Rhamnus nummularia* Burm f.

Vernacular name - Bor-Jari, Jhari-bor.

Regional and other names- *Bengali* –Kantaikolli; *English* –Wild jujube; *Marathi* –Janglar; *Tamil* -Korgodi; *Telugu* –Parimpalu.

Distribution - Throughout India.

Taxonomic description - Much-branched, prickly shrubs, with zigzag branches & orbicular-ovate leaves tomentose above. Flowers greenish-yellow, in axillary cymes. Drupes globose, red when ripe.

Flowering & fruiting - July-January.

Habitat ecology - Common in wastelands.

Regeneration - By seeds.

Field no. - DJ-54.

Parts used - Fruits.

Chemical constituents - Root bark contains Cyclopeptide alkaloids Nummularine A, B, C, Mucronine-D & Amphibine-H.

Biological activity - Sweet, sour, appetiser, stomachic, astringent, cooling.

Folk uses

Food: Fruits are edible. Fruits are sold in the market.

Medicine:

- Powder of roots is useful in diarrhoea and dysentery.
- Powder of roots is taken with water in curing the pus in ear.

Uses in literature

Root- Diarrhoea (Audichya *et al.,* 1983); Tonic (Shah *et al.,* 1983).

Bark- Dysentery (Rajwar, 1983).

Fruit- Indigetion (Rajwar, 1983); Edible (Maheshwari & Singh, 1984; Shah, 1984; Maji & Sikdar, 1982; Bhandari, 1974; Singh & Singh, 1981; Shah *et al.,* 1981; Singh & Maheshwari, 1985).

PLANTS USED IN FOLK-SONGS

Plants have been an integral part of life and cluture. Folk songs give an idea of plants growing in the surrounding environs of tribals and throw light on plant based practices in usage in various spheres of their lives, including the material benefits derived and recreational opportunities offered. They also reflect the most popular plants and the plants which tribals dislike for one reason or the other. Among the tribes of Ratlam district, chorus songs are more prevalent in comparison to solo songs and musical instruments and dancing is not always necessary. Some folk songs are exclusively on plants, based on single species or a number of species; while majority of songs related to agricluture, famine, love, marriage, religion, recreation, looting, liquor etc have reference of plants in the body of the song. Present survey revaled that *Ailanthus excelsa* (Aydu ba), *Bombax ceiba (*Semal*)*, *Syzygium cumini (*Jamun*)* and *Madhuca indica (*Mahuwa*)* are the main species on each of which several songs are prevalent among Bhil tribes of the district. Some beautiful examples are:

'Uncho mauwro (*Madhuca indica*) bhai daro undo bhai...' This song quite picturesquely shows that right from collecting the intoxicating mahua flowers to brewing the liquor, there is a suggestion of turning up side down. Hence whosoever drinks the mahua-wine turns tipsy and is ruined.

'Mahre angdi he lili limodi (*Azadirachta indica*)...'Bhils sing this song wherein it is explained that bounteous canopy of *Azadirachta indica* (Limdo) provides deep shade.

'Lilo ambo (*Mangifera indica*) **pili keri...**'Bhils sing this song wherin it is explained that bounteous canopy of *Mangifera indica* (Ambo) produces sweet fruits & provides shade.

'Hede himliye (*Bombax ceiba*) **ramto...**'. The song speaks of the young Bhils earnest wish to while away his time playfully under the beautiful, shady and flowery semal tree and his very many excuses not to go home.

'Dungri pitan aduo (*Ailanthus excelsa*)...'. Through this piece of folk song the daughter makes out her tragic situation at her in-laws. That her in-laws press her unduly as to collect ornaments like todi and so on from her fathers house or they will kick her out. Hence she is unwilling to go to her in-laws.

'Ghadi alo ghadi alo...'. Some of the tribal songs paint the famine stricken life and the young Bhils desperate desire to make use of bow and arrow and that way to rob and loot.

'**Amli** (*Tamarindus indica*) **ne dore morelu**...'Tree as a fine messenger to break the price of news to her parents that she has been gifted with a child and they have become grand parents. Hence, to observe the tradition.

'Phagun ro maino...'This song paints the eagerness to rush at the bhagoria fair and bathe in the holi colours. As to enjoy fully the beauty of the festivals there is a suggestion to harvesting and winnowing to grains (*Triticum aestivum*) and Grams (*Cicer arietinum*).

'Sai jamun (*Syzygium cumini)* **sai chayie...'**Bhils sing this song wherein it is explained that bounteous canopy of *Syzygium cumini* (Jambu) offer shade for Haats.

'Siksa vager ni revay...'This song paints learning as deep and shady as that of a banyan [Vadliya (*Ficus benghalensis*)] tree.

'Taro mati ramasyo...'In this song the tribal-wives through the jungle fruits timru (*Diospyros melanoxylon*), rice (*Oryza sativa*)and grams (*Cicer arietinum*)in a festive mood playfully dig at their in-laws.

Songs associated with the events of marriage also include several plants. References to *Curcuma longa* (Haldi), *Cicer arietinum* (Channa, Bhungna), *Mucuna pruriens* (Kevach), *Butea monosperma* (Khankhra), *Oryza sativa* (Chokha, Chawal, Dhan) are available in some folk songs of Bhils, viz. **'Haldi le re lodela piro ne panjdo janaye...'**; **'Bhai ne atri parja aavi...'** and **'Maro kavanada ghar dhedya dhase jhuni...'**. Besides this, majority of folk songs of tribals have reference of many plants in the body of song i. e. *Calotropis gigantea* (Ankdu), *Cannabis sativa* (Ganjo), *Cocos nucifera* (Narel), *Cuscuta reflexa* (Amervel), *Gossypium herbaceum* (Kapas), *Lawsonia inermis* (Mendi), *Musa paradisiaca* (Kero), *Phoenix sylvestris* (Khajor), *Tectona grandis* (Sagad), *Terminalia cuneata* (Kavdo), *Ziziphus mauritiana* (Bor) etc.

PLANTS IN PROVERBS

Folk proverbs having references to plants are indicative of the deep insight, common sense and practical wisdom of the common folk. Things which cannot be remembered or explained easily are learnt, taught and emphasised through proverbs. The words used in a proverb have many hidden meanings of wisdom. Some of the proverbs are invented on plants in general, some involve single species and rests two or more species.

Table 5: Proverbs based on plants.

Name of species with local name	Proverb in tribal's words and sense of proverb
Allium cepa (Kanda)	Yun hun kanda na photra cholira hai. It is not proper to expose others, as generally all have faults. 'Kanda' has been used as an example since it is made of scaly leaves covering one another. If these leaves are removed, nothing is left.
Brassica juncea (Rai) & *Sesamum orientale* (Til)	Na rai vade na til ghate. To held responsible equally 2 groups or person for quarrel, it is said that neither rai is small nor til is big.
Madhuca indica (Mawdi)	Mawdi mathe chadi gi. Success has gone to his head.
Oryza sativa (Chokha)	Handi me chokho ekaj hemare. A single rice speaks about the whole pot.
Cicer arietinum (Channa)	Channa ka jhad pe kim chadai ra. Why do you make me into a braggart.
Phoenix sylvestris (Tad)	Tad ko tad thai gyo, ne kaam kai ni ve. A tall tree bearing little fruit. OR Offcourse he holds a big portfolio but is of no use.
Calotropis gigantea (Ankdo)	Haji gande ankdo kyun lagave. It is unwisely for a healthy man to go to an apothecary.
Sorghum bicolor (Juwar)	Bole tini buri bike, ni bole tini juwar ni bike.

	Publicity sales husk or grains will go rot. OR Publicity sales coal or even diamonds go unsold.
Brassica juncea (Rai)	Rai no bhav ek rate. To feel the people ashamed for their act creating harm to others, as a merchant.
Butea monosperma (Khankhro)	Khakra ne khali, hun jane hag no hawak. A low breed person is ignorant of good pleasure, as a squirrel living on khankhro does not know the taste of mango fruit (hag).
Cicer arietinum (Channa)	Channa he to chabuwa wara koi ne. Where there is wealth, no offspring is there to enjoy it i. e. channa are there, but no one to eat them.
Cicer arietinum (Channa)	Khali channa vaje ghana. When an unknowledgeable person speaks more and presents himself as most efficient, such person is compared with an empty infested channa which produces more sound, but of no other use.
Cicer arietinum (Channa)	Eklo chano Bhad ni phode. One person alone cannot do a big work, as a single channa cannot break the furnace (Bhad).
Saccharum officinarum (Hanttha)	Hanttha ne bharose dhadhan pilayee jaye. One must keep away himself from the person who is in trouble if he is also likely to fall in trouble, otherwise his position will be like the culms of other species which are crushed with hanttha culms.

DISCUSSION & CONCLUSION

During the last half century, ethnobotany has more and more been recognized as a valid discipline that can play a very important role in the advancement of many aspects of scientific, sociological and historical studies. An increasing number of investigators have been devoting attention to the vast store of knowledge of plant properties and uses still intact in native cultures in several parts of the world.

In India about eighty per cent of total population resides in villages. A large number of people, particularly tribals, living in far remote areas of the country in thick forests depand on nature for their food and shelter. These people, especially belonging to primitive or aboriginal culture, have been found to possess, a good deal of informations, not only about traditional medicines but also on less known economical important plants (food plants, dyes and tannins etc.). The important scope and urgent need of the ethnobotanical studies for unearthing the secrets of nature through the collection of data from the rapidly disappearing primitive cultures have been recognised by various works.

The total belief on traditional medicines among primitive societies are based on their long felt experiences and it will be considerable significance if the value of such experiences be held in proper esteem and due importance. An understanding of ethnobotany is critical for assessing the ecological relations with human and ecosystem. Knowledge of resources and successful resource management harvesting system provide information about human adaptation to their social and natural environment. Ethnobotanical knwoledge is important for policy-makers who device solution to local and regional problems. According to Jain (1967a) for any ethnobotanical study, it is usually more convenient and useful to delimit the area of study. Keepping this is view the present studies were taken up in the study area, where there appears to be no record of any ethnobotanical survey. Villages of Ratlam district are very poor and in underdeveloped conditons. During the course of field study, it was observed that the rural and tribal, particularly Bhils, appeared more conservative and did not like to share their knowledge with others, while some local medicine men seemed proud of telling the secretes known to them. However, others required private consultation and great persuasion.

The present ethnobotanical work was carried out by doing periodical and extensive ethnobotanical survey, assessment of field and tribal knowledge bank of tribal inhabitated localities of Ratlam district from 2004-2007. A total number of 210 plant species of ethnobotanical importance [Medicinal- 186; Vegetables- 27; Wild edible fruits-36; Fodder/Forage - 18; Beverage and drinks-3; Gum and Resin-3; Magico-religious beliefs and offerings-25; Agriculture implements, Handle, Tools, Boats-5; Fibres-11; Detergent- 3; Dye-3; Tannin-12; Oil yielding plants-3; House, Building construction, Thatching and Furniture-11; Taboos and Totems-6; Musical instruments-11; Fuel-6; Biofencing-10; Fish poison-10; Other economic uses-7] belonging to 178 genera and 72 families have been including in the present study. Further, the dicots are represented by 180 species belonging to 150 genera and 59 families, and the remaining 30 species belong to 28 genera and 12 families of monocots. Statistically, 210 plant species of ethnobotanical interest belong to different habits, viz. 96 herbs, 55 trees, 37 shrubs, 16 climbers, 5 grasses and one angiospermic parasite plants. The Leguminous (30 species with 23 genera) plants belonging to 3 families are dominant, followed by Poaceae and Solanaceae (9-9 species) and Euphorbiaceae (8 species). Lamiaceae, Liliaceae, Malvaceae, Cucurbitaceae (7-7 species), Asteraceae and Apiaceae (6-6 species) and Convolvulaceae (5) are some co-dominant families of study sites. It is evident from the systematic enumeration that the largest genus is *Acacia* with 4 species *(Acacia catechu, A. leucophloea, A. nilotica* and *A. pennata)* which is followed by *Euphorbia, Ficus* and *Piper* with 3 species each. Following table shows the list of all the studied families with their genera and species.

Table 6. Ethnobotanically important families with their genera & species reported from tribal inhabited localities of Ratlam district.

S.N.	Families	Genus	Species
1.	Acanthaceae	4	4
2.	Amaranthaceae	5	5
3.	Amaryllidaceae	1	1
4.	Anacadiaceae	1	1
5.	Annonaceae	1	1
6.	Apiaceae	6	6
7.	Apocynaceae	3	3
8.	Araceae	1	1
9.	Arecaceae	2	2
10.	Asclepiadaceae	1	2
11.	Asteraceae	6	6
12.	Aizoaceae	1	1
13.	Bignoniaceae	1	1

14.	Bombacaceae	2	2
15.	Boraginaceae	2	2
16.	Brassicaceae	3	4
17.	Burseraceae	1	1
18.	Cactaceae	1	1
19.	Caesalpiniaceae	4	7
20.	Cannaceae	1	1
21.	Capparaceae	2	3
22.	Caricaceae	1	1
23.	Chenopodiaceae	1	1
24.	Combretaceae	3	4
25.	Commelinaceae	2	2
26.	Convolvulaceae	4	5
27.	Crassulaceae	1	1
28.	Cucurbitaceae	6	7
29.	Cuscutaceae	1	1
30.	Cyperaceae	1	1
31.	Dioscoreaceae	1	1
32.	Ebenaceae	1	1
33.	Euphorbiaceae	5	8
34.	Gentianaceae	1	1
35.	Lamiaceae	6	7
36.	Liliaceae	6	7
37.	Linaceae	1	1
38.	Lythraceae	2	2
39.	Malvaceae	5	7
40.	Martyniaceae	1	1
41.	Meliaceae	1	1
42.	Menispermaceae	2	2
43.	Mimosaceae	4	7
44.	Moraceae	3	5
45.	Moringaceae	1	1
46.	Musaceae	1	1
47.	Myrtaceae	3	4
48.	Nyctaginaceae	1	1
49.	Oleaceae	2	3
50.	Pandanaceae	1	1
51.	Papaveraceae	1	1
52.	Papilionaceae	15	16
53.	Piperaceae	1	3
54.	Plumbaginaceae	1	1

55.	Poaceae	9	9
56.	Portulacaceae	1	1
57.	Rhamnaceae	1	2
58.	Rubiaceae	1	1
59.	Rutaceae	3	3
60.	Sapotaceae	2	3
61.	Santalaceae	1	1
62.	Simaroubaceae	2	2
63.	Solanaceae	7	9
64.	Sterculiaceae	1	1
65.	Tiliaceae	3	3
66.	Trapaceae	1	1
67.	Typhaceae	1	1
68.	Ulmaceae	1	1
69.	Vitaceae	3	3
70.	Verbenaceae	4	4
71.	Zingiberaceae	2	3
72.	Zygophyllaceae	1	1
	Total	178	210

The present study has revealed some interesting plants used by the local tribal people (Bhil) in the study area for their treatment of common ailment in their daily life. It was observed that the treatment of cuts, wounds, burn, boils, leprosy, scabies, bodyachae, headache, respiratory diseases, digestive diseases, blood diseases, air born diseases, brain diseases, bone facture, women and men diseases, toothache and other various diseases are attempted by indigenous herbs from their natural environment. In the present study it was estimated that more than 70 types of human ailments are cured by the tribals of the district with 210 plant species, singly or in combination. The frequent health problems are Appendicitis (1); Asthma (11); Backache (2); Blood sugar (2); Bodyache (7); Boils (17); Bone fracture (6); Bronchitis (1); Burn/Burning sensation/ Sunstroke (3); Cervical pain (2);Cholera (1); Constipation (1); Corn (5); Cough & Cold (1); Cuts & Wounds/injury (14); Deafness (1); Diarrhoea/ Dysentery (34); Dogbite (11); Epilepsy (3); Eye & Ear diseases (17); Gastric trouble (4); Gonorrhoea (1); Hair tonic/Shampoo (2); Headache (14); Heel cracking (5); Hoarseness (1); Hysteria (2); Indigestion (1); Intermittent/Fever (20); Intsetinal worms/Anthelmintic (8); Itch/Eczema (9); Jaundice (17); Joint diseases (17); Kidney stone (4); Leprosy (2); Liver disorders (1); Loose motion (6); Men diseases (12); Mental weakness (2); Muscular pain (4); Night blindness(1); Paralysis(1); Piles (13); Pimples(1); Pneumonia(8); Psoriasis(1); Rickets (2); Ringworm (5); Sciatica (2); Scorpion sting (15); Small pox/Chiken pox (5); Snake bite (14); Stomach trouble (14); Swelling (7); Tetanus (1); Throat infection (3); Toothache

(19); Tuberculosis (3); Tumour (5); Typhoid (17); Ulcer (9); Urinary diseases (7); Urticaria (1); Vomiting (6); Women diseases (50). A comparative analysis regarding the medicinal activity of plant parts show that leaves (60), seeds (51) and root (42) have the higher medicinal values in comparison to other plant parts. There are also used 22 plant species for the treatment of animal diseases.

REFERENCES

Abraham, Z. 1981. Ethonobotany of the Todas, the Kotas and the lrulas of the Nilgiris. In S. K. Jain (Ed.) *Glimpses of Indian Ethnobotany*, pp. 308 – 320. Oxford and IBH. New Delhi.

Aditya, P., Kumar, A. &. Yadav, D.K. 2006. Herbal remedies for Dental care system in Gaya district, Bihar. *J. Econ. Taxon. Bot.* 30(Suppl.): 167-170.

Agrawal, S.R. 1981. Trees, Flowers and Fruits in Indian folk songs, folk proverbs and folk tales, In S.K. Jain (Ed.) *Glimpses of Indian Ethnobotany*, pp. 3-12. Oxford and IBH. New Delhi.

Agarwal, V.S. & Arya, M. 2005. *Alternanthera sessilis* DC. (Amaranthaceae): A new source for the treatment of rickets and marasmus. *Ethnobotany* 17: 200-201.

Ahluwalia, K.S. 1968. Medicinal plants of Kerala-VI. *Nagarjun* 11: 392-382.

Ahmad, A. 1995. Drugs of plant origin as used by certain tribes of eastern (Purvanchal) U.P. (Part -IV). *Int. J. Mendel.* 12(1-4):12 – 14.

Ahmad, A. 1996. Drugs of plant origin as used by certain tribes of eastern (Purvanchal) U.P. (Part -V). *Indian J. Appl. Pure. Biol.* 11:41-52.

Ainslie, W. 1813. *Materia Medica of Hindustan*. Neeraj Publishing House, New Delhi.

Ambasta, S.P. 1991. (Ed.). *Useful Plants of India*. Coun. Sci. Industr. Res., New Delhi.

Aminuddin, Girach, R.D. & Subhan Khan, A. 1993. Treatment of Malaria through herbal drugs from Orissa. *Fitoterapia* 64: 545 – 548.

Amit Tomar & Singh, H. 2006. Exotic medicinal plants from Baghpat district, Uttar Pradesh. *J. Non-Timber Forest Prod.* 13(4):273-280.

Anonymous. 1980. Medico-ethno-botanical survey of Surguja district, Madhya Pradesh. *Newslett.* (CCRAS) 2(21) : 1-7.

Anonymous. 1981. Medico-ethno-botanical and health and medicare research programme of Surguja. *Newslett.* (CCRAS) 3(6) : 1-6.

Anonymous 1984. All India *Coordinated Research Project on ethnobotany*. Annual Report, Department of Environment, New Delhi.

Anonymous. 1985. Ethnobotany of Baigas of Madhya Pradesh. *NBRI Newslett.* 12: 28-29.

Apparananthum, T. & Chelladurai, V. 1986. Glimpses on folk medicines of Dharmapuri forest division, Tamil Nadu, *Anc. Sci. Life* 5: 182 – 185.

Arinathan, V., Amuthavalli, A. & Malaiarasi, M. 2006. Studies of medicinal plants in Thoothukudi, Tamilnadu. *J. Econ. Taxon. Bot.* 30 (Suppl.): 353-357.

Arinathan, V., Visuvasam, J. Joseph & Muthu, R.G. 2006. Survey of medicinal plants in Kulathur, Tamilnadu. *J. Econ. Taxon. Bot.* 30(Suppl.): 370-374.

Arjariya, A. & Rawat, A. 2004. Ethnobotanical studies on *Chenopodium* from Central India. *Nat. Environ. And Poll. Tech.* 3(1): 71-72.

Arora, C.M. 1968. The botany of Bailadila, Bastar State, Madhya Pradesh. *Bull. Bot. Surv. India* 10 : 61-66.

Arora, R.K. 1981. Native food plants of the north-eastern tribals. In S. K. Jain (Ed.) *Glimpses of Indian ethnobotany*, pp. 91-106. Oxford & Indian IBH, New Delhi.

Arora, R.A. 1997. Ethnobotany and its role in the conservation and use of plant genetic resources in India. *Ethnobotany* 9: 6-15.

Arya, K.R. & Agrawal, S.C. 2001. Survey of herbal folk remedy for Piles. *Ethnobotany* 13(1 & 2): 142-145.

Arya, K.R. & Agrwal, S.C. 2006. Conservation of threatened medicinal and folklore plants through cultivaiton in Uttaranchal State. *Ethnobotany* 18: 77-86.

Audichya, K.C., Billore, K.V., Joseph, T.H. & Chaturvedi, D.D. 1983. Role of indigenous folk remedies for certain acute illnesses in primary health care. *Nagarjun* 26: 199 – 201.

Bajpai, A., Ojha, J.K. & Sant, H.R. 1995. Medicobotany of the Varanasi district, U.P. India. *Int. J. Pharmacog.* 33: 172-176.

Balaji Rao, N.S., Rajasekhar, D., Chengal Raju, D. & Nagaraju, N. 1996. Ethnomedicinal notes on some plants of Tirumala hills for dental disorders. *Ethnobotany* 8: 88-91.

Ballabh, Basant & Chaurasia, O.P. 2006. Ethnobotanical studies on Boto tribe in Ladakh. *Ethnobotany* 18: 87-95.

Balodi, B. & Singh, D.K. 1997. Medico-ethnobotany of Ladakh. *Ann. For.* 5: 189-197.

Banerjee, D.K. 1974. Magico-religious beliefs about plants among some adibasis of India. *J. Myth. Soc.* 65(3): 5-8.

Banerjee, A.K. & Banerjee, I. 1986. A survey of the medicinal plants in Shevaroy hills. *J. Econ. Taxon. Bot.* 8: 271-290.

Baruah, P. & Sarma, G.C. 1984. Studies on the medicinal uses of plants by the Boro tribals of Assam-II. *J. Econ. Taxon. Bot.*5: 599 – 604.

Baruah, Parukutty & Sarma, G.C. 1987. Studies on the medicinal uses of plants by the Northeast tribes – III. *J.Econ. Taxon. Bot.*11: 71– 76.

Baruah, A. & Nath, S.C. 2006. Ethnobotanical evaluation of *Cinnamomum* species used as spices and condiments in North-East India. *Ethnobotany* 18: 27-36.

Baruah, M. K., Choudhury, P. D. & Sarma, G.C. 2006. Ethnomedicinal plants used by Khasi tribe of Cachar district, Assam. *J. Econ. Taxon. Bot.* 30(Suppl.): 110-114.

Basak, S.K. 1997. Medicinal plants of Bankura (West Bengal) and their uses. *J. Nat. Bot. Soc.* 51: 61 – 68.

Basak, S.K. 2006. Medicinal plants of Bankura (West Bengal) and their uses. *J. Econ. Taxon. Bot.* 30(Suppl.): 217-224.

Bedi, S.J. 1978. Ethnobotany of the Ratan Mahal Hills, Gujarat, India. *Econ. Bot.* 32: 278 – 284.

Behura, N.K. 1962. Meals and food habits in rural India. *Bull. Anthro. Surv. India* 11: 111-137.

Bennet, S.S.R. 1985. Ethnobotanical studies in West Sikkim. *J. Econ. Taxon. Bot.*7: 317 – 321.

Bennet, S.S.R. 1987. *Name changes in flowering plants of India and Adjacent regions.* Trisesa Publishers, Dehradun.

Bhalla, N.P., Sahu, T.R., Mishra, G.P. & Dakwale, R.N. 1982. Traditional plant medicines of Sagar Distt., Madhya Pradesh, India. *J. Econ. Taxon. Bot.*3: 23 – 32.

Bhalla, S., Patel, J.R. & Bhalla, N.P. 1996. Ethnomedicinal observation on some Asteraceae of Bundelkhand region, Madhya Pradesh, *J. Econ. Taxon. Bot. Addl. Ser.* 12: 175-178.

Bhandari, M.M. 1974. Native resources used as famine food in Rajasthan. *Econ. Bot.* 28: 73-81.

Bhardwaj, A.K. & Rai, M.K. 2000. Documentation on traditional knowledge of medicinal plants of indigenous people in Bastar (Madhya Pradesh). In Rai, M.K., Verma, A. and Rajak, R.C. (Ed.) *Integrated Management of Plant resources,* pp. 97-102. Scientific Publisher, Jodhpur.

Bhardwaj, S. & Gakhar, S.K. 2005. Ethnomedicinal plants used by the tribals of Mizoram to cure cuts & wounds. *Indian J. Trad. Knowledge* 4(1): 75-80.

Bhargava, N. 1983. Ethnobotanical studies of the tribes of Andaman Nicobar Island, Indian Onge. *Econ.Bot.* 37(1):110-119.

Bhatt, R.P. &. Sabnis, S.D. 1987. Contribution to the ethnobotany of Khedbrahma region of north Gujarat. *J. Econ. Taxon. Bot.* 9: 139 – 145.

Bhatt, D.C., Jadeja, B.A., Odedra, N.K. & Lashkari, P.I. 2004. Enumeration of wild food plants of ethnobotanical significance in Barda hills of Gujarat. *J. Econ. Taxon. Bot.* 28(3): 669-674.

Bhattacharjee, S., Tiwari, K.C., Majumdar, R. & Mishra, A.K. 1980. Folklore medicine from district Kamrup (Assam). *Bull. Medico – ethnobot. Res.*1: 447 – 460.

Bhattacharya, 1950. Deathrite, funeral ceremonies and idea of life after death among the hill Sora of Orissa. *Bull. Dept. Anthrop.*1(2): 1-16.

Bhattacharya, A.K. & Patra, K. 2004. Impact of consumption of non-wood forest products (NWFP) on the nutriture of *Pahadi Korwa* primitive tribal group-A study from Sarguja district of Chhatisgarh. *J. Non-Timber Forest Prod.* 11(4): 254-261.

Bhattacharya, A.K., Tiwari, P., Patra, K. & Sinha, V.K. 2004. Seasonal availability and consumption pattern of non-wood forest products (NWFP) as food among Baiga tribe of Dindori district of Madhya Pradesh, India. *Ethnobotany* 16(1&2): 64-68.

Bhattarai, N. K. 1968. Folk medicinal use of plants for respiratory complaints in Nepal. *Fitterapia* 64.

Bhogaonkar, P.Y. & Devarkar, V.D. 2002. Some unique ethnobotanical plants of korkus of Melghat tiger reserve (Maharashtra). *Ethnobotany* 14(1&2): 16-19.

Bhowmick, P.K. & Chowdhary, B. 1966. Some aspects of the magico-religious beliefs and practices of the Mundas. *Folklore* 7 : 100-107.

Bhuyan, L.R. 2003. Some plants used as medicine by the nishi tribe of Arunachal Pradesh. *J. Econ. Taxon. Bot.* 27(2): 447-450

Billore, K.V. & Audichya, K.C. 1978. Some oral contraceptives: family planning tribal way. *J. Res. Indian Med. Yoga & Homeo.*13: 104.

Billore, K.V., Joseph, T.G. & Dave, S.K. 1998. Interesting folk remedies by the 'Lok Vaidyas' of Rajasthan for 'Swas Roga'. *Ethnobotany*10: 42 – 45.

Binu, S., Nayar, T.S. & Pusphpangadan, P. 1992. An out line of ethnobotanical research in India. *J. Econ. Taxon. Bot.* (Add. Ser.) 10: 405-428.

Biswas, K. 1956. Vegetable drugs in the treatment of Leucoderma. *J. Asiat. Soc.*22: 61-66.

Bodding, P.O. 1898. On taboos and customs connected there with amongst the Santhal. J. *Asiat. Soc. Bengal* 67(3): 1-24.

Bodding, P.O. 1925. Studies in Santal medicine and connected folklore-I. Santhals and disease. *Mem. Roy. Asiatic. Soc. Bengal* 10(1): 1-132.

Bodding, P.O. 1927. Studies in Santal medicine and connected folklore-II. Santhals and disease. *Mem. Roy. Asiatic. Soc. Bengal* 10(2): 133-426.

Bodding, P.O. 1940. Studies in Santal medicine and connected folklore. *Mem. Roy. Asiatic. Soc. Bengal.* 10: 427-502.

Bodhe, P.D., Sharma, L.K. & Pandey, V.K. 1997. Rare folk medicine practiced by Gond tribes of wadsa forest division, Gadchirol district, Maharashtra. *Bull. Medico-Ethnobot. Res.* 18(1&2): 12-17.

Bondya, S. L. & Sharma, H.P. 2004. Ethnobotanical studies on plants used in diabetes (Madhumeha) under the Baharagora Block of Jharkhand. *Ethnobotany* 16: 139-140.

Borthakur, S.K. 1976. Less known medicinal uses of plants among the tribals of Mikir Hills. *Bull. Bot. Surv. India* 18: 166 -171.

Borthakur, S.K. 1981. Certain plants in folklore & folklife of Karbis (Mikirs) of Assam. In Jain, S.K. (ed.) *Glimpses of Indian ehtnobotany*, pp. 170-181. Oxford and IBH, New Delhi.

Borthakur, S.K. 1993. Native phytotherapy for child and women diseases from Assam. *Ethnobotany* 75: 87-89.

Borthakur, S.K. & Goswami, N. 1995. Herbal remedies from Dimoria of Kamrup district Assam in Northeastern India. *Fitoterapia* 66: 333-340.

Borthakur, S.K. 1996. Postnatal care of women in traditional system in Assam. *Ethnobotany* 8: 51-55.

Borthakur, S.K., Nath, K. & Gogoi, P. 1996. Herbal remedies of the Nepalese of Assam. *Fitoerapia* 67: 231 – 237.

Borthakur, S.K., Choudhury, B.T. & Gogoi, R. 2004. Folklore hepato-protective herbal recipes from Assam in northeast India. *Ethnobotany* 16(1&2): 76-82.

Brij Lal, Upreti, D.K. & Kalakoti, B.S. 1985. Ethnobotanical utilization of lichens by the tribals of Madhya Pradesh (India). *J. Econ. Taxon. Bot.* 7 : 235-204.

Brij Lal, 1988a. Traditional remedies for bone fracture among the tribals of Madhya Pradesh, India. *Aryavaidyan* 1(3) : 190-195.

Brij Lal. 1988b. Ethnobotanical studies of Baigas in Baigachak region in Madhya Pradesh. *Bull. TDRI.* 16(1-2): 29-35.

Brij Lal. 1993. Ethonobotany of Baigas of Madhya Pradesh- a preliminary report. *Arunachal Forest News* 11(1): 17-20.

Brij Lal & Maheshwari, J.K. 1993. Prospects of plant-based cottage industries in tribal areas of Madhya Pradesh. *J. Econ. Taxon. Bot.* 17(1) : 235-238.

Brij Lal, Vatsa, S.K., Singh, R.D. & Gupta, A.K. 1996. Plants used as ethnomedicine and supplementary food by Gaddis of Himachal Pradesh. In S.K. Jain (ed.) *Ethnobiology in human welfare, pp.* 384-387. Deep Publ, New Delhi.

Caius, J. F. 1998. *The Medicinal and Poisonous plants of India.* Scientific Publishers, Jodhpur.

Chaghtai, S.A. & Javed, Ahmad 1978. An ethnobotanical study of Baghrendah (*Jatropha curcas* L.). *J. Res. Indian Med. Yoga & Homeo.* 13: 134 – 136.

Chakraborty, M.K. & Bhattacharya, A. 2004. Magico-religious beliefs among the tribal of Purulia district, West Bengal. *J. Econ. Taxon. Bot.* 28(2): 510-512.

Chakraborty M.K. & Pal, D.C. 2004. Plant conservation through tribal deities in some sacred groves of Purulia district, West Bengal. *Ethnobotany* 16(1&2): 99-102.

Chandra, K., Pandey, B.N., Sudha, G.N. & Pandey, P. 1989. Medicinal plants wealth of the Rajgir, Bihar. *Bull. Med. Ethnobot. Res.* 1: 124-161.

Chandra, K. 1995. An ethnobotanical study on some medicinal plants of District Palamau (Bihar). *Sachitra Ayurved.* 48: 311 – 314.

Chatterjee, S. & Bhatterjee, A. 2006. Some ethnomedicinal plants of North 24 Paragana, West Bengal. *J. Econ. Taxon. Bot.* 30(Suppl.): 281-282.

Chattopadhaya, K.P. 1954. A short note on cultivation among tribal folk in India. *Vanyajati* 2(4) : 110-112.

Chattopadhaya, K.P. 1955. A torsion trap of Korkus. *Vanyajati* 3(2) : 58-59.

Chaudhuri, Rai H.N., Pal, D.C. & Tarafder, C.R.. 1975. Less known uses of plants from the tribal areas of Orissa. *Bull. Bot. Surv. India* 17: 132 – 136.

Chaudhuri, Rai H.N. & Pal, D.C. 1976. Preliminary observations on ethnobotany of Midnapur district, West Bengal. *India. Mus. Bull.*11: 51 – 53.

Chaudhuri, Rai H.N., Dam, S.N., Pal, D.C., Tarafder, C.R. & Banerjee, K.K. 1980. Edible jungel leaves. *Folklore* 18: 161-163.

Chaudhuri, Rai H.N., Molla, H.A., Pal, D.C. & Roy, B. 1982. Plants used in traditional medicine by some tribals of Jalpaiguri district, West Bengal. *Bull. Bot. Surv. India* 24: 87 – 90.

Chaudhuri, Rai H.N., Pal, D.C., Saha, N.C. & Roy, B. 1985. Some wild edible plants in Calcutta markets. *Man & Life* 11: 45 – 58.

Chaudhary, D. & Neogi, B. 1999. Ethonobotany of Khasi and Chakma tribes of North-East India. *J. Econ. Taxon. Bot.* 23(2): 583-589.

Chauhan, Mayarani. 1982. *Ratlam jile me Sinchai: Ek Bhaugolik adhyayan.* Laghu Sodh Prabandha. (Hindi).

Chauhan, V. & Chauhan, N.S. 1988. Ethno-botany of Trans-Giri area of Sirmour district of Himachal Pradesh, *Bull. Med. Ethnobot. Res.* 9: 106-122.

Chelvan, P.T. 1998. Traditional phytotherapy among the migrant Tamilian settlers in South Gujarat. *Biojournal*10: 9 – 14.

Chhetri, D.R. 2005. Ethnomedicinal plants of the Khangchendzonga National Park, Sikkim, India. *Ethnobotany* 17: 96 – 103.

Chetty, K.M. & Rao, K.N. 1989. Ethnobotany of Sarakulla & adjacent areas of Chittoor district, Andhra Pradesh. *Vegetos* 2: 51-58.

Chetty, K.M., Chetty, M.L., Sudhakar, A. & Ramesh, C. 1998. Ethno-medico-botany of some aquatic angiosperm in Chittoor district of Andhra Pradesh, India. *Fitoterapia* 69: 7-12.

Chopra, R.N., Nayar, S.L. & Chopra, I.C. 1956. *Glossary of Indian Medicinal Plants.* CSIR., New Delhi.

Chopra, R.N., Badhwar, R.L. & Ghosh, S. 1965. *Poisonous Plants of India,* Vol. I & II. ICAR, New Delhi.

Dabas, Kiran, Parabia Minoo & Jayaprakash, Dabas 1990. Cultivation and food habits of tribals of Dangs Distt. in Gujarat – A case study. *Ethnobotany* 2(1-2): 61-65.

Dagar, H.S. 1986. Ethnobotany of the Nicobarese tribals. *Indian Forester* 112(2): 174-179.

Dagar, H.S. 1989. Plants folk medicines among Nicobarese tribals of Car Nicobar Island, India. *J. Econ. Taxon. Bot.* 43(2): 215-224.

Dagar, H.S. & Chaghtai, S.A. 1991. Some plant lore of the Nicobarese in the treatment of domestic animals ailments. *J. Econ. Taxon. Bot.* 15(3): 603-605.

Dagar, H.S. & Dagar, J.C. 1992. Some ethnobotanical observations amongst the Nicobarese. *J. Econ. Taxon. Bot.* 10: 1-5.

Dagar, H. S. & Dagar. J.C. 1996. Ethnobotanical studies of Nicobarese of Chowra Island of Nicobar Group of Islands. *J. Econ. Taxon. Bot. Addl. Ser.* 12: 381-388.

Dagar, J. C. & Dagar, H.S. 1999. *Ethnobotanical aspect of Aborigins of Andaman & Nicobar Islands.* Surya Intern.Publ., Dehradun.

Dar, G.H., Vir jee, Kachroo, P. & Buth, G.M. 1984. Ethnobotany of Kashmir – I. Sind Valley. *J. Econ. Taxon. Bot.*5: 668 – 675.

Das, M.B. & Sharma, P.C. 1958-59. Traditional methods of treatment of leprosy among Mikirs. *Adibas.* 3: 25-30.

Das, A.K. 1959. Food habits and dietaries of the Oraons and their nutritional efficiency. *Vanyajati* 7: 53-58.

Das, A.K. & Sarkar, B.K. 1963. A note on bidi leaf *(Diospyros melanoxylon). Ind. Forester* 89(3): 39-45.

Das, S.N., Janardhanan, K.P. & Roy, S.C. 1983. Some observations on the ethnobotany of the tribes of Totopara and adjoining areas in Jalpaiguri district, West Bangal. *J. Econ. Taxon. Bot.*4: 453 – 474.

Das, S.R. 1995. Some beneficial weeds of cultivation from Nadia district, West Bengal. *Bull. Med. Ethnobot. Res.*16: 74 – 81.

Das, A. K. 1997. Less-known uses of plants among the Adis of Arunachal Pradesh. *Ethnobotany* 9: 90-93.

Das, A. K., Dutta, B. K. & Sharma, G. D. 2006. Study of medicinal plants used by the different tribes/communities of Cachar district Assam. *J. Econ. Taxon. Bot.* 30(Suppl.): 294-309.

Das, S., Sheeja, T. E. & Mandal, A. B. 2006. Ethnomedicinal uses of certain plants from Bay Islands. *Indian J. Trad. Knowl.* 5(2):207-211.

Datta, M.L. & Nath, S.C. 1998. Ethno-medico-botany of the Deories of Assam, India. *Fltoterapia* 69: 147-154.

Deka, D., Sarma, G. C., Devi, N. & Pathak, N. 2006. Indigenous herbal medicines used against Malaria in Goalpara and Morigaon district of Assam. *J. Econ. Taxon. Bot.* 30(Suppl.): 177-183.

Deokule, S. S. 2006. Ethno-medicinal plants of Baramati region of Pune district, Maharashtra. *J. Econ. Taxon. Bot.* 30(Suppl.): 59-69.

Devbarman, M. 1976. Chuak-A drink for tribals in Tripura. *Folklore* 17: 347-351.

Devi, Maya 2003. Wild edible plants of Sonitpur district, Assam. *J. Econ. Taxon. Bot.* 27(2): 396-409.

Dhar, U. & Siddique, M.A.A. 1993. Ethnobotanical studies and biological diversity of Suru Valley (Zanskar). *Glimpses Plant Res.* 10: 31-39.

Dobriyal, R.M., Singh, G.S., Rao, K.S. & Saxena, K.G. 1997. Medicinal plant resources in Chhakinal watershed in the North-Western Himalaya, *J. Herb. Spices. Med. Plant* 5: 15-27.

Dubey, G., Sahu, P. & Sahu, T.R. 2000. Role of plant in different religious ceremonies common to Buldelkhand region of Madhya Pradesh. *Proc. Nat. Sem. CIMAP,* pp. 542-545.

Duthie, J. F. 1902-29. *Flora of the Upper Gangatic plain.* Bishen Singh Mahendra Pal Singh, Dehradun.

Dutta Choudhury, M. & Shil, S. 2006. Some traditional antihepatitis plants used by the People of Manipur. *J. Econ. Taxon. Bot.* 30(Suppl.): 289-293.

Dwivedi, S. N., Dwivedi, S. & Patel, P.C. 2005. Herbal remedies used by the rural people in the treatment of joint diseases. *Ethnobotany* 17: 193-196.

Elvin, V. 1947. *Maria and their Ghotul.* Oxford University Press, London.

Elvin, V. 1950. *Maria murder and suicide.* Oxford University Press, (1st Ed., 1943).

Elwin, V. 1953. Toys of tribal children. *Illustrated Weekly of India,* Oct.18, 74(42): 28-29.

Elwin, V. 1955a. The musical instruments of tribal India (ii) *Illustrated Weekly of India,* Nov. 27, 76(49): 26-27.

Elwin, V. 1955b. The musical instruments of tribal India (ii) *Illustrated Weekly of India,* Dec.4, 76(49): 48-49.

Elwin, V. 1965. Field work in tribal world. *Folklore* 6 : 480-481.

Gadgil, M. & Vartak, V.D. 1973. Groves dedicated to Gods. *Illustrated weekly of India* 35: 19-21.

Ganesan, S. & Kesavan, L. 2003. Ethnomedicinal plants used by the ethnic group Valaiyans of Vellimalai hills (Reserved forest), Tamilnadu, India. *J. Econ. Taxon. Bot.* 27(3): 754-766.

Ganesan S., Venkateshan, G. & Banumathy, N. 2006. Medicinal plants used by ethnic group Thottianaickans of Semmalai hills (reserved forest), Tirucira-ppalli district, Tamil Nadu. *Indian J. Trad. Knowl.* 5(2): 245-252.

Gantait, S., Kundu, S. K. & Giri, G. S. 2006. Some important medicinal plants of North-East India. *J. Econ. Taxon. Bot.* 30(Suppl.): 283-288.

Gaur, R.D. 1997. Wild edible fruits of Garhwal Hills. *The Himalaya* 1: 65-70.

Gaur, R.D., Sharma, M.P. & Semwal, J.K. 1980. Ethnotoxic plants of Garhwal hills in India. *East. Anthrop.*33: 159 – 163.

Gera, M., Bisht, N.S. & Rana, A.K. 2003. Ex situ conservation of medicinal and aromatic plants in India with special reference to Madhya Pradesh. *Indian For.* 123(1): 102-108.

Ghate, Vinaya S. 1998. Plants in patra-pooja, notes on their identity in West Bengal. *Indian J. Trad. Knowledge* 2(4): 393-396.

Ghosh, S. & Sensarma, P. 1997. Ethnomedicine to modern medicine: an observational study in some villages of West Bengal. *Ethnobotany* 9: 80-84.

Ghosh, A. 2006. Medicinal plants used for treatment of diabetes by the tribals of Bankura, Purulia and Medinipur of West Bengal. *J. Econ. Taxon. Bot.* 30(Suppl.): 233-238.

Girach, R.D., Aminuddin & Ahmad, M. 1998. Folk veterinary herbal medince of Bhadark district, Orissa. India. *Ethnobotany*10: 85-88.

Goel, A.K. & Bhattacharya, U.C. 1981. A note some plants found effective in treatment of jaundice (Hepatitis). *J. Econ. Taxon. Bot.* 2: 157-159.

Goel, A.K., Sahoo, A.K. & Mudgal, V. 1984. *A contribution to the ethnobotany of Santal Pargana, Bihar.* BSI, Calcutta.

Goel, A.K. & Mudgal, V. 1988. A survey of medicinal plants used by the tribals of Santal Pargana (Bihar). *J. Econ. Taxon. Bot.*12: 329 – 335.

Gogoi, P. & Boissya, C.L. 1984. Informations about a few herbal medicines used by the people of Assam (India) against Jaundice. *Himal. Res. Dev.*2: 41 – 44.

Gohain, B.C. 1971. *Annexure to the Tribal Map of tribes.* Anthr. Surv. India, Calcutta.

Gonoi, K.A. & Nawchoo, I.A. 2003. Taditional treatment of toothache by the Gujjar and Bakerwal tribes of Kashmir in India. *J. Econ. Taxon. Bot.* 27(1): 105-120.

Gonzalez, J. 1980. Medicinal plants in Columbia. *J. Ethnopharmacol.* 2: 43-47.

Gopakumar, K., Vijayalakshmi, B., Shantha, T. R. & Yoganarasimhan, S.N. 1991. Plants used in Ayurveda from Chikmangalur district, Karnataka – II. *J. Econ. Taxon. Bot.*15: 379 – 389.

Gopal, G.V. & Shah, G.L. 1985. Some folk medicinal plants used for jaundice in Gujarat, India. *J. Res. Edu. Indian Med.*4: 45 – 49.

Grigson, W.V. 1949. *The Maria Gonds of Bastar.* Oxford Universtiy Press, (1st Ed. 1938, Revised Ed. 1949).

Guha, B.S. 1939. Simphak the bark cloth of the Garos of Assam. *J. Asiat. Soc.* 6: 174.

Gunjatkar, N. & Vartak, V. D. 1982. Enumeration of wild edible legumes from Pune district, Maharashtra State. *J. Econ. Taxon. Bot.*3: 1-9.

Gupta, S.P. 1981. Native medicinal uses of plants by the Asurs of Netarhat plateau (Bihar). In S.K. Jain (Ed.) *Glimpses of Indian Etnobotany*, pp. 218 – 231. Oxford and IBH publishing Co., New Delhi.

Gupta, S.P. 1981. Folklore about plants with reference to Munda Culture. In S.K. Jain (Ed) *Glimpses of Indian ethnobotany,* pp. 199-207. Oxford and IBH publishing Co., New Delhi.

Gupta, Rajendra 1981. Plants in folk medicine of the Himalayas. In S.K. Jain (Ed.) *Glimpses of Indian Ethnobotany*, pp. 83 – 90. Oxford and IBH publishing Co., New Delhi.

Gupta, S.P. 1987. Study of plants during ethnological research among the tribals. In S.K. Jain (Ed.) *A Manual of Ethnobotany,* pp. 12-22. Scientific publishers, Jodhpur.

Gupta, R. 1993. Conservation and utilization of Indian medicinal plants. *Indian J. Plant Genetic Resources* 6(2): 131-137.

Gupta, V.C., Hussain, S.J. & Imam, S. 1997. Medico – ethonbotanical survey at Paderu forests of Araku Valley, Andhra Pradesh, *Fitoterapia* 68: 45 – 48.

Gupta, A.K., Mishra, S.K. & Khan, A.A. 1999. Ethnobotanical notes on some herbs from Chhattisgarh region of Madhya Pradesh. *Advances in Plant Sci.* 12(1) : 163-166.

Gupta, A.K. & Tandon, N. 2004 (Ed). *Reviews on Indian Medicinal Plants, Vol.* I, II, III & IV. Indian Council of Medical Research. New Delhi.

Gupta, A.K. 2005 (Ed.). *Quality standards of Indian Medicinal Plants, Vol.* I, II & III . Indian Council of Medical Research. New Delhi.

Gupta, V. 2006. Traditional medicinal plants of the Bangnis of East Kameng district, Arunachal Pradesh. *J. Econ. Taxon. Bot.* 30(Suppl.):310-319.

Hajra, P.K. & Baishya, A.K. 1981. Ethnobotanical notes on the Miris (Mishings) of Assam. In S.K. Jain (Ed.) *Glimpses of Inidian Ethnobotany,* pp. 161-169. Oxford and IBH publishing Co., New Delhi.

Harshberger, J.W. 1896. The purpose of Ethnobotany. *Bot. Gaz.* 21: 146-158.

Hema, Sane & Ghate, Vinaya 2006. Sacred conservaiton practices at species level through tree worship. *Ethnobotany* 18: 46-52.

Hema, E.S., Sivadasan, M. & Anilkumar, N. 2006. Studies on edible species of Amaranthaceae and Araceae used by Kuruma and Paniya tribes in Wayanad district, Kerela, India. *Ethnobotany* 18: 122-126.

Hemadri, K., Raj, P.V., Rao, S.S. & Sarma, C.R.R. 1980. Folklore claims from A.P.-I. *J. Sci. Res. Plant Med.* 1: 37-49.

Hemadri, K. 1981. Rheumatism: Tribal medicine. *Ancient Sci. Life* 1: 117-120.

Hemadri, K. & Rao, S.S. 1983. Leucorrhoea and menorrhagia : Tribal medicine. *Anc. Sci. Life* 3: 40 – 41.

Hemadri, K. & Rao, S.S. 1984. Jaundice: Tribal medicine. *Anc. Sci. Life* 3: 209–212.

Hemadri, K. & Rao, S.S. 1989a. Folk medicine of Bastar. *Ethnobotany* 1: 61-66.

Hemadri, K. & Rao, S.S. 1989b. Folk-lore claims of Koraput and Phulbani district of Orissa State. *Indian Med.* 1(4): 11-13.

Hemadri, K. & Rao, S.S. 1990. Folklore claims of Koraput and Phulbani district of Orissa State. *Indian Med.* 2(1): 4-6.

Hemadri, K. 1992. Tribals of Andhra Pradesh: their knowledge in nutritional and medicinal herbs. *Indian Med.* 4(1):1– 6.

Hembrom, P.P. & Goel, A.K. 2005. Horopathy : Ethnomedicine of Mundas. *Ethnobotany*17: 89 – 95.

Hosagoudar, V.B. & Henry, A.N. 1993. Plants used in birth control and reproductive ailments by Soligas of biligiri rangana betta in Mysore district of Karnataka. *Ethnobotany* 5: 117-118.

Issar, R.K. 1981. Traditional important medicinal plants and folklore of Uttaranchal Himalaya for animal treatments. *J. Sci. Res. Pl. Med.* 2: 61-66.

Jadeja, B.A., Odedra, N.K. & Choudhary, S.I. 2006a. Ethnomedicinal plants used by the tribals of Gujarat to cure digestive disorders. *J. Econ. Taxon. Bot.* 30(Suppl.): 115-123.

Jadeja, B.A., Odedra, N.K., Mori, H.N. & Bhati, D.C. 2006b. Medicinal plants of Gujarat used in Gout. *J. Econ. Taxon. Bot.* 30(Suppl.): 225-232.

Jadeja, B.A., Odedra, N.K. & Gajera, M.R. 2006c. Plants used in traditional phytotherapy for hair care in Gujarat. *J. Econ. Taxon. Bot.* 30(Suppl.):258 - 268.

Jain, A.K. 1992. Clans of Sor tribals of Madhya Pradesh and their role in conservation. *Ethnobotany* 4: 67-69.

Jain, S.K. 1963a. Magico religious beliefs about plants among the tribals of Bastar, Madhya Pradesh. *Q. J. Myth. Soc.* 54(3): 73-94.

Jain, S.K. 1963b. Studies in Indian Ethnobotany-I : Wild plant foods of the tribals of Bastar, Madhya Pradesh. *Proc. Nation. Inst. Sci. India* 30B: 3.

Jain, S.K. 1963c. Observations on ethnobotany of tribals of Madhya Pradesh. *Vanyajati* 11: 177-183.

Jain, S.K. 1963d. Studies in Indian ethnobotany (1) less known uses of 50 common plants from tribal areas of Madhya Pradesh. *Bull. Bot. Surv. India* 5: 223-226.

Jain, S.K. 1963e. Studies in Indian ethnobotany II- Plants used in medicine by tribals of Madhya Pradesh. *Bull. Regional Research Lab. Jammu* 1(1 & 2): 126-128.

Jain, S.K. 1963f. Studies in Indian ethnobotany III- The origin and utility of some vernacular plant names. *Proc. Nat. Acad. Sci. India* 33(4): 525-530.

Jain, S.K. & Tarafder, C.R. 1963. Native plant remedies for snakebite among the Adivasis of Central India. *Ind. Med. J.* 57: 307-369.

Jain, S.K. 1964a. An indigenous water bottle. *Indian Forester* 90: 709.

Jain, S.K. & Tarafder, C.R. 1964b. Observation on *Nyctanthes arbor-tristis* L. *J. Soc. Indian Forester* 4(4): 97-101.

Jain, S.K. 1964c. The role of botanist in folkore research. *Folklore* 5: 145-150.

Jain, S.K. 1964d. Medicinal plantlore of the tribals of Bastar. *Econ. Bot.* 19: 236-250.

Jain, S.K. 1964e. Plant resources in tribal areas of Bastar. *Khadi Gramodyog* 10: 557-561 & *Vanyajati* 12: 174-178.

Jain, S.K. 1964f. Wild plant foods of tribals of Bastar. Madhya Pradesh. *Proc. Nat. Inst. Sci. India* 30B: 56-80.

Jain, S.K. & De, J.N. 1964. Some less known plant foods among the Tribals of Purulia, West Bengal. *Sci. and Cult.* 30: 385-386.

Jain, S.K. 1965a. Medicinal plantlore of the tribals of Bastar, Madhya Pradesh. *Econ. Bot.* 19: 236-250.

Jain, S.K. 1965b. On the prospects of some new or less known medicinal plant resources. *Indian Med. J.* 59: 270-272.

Jain, S.K. 1965c. Wooden musical instruments of the Gonds of Central India. *Ethnobotany* 9(1): 39-42.

Jain, S.K. & De, J.N. 1966. Observations on ethnobotany of Purulia district, West Bengal. *Bull. Bot. Suv. India.*8: 237 – 251.

Jain, S.K. 1967a. Ethnobotany : its Scope and study. *Indian Mus. Bull.* 2(1): 39-43.

Jain, S.K. 1967b. Plants in Indian medicine and folklore associated with healing of bones. *Indian J. Ortho.* 1: 95-105.

Jain, S.K. 1968. *Medicinal Plants.* National Book Trust, New Delhi.

Jain, S.K. & Tarafder, C.R. 1970. Medicinal plant lore of the Santals. A revival of P. O. Boddings's work. *Econ. Bot.* 24: 241 – 278.

Jain, S.K., Banerjee, D.K. & Pal, D.C. 1973. Medicinal plants among certain Adibasis in India. *Bull. Bot. Suv. India* 15: 85 – 91.

Jain, S.K., Hanif, M.L. & Tarafder, C.R. 1973. *Bauhinia vahlii* – A multipurpose plant in tribal area. *Vanyajati* 21(4): 106-108.

Jain, S.K. & Banerjee, D.K. 1974. Preliminary observations on the ethnobotany of the genus *Coix. Econ. Bot.* 28: 38-42.

Jain, S.K. 1977. Utilisation of medicinal plant from wild sources. *Cult. Util. Med. Arom. Pl. Jammu* 461-467.

Jain, S.K. & Dam, N. 1979. Some ethnobotanical notes from north – eastern India. *Econ. Bot.*33: 52 – 56.

Jain, S.K. & Borthakur, S.K. 1980. Ethnobotany of the Mikirs of India. *Econ. Bot.*34: 264 – 272.

Jain, S.K. 1981a. Ethnobotanical research unfolds new vistas of traditional medicine. *Glimpses of Indian ethnobotany,* pp.13: 36. Oxford and IBH publishing Co., New Delhi.

Jain, S.K. 1981b. Observations on ethnobotany of the tribals of Central India. In S.K. Jain (Ed.) *Glimpses of Indian ethnobotany,* pp. 193-198. Oxford and IBH publishing Co., New Delhi.

Jain, S.K. 1981 (Ed.). *Glimpses of Indian ethnobotany.* Oxford and IBH publishing Co., New Delhi.

Jain, S.K. & Rao, R.R. 1983. (Ed.) *An Assessment of Threatened plant of India.* Bot. Surv. India, Calcutta.

Jain, S.K. 1984. Ethnobotany of Morni and Kalesar (Ambala – Haryana). *J. Econ. Taxon. Bot.*5: 809 – 813.

Jain, S.K., Mudgal, V., Banerjee, D.K., Guha, A. & Das, D. 1984. *Bibliography of Ethnobotany*. BSI. Calcutta.

Jain, S.K. 1986. Ethnobotanical studies on medicinal plants. *Curr. Res. Med. Pl. in India,* INSA, pp. 1-5.

Jain, S.K. 1987. Plants in Indian medicine and folklore associated with healing of bones. *Indian J. Orthopaed.* 1: 95-104.

Jain, S.K. & Sinha, B.K. 1988. Ethnobotanical aspects of life support species. Some emergency and supplementary foods among aboriginals in India. In *Life support species: Diversity and Conservation,* pp. 173-180. National Bur. Pl. Gen. Res. New Delhi.

Jain, A.K. 1988. Tribal clans in Central India and their role in conservation. *Environ. Conserve.* 15: 368.

Jain, S.K., Sinha, B.K. & Saklani, Arvind 1989. Some interesting medicinal plants known among several tribal societies of India. *Ethnobotany* 1: 89-10.

Jain, S.K. 1989a. Medicinal plants and Ethnobotany In Johri. B.M. (ed.) *Progress of Botany (1910-1985) Retrospect & prospect.* Indian Sci. Con. Assoc. Calcutta.

Jain, S.K. 1989b. *Method and Approaches in ethnobotany*. (I Repr. 1994), SEB, Lucknow.

Jain, A.K. & Sharma, H.O. 1990. Certain medicinal plants used by Sahariya tribals of northern Madhya Pradesh. *Proc. Nat. Sem. Med. & Arm. Plants* Jabalpur 2: 155-165.

Jain, S.K. 1991a. *Dictionary of Indian Folkmedicine and Ethnobotany*. Deep Publications, New Delhi.

Jain, S.K. 1991b. Observations on ethnobotany of the tribals of Central India. In Jain S.K. (Ed.) *Glimpses of Indian Ethnobotany,* pp. 192-198. Oxford and IBH, New Delhi.

Jain, P. & Sahu, T.R. 1993. An ethnobotanical study of Noradehi Sanctuary Park of Madhya Pradesh, India: Native plant remedies for scorpion sting and snake bite. *J. Econ. Taxon. Bot.*17: 315 – 328.

Jain, S.P. & Singh, S.C. 1994. *Ethno-medico-botanical survey of Ambikapur, Madhya Pradesh*, IV. ICE, Lucknow, Nov. 17-21, 1994.

Jain, A.K. & Sharma, H.O. 1996. Ethnobiological studies of by Sahariya tribes of Central India. In S.K. Jain, (ed.) *Ethnobiology In Human Welfare,* pp. 397-399. Deep Publ., New Delhi.

Jain, S. K. 1997. Observation on Ethnobotany of the tribals of Central India. In S. K. Jain (Ed) *Contribution to Indian Ethnobotany,* pp 61-65. Scientific Publishers, Jodhpur.

Jain, S.P. & Singh, S.C. 1997. An Ethno-medico-botanical survey of Ambikapur district Madhya Pradesh. In S.K. Jain (Ed.) *Contribution to Indian Ethnobotany,* pp. 83-91, Scientific Publishers, Jodhpur.

Jain, S.K. & Sikarwar, R.L.S. 1998. Bharat ki Janjyotiyon ki Kuchha Aparichit Vanaushdhiyan -2, Madhya Pradesh. *Sachitra Ayurved* 51(5): 348-350 (In Hindi).

Jain, S.K., Shrivastava, S. & Kumar, V. 2000. Some observations of the life support species in Surguja forest of Central India: A report. *MFP NEWS.*10, (Nov.4) 2000: 8(Short Communication).

Jain, S.K. 2001. Ethnobotany in modern India. *Phytomorphology.* Golden jublee issue: 39-54.

Jain, S.K. 2002. *Bibliography of Indian Ethnobotany.* Scientific Publishers, Jodhpur. 1-144.

Jain, S.K. 2004 (Ed.). *A manual of ethnobotany.* Scientific Publishers, Jodhpur.

Jamir, N.S. 1990. Some interesting medicinal plants used by Nagas. *J. Res. Edu. Indian Med.* 9: 81-87.

Janaki Ammal, E.K. & Nagendra Prasad, P. 1984. Ethnobotanical findings of *Costus speciosus* (Koen) Sm. among the Kanikkars of Tamil Nadu., *J. Econ. Taxon. Bot.* 5: 129-133.

Janardhanan, K.P. 1963. An enumeration of the medicinal plants of Khed Taluka, Maharashtra State. *Bull. Bot. Surv. India.*5: 363 – 374.

Janmeda, B.S., Singh, L. & Vats, Pritti 2006. Folk claims on monocotyledonous plants in East Sikkim. *J. Non-Timber Forest Prod.*13(2): 155-160.

Jha, P.K.,Choudhary, R.S. & Choudhary, S.K. 1997. Studies of medicinal plants of Palamau (Bihar) (Part – II). *Bijournal* 9: 21 – 38.

Jha, V. & Goel, A.K. 2004. Botanical rosaries and non-flower garlands in folklore life of Mithila region (North Bihar). *Ethnobotany* 16(1&2): 27-32.

Johes, Volluey H. 1941. The nature and scope of ethnobotany. *Chronica Botanica* 6(10): 214-221.

John, D. 1981. One hundred useful drugs of the Kani tribes of Trivendrum forest division, Kerela. *Indian J.Crude Drug. Res.*22: 17-39.

Joshi, M.C., Patel, M.B. & Mehta, P.J. 1980. Some folk medicines of Dangs, Gujarat Satate. *Bull. Medico – ethnobot. Res.* 1: 8 – 24.

Joshi, P. 1982. An ethnobotanical study of Bhils– A preliminary survey. *J. Econ. Taxon. Bot.*3: 257 – 266.

Joshi, M.C. 1987. Conservation and cultivation of medicinal plants in Gujarat. *Bull. Med. Ethno. Bot. Res.* 8 (3-4): 183-187.

Joshi, G.C. & Tiwari, K.C. 2000. Wild edible plants diversity in Uttar Pradesh Himalaya. *J. Econ. Taxon. Bot.* 24(2): 433-443.

Jubilee, P. & Subhan, C. N. 2006. Biological activities of ethnomedicinal claims of some plant species of Assam. *Indian J. Trad. Knowl.* 5(2):229-236.

Kadel, C. & Jain, A. 2008. Folklore claims on snakebite among some tribal communities of Central India. *Indian J. Trad. Knowl.* 7(2):296-299.

Kadamban, D., Ramanujam, M.P. & Balachandran, N. 2004. Plants used by the traditional medical practitioners for treating hepatic disease in Pondicherry region. *J. Econ. Taxon. Bot.* 28(2): 497-501.

Kadavul, K. & Jisha, P. 2006. Medicinal plants and their conservation in Mahe region, U. T. of Pondicherry, Malabar Coast, South India. *J. Econ. Taxon. Bot.* 30(Suppl.): 203-207.

Kamble, S.Y. & Pradhan, S.G. 1980. Some ethnobotany of the Korkus in Maharashtra. *Bull. Bot. Surv. India* 22(1&4): 201–202.

Kanwar, P. & Yadav, D. 2005. Indigenous animal healthcare practices of Kangra district, Himachal Pradesh. *Indian J. Trad. Knowl.* 4(2): 164-168.

Kapoor, S.L. & Kapoor, L.D. 1980. Medicinal plant wealth of the Karimnagr district of Andhra Pradesh. *Bull. Med. Ethnobot. Res.* 1: 120-144.

Kapur, S.K. 1993. Ethno – medico plants of Kangra Valley (Himachal Pradesh). *J. Econ. Taxon. Bot.*17: 395 – 408.

Karnick, C.R., Tiwari, K.C., Majumdar, R. & Bhattacharjee, S. 1981. Newer ethnobotanical and folklore studies of some medicinal plants of Gauhati and surrounding areas. *Nagarjun* 24: 240 – 245.

Kasera, P.K., Jitendra, K., Shukla, Jaiprakash, Priya Saharan & Chawan, D.D. 2000. Biology conservation and mediculture of important medicinal plants from Indian Thar Desert. *Proc. Nat. Sem. CIMAP,* pp. 342-443.

Kataria, R. L. & Gauchan, D. P. 2006. Ethnomedicinal plants used in dysentery and diarrhoea in Hanumangarh district, Rajasthan. *J. Econ. Taxon. Bot.* 30(Suppl.): 128-132.

Katewa, S.S. & Arora, A. 1997. Some plants in folk medicines of Udaipur district, Rajasthan. *Ethnobotany* 9: 48 – 51.

Katewa, S.S. & Sharma, R. 1998. Ethnomedicinal observations from certain watershed areas of Rajasthan. *Ethnobotany* 10: 46 – 49.

Katiyar, P. & Kolhe, S.S. 2000. Medicinal weeds of Raipur region in Chhattisgarh. *JMAPS* (22/A & 23/1A): 455-460.

Kaufmann, W. 1961. The musical instruments of the Hill Maria, Jhoria and Bastar Mauria Gond tribes. *Ethnomusicology* 5: 1–9.

Kaul, M.K. & Atal, C.K. 1983. Studies on *Holarrhena antidysenterica* Wall-I. Botany, Medico-ethnobotany and distribution. *J. Ethno-pharmacol.* 8: 349-356.

Khan, S.S. & Chaghtai, S.A. 1982. Ethnobotanical study of some plants used for curing skin affliction. *Anc. Sci. Life* 1(4): 236-238.

Khan K., Shukla, M. L. & Khan, I.M. 2000. Enumeration of wild food plants of ethnobotanical significance in Central India. *Adv. in Plant Sci.* 13(1): 277-281.

Khanna, K.K. & Mudgal, V. 1992. Ethnobotanical data from the herbarium of Botanical Survey of India. Central circle, Allahabad. *Bull. Bot. Surv. India* 34(1-4): 112-135.

Khanna, K.K., Dongarwar, N.M. & Makde, K.H. 2005. Unreported ethnomedicinal uses of *Canscora decussata* (Gentianaceae) among the tribals of Maharashtra. *Ethnobotany* 17: 187-188.

Khare, P.K. & Khare, L.J. 1999. Plants used in rheumatism by rural people of Chhatarpur district (Madhya Pradesh). *J. Econ. Taxon. Bot.* 23(2): 301-304.

Kharkongor, P. & Joseph, J. 1981. Folklore medico – botany of rural Khasi and Jaintia tribes in Meghalaya. In S.K. Jain (Ed.) *Glimpses of Indian Ethnobotany*, pp. 124 – 136. Oxford and IBH, New Delhi.

Khumbongmayum, A.D., Khan, M.L. & Tripathi, R.S. 2005. Ethnomedicinal plant in the sacred groves of Manipur. *Indian J. Trad. Knowledge* 4(1): 21-32.

Khory, J.K. 1981. *Materia medica of India & their Therapeutics.* Neeraj Publishing house, New Delhi.

Kirtikar, K.R. & Basu, B.D. 2004. *Indian Medicinal Plants*, Vol. I ,II,III &IV(Reprint). Bishen Singh Mahendra Pal Singh, Dehradun.

Kiruba, S., Jeeva, S., Venugopal, N., Das, S. S. M., Regini, G. S., Laloo, R. C. & Mishra, B. P. 2006. Ethnomedicinal herbs of Koonthakulam water Sanctuary, Nellai, Tamilnadu, India. *J. Non-Timber Forest Prod.* 13(1):25-27.

Kohli, R.K., Jerath, N. & Batish, D. 1996. *Some facets of biodiversity.* Society Env. Sci. & Punjab State Council for Sci. & Tech.

Krishna, B. & Singh, S. 1987. Ethnobotanical observations in Sikkim. *J. Econ. Taxon. Bot.* 9: 1 – 7.

Kshirsagar, R.D. & Singh, N.P. 2003. Less known herbal remedies from Soligas of Mysore, southern India. *J. Econ. Taxon. Bot..* 27(3): 656-661.

Kulkarni, D.K., Kumbhoj Kar, M.S., Anuradha, S.,Upadhye & Bonde, S. D. 2001. Ethnobotany of *Aloe vera* from Western Maharashtra. *Proc. Nat. Sem. CIMAP,* pp. 479-481.

Kumar, Y., Haridasan, K. & Rao, R.R. 1980. Ethnobotanical notes on certain medicinal plants among some Garo people around Balphakram Sanctury in Meghalaya. *Bull. Bot. Surv. India* 22: 161 – 165.

Kumar, Y., Fancy, S. & Rao, R.R. 1987. Further contribution to the ethnobotany of Meghalaya: Plants used by War Jaintia of Jaintia Hill district. *J. Econ. Taxon. Bot.* 11: 65 – 70.

Kumar, M. 1992. Ethnomedicinal studies of plants of Chotanagpur plateau, Bihar. *Int. J. Mendel.*9: 137 – 139.

Kumar, A. 1996. Some ethnomedicinal plants of Murias of the Indravati Tiger Reserve, Bastar (Madhya Pradesh) *J. Econ. Taxon. Bot.* 12: 201–205.

Kumar, V. & Jain, S.K. 1998. A contribution to ethnobotany of Surguja district in Madhya Pradesh. *Ethnobotany.*10: 89 – 96.

Kumar, V. 1999. Some indigenous tools of Surguja district, Madhya Pradesh, India. *Ethnobotany* 11: 135-137.

Kumar, V. & Jain, S.K. 1999. Some less known ethnomedicine among the tribals of Surguja district in Madhya Pradesh, India. *J. Non-Timber Forest Prod.* 6(3/4): 110-113.

Kumar, V. 1999. Madhya Bharat ke Surguja Janpad Main Samanya paudhon ke kuch Aparichit Aushdiyha Gun. *Sachitra Ayurved* 51(11): 837-838 (In Hindi).

Kumar, V. & Jain, S.K. 2002. Plant products in some tribal markets of Central India. *Econ. Bot.* 56(3): 242-245.

Kumar, V. & Rao, R.R. 2002. *Elephantopus scaber* L. An important medicinal plant used by tribals of Surguja. *Tropical medicinal Plants,* Malaysia 3(2): 219-221.

Kumar, V. & Sikarwar, R.L.S. 2002. Observations on some rare and endangered plants of Chhattisgarh State, India. *Phytotaxonomy* 2: 135-142.

Kumar, V. & Sikarwar, R.L.S. 2003. Plants used as fish poison by tribals of Surguja district in Chhattisgarh State. *Ethnobotany* 15: 87-89.

Kumar, A., Tiwari, D.D. & Pandey, Y.N. 2003. Ethnophytotherapeutics among Thaus of Beerpur Semra Forest range of Balarampum district (U.P.). *J. Econ. Taxon. Bot.* 27(4): 839-844.

Kumar, D. & Yadav, A. 2003. Ethnoboanical investigation of some medicinal plants availed by Gond tribes of Noradehi wildlife sanctuary, Madhya Pradesh. *Anthropologist* 5(3): 201–202.

Kumar, V. 2003. Wild edible plants of Surguja district of Chhatisgarh State, India. *J. Econ. Taxon. Bot.* 27(2): 272-282.

Kumar, V. & Sikarwar, R.L.S. 2004. Role of tribals in conservation of plant diversity in Central India. *Vanyajati* XLXII(2): 7-9.

Kumar, A. & Yadav, D.K. 2004. Significance of sacred plants in *Shraddh* ritual *(Pindadan)* in Gaya, Bihar. *Ethnobotany* 16(1&2): 103-107.

Kumar, S. & Chauhan, A.K.K. 2005. Medicinal plants used by local inhabitants in Bharatpur district, Rajasthan. *Ethnobo.*17: 179 – 183.

Kumar, A., Singh, A., Sharma, Y. & Rana, J. C. 2006. Ethno-medico-botany of some important plants in Mandhala watershed of Himachal Pradesh. *J. Econ. Taxon. Bot.* 30(Suppl.): 145-150.

Kumar, A. & D. K. Yadav. 2006. Important Ethno-medicinal plants of family Poaceae in Gaya district, Bihar. *J. Econ. Taxon. Bot.* 30(Suppl.): 171-176.

Kurup, A.M. 1962. Glimpses of life and culture of the tribes of Madhya Pradesh. *Vanyajati* 10(4): 134-145.

Kurup, A.M. 1963. Impact of industrialization on tribal life. *Vanyajati* 11: 71-75.

Kurup, A.M. 1970. Tribal festivals of Central India. *Folklore* 11(5): 159-165.

Kutty, M.K. 1995. Adivasi medicine: a study. *Aryavaidyan* 9(1): 23-27.

Lakshamanan, K.K. & Sankaranarayanan, A.S. 1988. Some folklore medicines in the remote hemlets, Dhoomanoor & Champukarai of Anai Katty hills, Coimbatore, Tamil Nadu. *Indian J. Forestry* 11: 217 – 219.

Lakshamanan, K.K., Subramanian, M. S., Kalyani, K. & Sankaranarayanan, A.S. 1986. Native uses of plants in the Nilgiri. *Proc. Nat. Acad. Sci. India.* 56 (B): 258-265.

Lal, S.D. & Lata, K. 1980. Plants used by the Bhat community for regulating fertility. *Econ. Bot.*34: 273 – 275.

Lal, S.D. & Yadav, B.K. 1983. Folk medicines of Kurukshetra district (Haryana), India. *Econ. Bot.*37: 299 – 305.

Lal, B. 1993. Ethnobotany of Baigas of Madhya Pradesh– A preliminary survey. *Arunachal Forest News*. 11(1): 17-20.

Lal, D., Vats, S.K., Singh, R.D. & Gupta, A.K. 1996. Plants used as ethnomedicine and supplementary food by Gaddis of Himachal Pradesh. In Jain, S.K. (ed.) *Ethnobiology in human welfare,* 384-387. Deep Publ., New Delhi.

Lalramnghinglova, H. 2002. Ethnobotanical study on the edible plants of Mizoram. *Ethnobotany* 14(1&2): 23-33.

Lindley, John. 1981. *Flora medica.* Ajay book service, New Delhi.

Maheshwari, J. K. 1970. New vistas in ethnobotany. *J. Econ. Taxon. Bot.*.(Add. Ser.): 1-11.

Maheshwari, J. K., Singh, K. K. & Saha, S. 1980. Ethnomedicinal uses of plants by Tharus in Kheri district, U. P. *Bull. Medico-ethnobot. Res.* 1:318-337.

Maheshwari, J.K., Singh, K.K. & Saha, S. 1981. *The Ethonbotany of the Tharus of Kheri district. Uttar Pradesh.* Nation. Bot. Res. Inst. Lucknow.

Maheshwari, J.K. 1983. Developments in ethnobotany. *J. Econ. Taxon. Bot.* 4(1): 1-5.

Maheshwari, J.K. & Singh, J.P. 1984. Contribution to the ethnobotany of Bhoxa tribe of Bijnor & Pauri Garhwal district, U.P. *J. Econ. Taxon. Bot.*5: 251 – 259.

Maheshwari, J.K. & Dwivedi, R.P. 1985. Ethnobotany of Abujhmarhia tribe of Bastar district, Madhya Pradesh. *J. Indian Bot. Soc.* Abstracts 64 suppl: pp-53.

Maheshwari, J.K., Kalakoti, B.S. & Brijlal. 1986. Ethnomedicine of Bhil tribe of Jhabua district, Madhya Pradesh. *Anc. Sci. Life* 5: 255 – 261.

Maheshwari, J.K., Singh, K.K. & Saha, S. 1986. *Ethnobotany of tribals of Mirzapur district. U.P.* Nation. Bot. Res. Inst. Lucknow.

Maheshwari, J.K. 1987. Tribal ecosystem- an overview. *Tribal Res. Bull.* 9: 1-4,14.

Maheshwari, J.K. & Singh, J.P. 1987. Traditional phytotherapy amongst the Kol tribe of Banda district, Uttar Pradesh. *J. Econ. Taxon. Bot.*9: 165 – 171.

Maheshwari, J.K. & Singh, H. 1988. Ethnobotanical observation on the Saharia tribe of Lalitpur district, Uttar Pradesh. *Vanyajati* 36(3): 23-34.

Maheshwari, J.K. 1989. Case studies of three primitive tribes (Abujhmarhia, Baiga and Bharia) of Central India. In S.K. Jain (ed.) *Methods and Approaches in Ethnobotany,* pp. 187-188. SOE, Lucknow.

Maheshwari, J.K. 1990a. Interaction of tribals with the forest. *Trop. For. Ecosyst. Conserve. Develp. S. & SE Asia,* pp. 115-126. Proc. MAB. Regional Workshop.

Maheshwari, J.K. 1990b. Recent ethnobotanical research in Madhya Pradesh, *SEB Newsletter* 9(1-3): 5.

Maheshwari, J.K. & Painuli, R.M. 1990. Plants used in tribal crafts by Sahariyas of Madhya Pradesh. An account of a few trees in their life. *Folklore* 31: 239-243.

Maheshwari, J.K., Painuli, R.M. & Dwivedi, R.P. 1990. Notes on ethnobotany of the Oraon and Korwal tribes of Madhya Pradesh. In Jain, S.K. (ed.). *Contribution to Indian ethnobotany,* pp. 75-90. Scientific Publishers, Jodhpur.

Maheshwari, J.K. 1996. Ethnobotanical documentation of primitive tribes of Madhya Pradesh, India. *J. Econ. Taxon. Bot. Addl. Ser.* 12: 206-213.

Mahlawat, S. & Kumar, A. 2004. Ethnobotanical and pharmacognostical studies of medicinal plant *Argemone mexicana* L. *Int. J. Mendel.* 21(3-4): 81-82.

Malhotra, S.K. & Moorthy, S. 1973. Some useful and medicinal plants of Chandrapur district (Maharasthra State). *Bull. Bot. Surv. India*15: 13 – 21.

Malhotra, C.L. & Balodi, B. 1984. Medicinal plants of Alpine region of Gori Valley, Kumaon. *J. Sci. Res. Plant Med.* 5(1&2): 1-8.

Mani, S.B. 1964. Bows and arrows in Tribal India. *Vanyajati* 12: 40-45.

Masih, S.K. 1997. Antidiabetic plants of Madhya Pradesh. *Indian J. Appl. Pure. Biol.* 12(1): 11-14.

Megoneitso & Rao, R.R. 1983. Ethnobotanical studies in Nagaland – 4. Sixty two medicinal plants used by the Angami Nagas. *J. Econ. Taxon. Bot.*4: 167 – 172.

Mishra, R. & Dixit, R.D. 1973. Vajradanti – An ethnobotanical study. I. *J. Res. Indian Med.*8: 89 – 91.

Mishra, R. & Billore, K.V. 1983. Some ethnobotanical lores from Banswara district. Rajasthan. *Nagarjun* 26: 229 – 231.

Mishra, D.P. & Sahu, T.R. 1984. Euphorbiaceous plants used in medicine by the tribals of Madhya Pradesh. India. *J. Econ. Taxon. Bot.* 5(4): 191-194.

Mishra, D.P. & Sahu, T.R. 1984. Euphorbiaceous plants used in medicine by the tribals of Madhya Pradesh. India. *J. Econ. Taxon. Bot.* 5: 791-794.

Mishra, O.P. & Naquvi, S.M.A. 1995. Ethno-medico-botany from tribes of Madhya Pradesh. *Bull. Medico-Ethno-bot. Res.* 16 (1-2): 17-26.

Mishra, S.K. & Verma, R.M.P. 1995. Medicinal plants of Champaran (Bihar) used in various diseases. *Flora Fauna.*1: 75 – 76.

Mishra, S., Jha, Vidyanath & Jha, Shaktinath 1996. Plants in Ethnoveterinary practices in Darbhanga (North-Bihar). In Jain, S.K. (Ed.) *Ethnobiology In Human Welfare,* pp. 189-193. Deep Publicatins, New Delhi.

Mishra, B. & Vaisya, R. 2004. *Bhavaprakasa,* Vol. I & II. Chaukhambha Sanskrit Sansthan, Varanasi.

Mishra, D.K. & Agrawal, S.C. 2006. Threatened plants of North Western Ghats: Need to be included in the Red Data Book of Indian Plants. *J. Non-Timber Forest Prod.* 13(1): 73-76.

Mohanty, P.C. 1967. Food habit of a tribal community *Adibasi* 9: 20-25.

Mohanty, R.B., Padhy, S.N. & Dash, S.K. 1996. Traditional phytotherapy for diarrhoeal diseases in Ganjan and Phulbani district of South Orissa, India. *Ethnobotany* 8: 60 – 65.

Mokat, D.N. & Deokule, S.S. 2004. Plants used as veterinary medicine in Ratnagiri district of Maharashtra. *Ethnobotany* 16(1&2): 131-135.

Mokat, D. N. & Deokule, S. S. 2006. Ethno-medico-botanical survey of Ratnagiri district of Maharashtra State. *J. Econ. Taxon. Bot.* 30 (Suppl.): 140-144.

Molla, H.A. & Roy, B. 1985. Traditional uses of some medicinal plants by the Rabha tribals in Jalpaiguri district, West Bengal. *J. Econ. Taxon. Bot.*7: 578 – 580.

Mudgal, V. & Jain, S.K. 1980. *Coptis teeta* Wall. Local uses distribution and cultivaiton. *Bull. Bot. Surv. India* 22: 179-180.

Mudgal, V. & Pal, D.C. 1980. Medicinal plants used by tribals of Mayurbhanj (Orissa). *Bull. Bot. Surv. India.* 22: 59 – 62.

Mudgal, V., Khanna, K.K. & Hajra, P.K. 1997. *Flora of Madhya Pradesh,* Vol. II. BSI. Calcutta.

Mukerji, B. 1953. *The Indian Pharmaceutical.* Codex, New Delhi.

Mukherjee, A., Mukherjee, S. & Mukherjee, M. 1986. Less known medicinal plants of Rubiaceae in Darjeeling Hills, West Bengal (India). *Indian J. Appl. & Pure. Bot.*1: 86 – 91.

Murthy, K.S., Sharam, P.C. & Prem Kishore. 1986. Tribal remedies for snake bite from Orissa. *Anc. Sci. Life* 6: 122 – 123.

Nadkarni, K.M. 1927. *Indian Materia Medica.* Popular Book Depot. Bombay, (3 ed.).

Nagarjun, N. & Rao, K.N. 1989. Folk - medicine for diabetes from Rayalaseema of Andhra Pradesh, *Ancient. Sci. Life* 9: 31 – 35.

Nagendra Prasad, P. & Abraham, Z. 1984. Ethnobotany of Nayadis of North Kerala. *J. Econ. Taxon. Bot.*5: 41 – 48.

Nagar, M. 1985. The use of wild plant foods by aboriginal communities in Central India. In Mishra, V.M. & Belwood, P. (Ed.) *Recent Adv. In Indo-Pacific Prehist.,* pp. 337 – 342. Oxford & IBH, New Delhi.

Nair, K.V., Nair, A.R. & Nair, C.P.R. 1995. Biodiversity of medicinal plants based on Ayurveda. *Sem. On Res.In Ayur. & Siddha.* pp. 79. CCRAS, New Delhi.

Namjoshi, A.N. 1960. *Studies in pharmacognosy of Ayurvedic drugs.* Board of Research on Ayurveda, Bombay.

Narain, S. & Singh, U. 2006a. Contribution to the Ethnobotanical plants of Mirzapur district (U. P.). *J. Econ. Taxon. Bot.* 30(Suppl.): 14 – 17.

Narain, S. & Singh, U. 2006b. Contribution to the Ethnobotanical plants of Sonbhadra district (U. P.). *J. Econ. Taxon. Bot.* 30(Suppl.): 18 – 20.

Nath, P.C., Sarmah, R., Arunachalam, A. & Barbhuiya, A. 2006. NTFPs: Potential and their conservation in Dihing-patkai wildlife santuary of Assam. *J. Non-Timber Forest Prod.* 13: 17-23.

Nautiyal, S. 1981. Some medicinal plants of Garhwal hills – A traditional use. *J. Sci. Res. Pl. Med.*2: 12 – 18.

Negi, K.S., Tiwari, J.K. & Gaur, R.D.1985. Economic importance of some common trees in Garhwal, Himalaya: An ethnobotanical study. *Indian J. Forestry* 8: 276 – 289.

Negi, C.S. 2005. Socio-cultural and ethnobotanical value of a sacred forest, Thal Ke Dhar, Central Himalaya. *Indian J. Trad. Knowledge* 4(2): 190-198.

Negi, K.S. 2006. *Allium* species in Himalayas and their uses, with special reference to Uttaranchal. *Ethnobotany* 18: 53-66.

Neher, R.T. 1968. The ethnobotany of *Tagetes*. *Econ. Bot.* 23: 317-325.

Nonhare, B.P. 1996. *Conserving the biodiversity in Bastar district - An ethnobotanical approach.* Nat Conf. on Relevance of Tax. in Cons of Biodiversity, Abs. Sagar, July 15.

Oommachan, M. 1977. *The flora of Bhopal.* J. K. Jain brothers, Bhopal.

Oommachan, M. 1986. Observation on certain plants used in human skin disease in Central India. *Indian J. Appl. Pure Biol.* 1(1): 23-27.

Oommachan, M. & Masih, S.K. 1987. Multiferous uses of plants by the tribals of Madhya Pradesh 1: Medicinal plants. *J.Appli. Pure Biol.* 2(2): 55-63.

Oommachan, M. & Masih, S.K. 1988. Multiferous uses of plants by the forest tribals of Madhya Pradesh: Wild edible plants. *J. Tropical Forestery* 4(II): 163-164.

Oommachan, M. & Masih, S.K. 1989. Ethnobotanical observations on certain forest plants of Jabalpur, Madhya Pradesh. *Indian J. Applied and Pure Biology* 4(2) : 73-78.

Oommachan, M., Masih, S.K. & Shrivastava, J.L. 1989. Ethnobotanical studies in certain forest areas of Madhya Pradesh. *Trop. Forestry* 5(3): 182-196.

Oommachan, M. & Masih, S.K. 1991. Ethnobotanical and conservation aspects of medicinal plants of Madhya Pradesh. *Indian J. Apppl. Pure. Biol.* 39: 44.

Oommachan, M. & Masih, S.K. 1993. Ethnobotanical observations on certain plants of the tribal regions of Madhya Pradesh. *Biome* 6(1): 59-64.

Oudhia, P. 2000. Medicinal weeds in kodomillet fields. A source of an additional income for Chhattisgarh framers. *Ecol. Env. and Cons.* 6(2): 171-174.

Oudhia, P. & Dixit, A. 1994. Weeds in Ambikapur region (Madhya Pradesh) and their traditional use. *Weeds News* 1(2): 19-21.

Oudhia, P.,Tripathi, R.S., Puri, Sunil & Chandel, D.S. 1999. Traditional knowledge about medicinal weeds in Chhattisgarh. *Agric.Vasundhara The Earth.* 1(1): 12-13.

Oudhia, P., Pal, A.R. & Pali, G.P. 2002. Traditional medicinal knowledge about common crop weeds, in Bagbahera (India) region: A survey. *Agric. Sci. Digest.* 22(1): 53-54.

Pal, D.C. & Banerjee, D.K. 1971. Some less known plants foods among the tribals of Andhra Pradesh and Orissa State. *Bull. Bot. Surv. India* 13(3-4):221-223.

Pal, D.C. 1973. Tribal folklore about some plants associated with eye treatment. *Folklore* 14: 446-447.

Pal, D.C. & Banerjee, D.K. 1974. A note on the plants used by the tribals in India for hair and scalp preparations *Adibasi* 15(1): 28-32.

Pal, D.C. & Srivastava, J.N. 1976. Preliminary notes on ethnobotany of Singhbhum district, Bihar. *Bull. Bot. Surv. India* 18:247 – 250.

Pal, D.C. 1980. Observations on folklore about plants used in veterinary medicine in Bengal, Orissa & Bihar. *Bull. Bot. Surv. India.*22: 96 – 99.

Pal, D.C., Soren, A.M. & Sen, R. 1984. A curative fungus from tribal areas. *Vanyajati* 32(3): 16-17.

Pal, G.D. 1984. Observations on ethnobotany of tribals of Subansiri. Arunachal Pradesh. *Bull. Bot. Surv. India* 26: 26 - 37.

Pal, D.C., & Mudgal, V. 1985. Conservational practices of tribals of Purulia and their relevance in ecodevelopment programmes. *Proc. Eco-development Camp.,* pp. 37-42. Baghmundi (Purulia). Calcutta. School of Fundamental Research.

Pal, D.C. & Jain, S.K. 1989. Notes on Lodha medicine in Midnapur district. West Bengal, India. *Econ. Bot.*43: 464 – 470.

Panda, T., Panigrahi, S.K. & Padhy, R. 2005. A sustainable use of phytodiversity by the Kandha tribe of Orissa. *Indian J. Trad. Knowledge* 4(2): 173-178.

Pande, P.C. & Pande, H.C. 2004. Traditional usage of 'urad' *(Vigna mungo)* in Kumaon Himalaya, Uttaranchal. *Ethnobotany* 16(1&2): 96-98.

Pandey, G.S. 1969. *Bhavaprakash Nighantu* (With Commentary by Dr. K.C. Chunekar). Chowkhamba Sanskrit Sansthan, Varanasi.

Pandey, R.K. & Shrivastava, J.L. 1993. Forest and tribal: an ethnobotanical observations on wild medicinal plants used by Bharia tribes of Patalkot in Tamia forest of Madhya Pradesh. *J. Tropical Forestry.* 9(III).

Pandey, H.P. & Verma, B.K. 2002. Plants in oral healthcare among the aborgines of Gonda and Balrampur districts, U.P., India. *Ethnobotany* 14(1&2): 81-86.

Pandey, A.K. & Rout, S.D. 2006. Ethnobotanical uses of plants of Similipal Biosphere Reserve (Orissa). *Ethnobotany* 18: 102-106.

Pandey, R.K. & Saini, S.K. 2007. Edible plants of tropical forests among tribal communities of Madhya Pradesh. *Indian J. Trad. Knowl.* 6(1): 185-190.

Pandit, B.R., Kotiwar, O.S., Oza, R. A. & Kumar, R.M. 1996. Ethno - medicinal plant lore from Gir forest, Gujarat. *Adv Plant Sci.* 9: 81 – 84.

Panigrahi, A.K. & Alka, S. 2000. *Glossary of useful and economically important plants.* New Central Book agency, Calcutta.

Pant, S.C. & Pandey, G. 1995. Ethnoboanical studies on medicinal flora in Tharu tribal pockets in Kumaon region in U.P. *Bull. Medico-Ethnobot. Res.* 16(12): 1-10.

Paranjpe, P. 2005. *Indian medicinal plants: forgotten healers.* Chaukhamba Sanskrit Pratisthan, Delhi.

Parmar, S. 1978. A note on folk musical instruments of Madhya Pradesh. *Folklore* 19: 26-27.

Patil, S.H. & Merat, M.M. 2003. Ethnoveterinary practices in Satpudas of Nandurbar district of Maharashtra. *Ethnobotany* 15: 103-106.

Patil, M.V., Pawar, S. & Patil, D.A. 2004. Ethnobotanical plants used for dental and oral healthcare in north Maharashtra. *J. Non-Timber Forest Prod.* 11(3): 208-210.

Paul, C.R. & Verma, N.K. 2004. Botany and ethnobotany of *Diospyros melanoxylon* Roxb. (Ebenaceae). *J. Econ. Taxon. Bot.* 28(3): 599-601.

Pawar, S. & Patil, D.A. 2004. Herbal folk remedies for urinogenital complaints in Jalgaon district of Maharashtra. *Ethnobotany* 16(1&2): 86-90.

Pawar, S., Patil, M. V. & Patil, D. A. 2006. Ethnoboanicinal uses of Bel [*Aegle marmelos* (L.) Corr.] in Khandesh region of Maharashtra. *J. Non-Timber Forest Prod.* 13(2): 131-133.

Pereira, N. 2006. Ethnobotanical relevance in tribal life: A study on Warli tribe of Thane district, Maharashtra. *J. Econ. Taxon. Bot.* 30(Suppl.): 31-39.

Peter Paul Hembrom & Goel, A. K. 2005. Horopathy: Ethnomedicine of Mundas. *Ethnobotany*17:89-95.

Porte, S.S. 1965. Dahi cultivation in Bindra Nawagarh tahsil of Raipur dist. *Vanyajati* 32(3): 93-101.

Prajapati, V.K. & Verma, B.K. 2004. Ethnoveterinary plants of district Mahoba. U.P. *J. Econ. Taxon. Bot.* 28(3): 623-626.

Prakash, A. & Singh, K. K. 2006. Useful medicinal plants for general tonic, sexual vigour, strength, vitality and venereal diseases among the tribals of Uttar Pradesh and Uttaranchal.. *J. Econ. Taxon. Bot.* 30(Suppl.): 326-331.

Prance, G.T. 1983. The ethnobotany of Amazan Indians: A rapidly disappearing source of botanical knowledge for human welfare. *Bull. Bot. Surv. India* 29: 148-159.

Prasad, S.S., Mohan, K. & Singh, A.K. 1990. Ethnobotanical exploration of the Tharu belt of Champaran (North Bihar), Part-I: medicinal plants of the area. *Bio. Journal.* 2(2): 203-206.

Prasad, K.V., Rajgaopal, T., Kant, Yogesh & Badrinath, K.V.S. 1999. Food plants of Konda reddis of Rampa agaence East Godavari district, Andhra Pradesh- a case study. *Ethnobotany* 11: 92-96.

Printya, J.V. 1963. *The Sushruta Samhita,* Vol. I&II, Motilal Banarsidas. Delhi.

Punjani, B.L. 2004. Ethnomedicinal uses of Vitaceae among the herbal healers in central India. *Indian Forester* 83(1&2): 83-85.

Punjani, B.L. 2006. Ethno-medico-botanical study of Kathodi tribe of Sabarkantha in Gujarat. *Ethnobotany.* 18: 135-138.

Puri, H.S. 1983. Medicinal plants of Tejpur (Assam). *Bull. Medico-ethnobot. Res.*4: 1 – 13.

Purkayastha, J. & Subhan, C.N. 2005. Biological activitis of ethnomedicinal claims of some plant species of Assam. *Indian J. Trad. Knowledge* 5(2), 229 – 236.

Purohit, V.P., Silas, R.A. & Gaur, R.D. 1985. Ethnobotanical studies of some medicinal plants used in skin diseases from Raath (Pauri), Garhwal Himalaya. *J. Sci. Res. Pl. Med.*6: 39 – 47.

Radhakrisnan, K., Pandurangan, A.G., Pusphangadan, P. & Sasidharan, A. 1996. Less known ethnomedicinal plants of Kerala State and their conservation. *Ethnobotany* 8: 82-84.

Raghavaiah, V. 1956. How to approach the Adivas,. *Vanyajati* 4: 31-38.

Rai Bahadur, H. 1925. Some notes about marriage, food, drink and occupations of castes affecting social status in Central Provinces. *Man in India* 5(1&2): 56-68.

Rai Chaudhuri, H.N., Banerjee, D.K., & Guha, A. 1977. 'Ranu' tribal tablet. *Vanyajati* 25(2): 9-13.

Rai Chaudhuri, H.N., Dam, N.S.,Pal, D.C., Tarafder, C.R. & Das, K.K. 1977. Edible jungle leaves. *Folklore* 18(5): 161-163.

Rai, M.K. 1985. Plants used as medicine by tribals of Chhindwara district (Madhya Pradesh). *J. Econ. Taxon. Bot.*7: 385 – 387.

Rai, M.K. 1988. Ethnomedicinal survey of Patalkot and Tamia (Distt. Chhindwara): I-plants used against skin diseases and liver disorders. *J. Econ. Taxon. Bot.* 12(2): 337-339.

Rai, B.K. & Rai, A. 1992. Use of fungus in folk medicine for family planning among Baiga. *Folklore* 33: 65.

Rai, B.K., Shiv Shanker & Rai, A.A. 1996. A note on ethno-mycomedicines from Central India. *J. Econ. Taxon. Bot. Addl. Ser.* 12: 186-191.

Rai, R. 2004. Forest products and their ethnobotanical uses among tribals of Madhya Pradesh. *Indian J. Trop. Biodiversity* 12 (1 & 2):57-60.

Rai, R. 2007. Studies on indigenous herbal remedies in cure of fever by tribals of Madhya Pradesh. In Neeru Singh (Ed.) *Tribal health,* pp 177-182. Regional Medical Research Centre for Tribals.

Rai, R. 2008a. Herbal medicines in cure of prevalent diseases in tribal pockets of Madhya Pradesh. In Trivedi, P. C. (Ed.) *Indian Medicinal Plants,* pp 175-183. Avishkar Publishers & distributers, Jaipur.

Rai, R. 2008b. Herbal plants in ethnomedicinal uses by Bharia tribes in Satpura plateau of Madhya Pradesh. In Trivedi, P. C. (Ed.) *Indian Medicinal Plants,* pp 196-211. Avishkar Publishers & distributers, Jaipur.

Rai, R. 2008c. Indigenous and herbal medicines prevalent among tribal communities in Madhya Pradesh. *J. Non-Timber Forest Prod.*15 (3):183-191.

Rai, R. 2008d. Tribes and their indigenous knowledge on wild plants used as food and their medicinal uses in Madhya Pradesh. *J. Non-Timber Forest Prod.*15 (4):255-263.

Raizada, A. 1984. *Tribal Development in Madhya Pradesh. A Planning perspective.* Inter-India Publication, New Delhi.

Rajendranan, A., Ravi Kumar, K. & Henry, A.N. 1997. Utilization of wild plant resources for food consumption in Tamilnadu. *J. Non. Timber. For Prod.,* 4(3-4): 93-95.

Rajendran , K. & Rengamani, S. K. 2006. Medicinal plants and their utilization by villagers in Southern districts of Tamilnadu. *J. Econ. Taxon. Bot.* 30(Suppl.): 208-216.

Rajwar, G.S. 1983. Low altitude medicinal plants of South Garhwal. *Bull. Medico-ethnobot. Res.*4: 14 – 28.

Ramachandran, V.S. & Nair, N.C. 1981a. Ethnobotanical studies in Cannanore district, Kerala state (India). *J. Econ. Taxon. Bot.*2: 65 - 72.

Ramachandran, V.S. & Nair, N.C. 1981b. Ethnobotanical observations on 'lrulars' of Tamil Nadu (India). *J. Econ. Taxon. Bot.*2: 183 – 190.

Ramdas, R.,Sagari, Ghote Nitya Sambamurti, Ashalate, S., Mathur P.,Nandini, Broome, Vivek Gour & Rao,Sanyasi.2000. Ethnoveterinery remedies used in common surgical conditions in some districts of Andhara Pradesh and Maharasthtra, India. *Ethnobotany* 12: 100-112.

Rana, T.S., Datt. & Rao, R.R. 1996. Strategies for sustainable utilization of plant resources by the tribals of the Tons – Valley, Western Himalayas. *Ethnobotany* 8: 96 – 104.

Ranade, R.M. 1956. Oil pressing as home industry among Bhils & Bailalas. *Vanyajati* 4(1): 40-41.

Rao, R.R. & Neogi, B. 1980. Observations on the ethnobotany of the Khasi & Garo tribes in Meghalaya. *J. Econ. Taxon. Bot.*1: 157 – 162.

Rao, M.K.V. & Shanpru, R. 1981. Some plants in the life of Garos of Maghalaya. In Jain, S.K. (ed.) *Glimpses of Indian Ethnobotany,* pp. 153-160. Oxford and IBH, New Delhi.

Rao, R.R. 1981a. Ethnobotanical studies on the flora of Meghalaya-some interesting reports of herbal medicines. In Jain, S.K. (ed.) *Glimpses of Indian Ethnobotany,* pp. 137-148. Oxford and IBH, New Delhi.

Rao, R.R. 1981b. Ethnobotany of Meghalaya - medicinal plants used by Khasi & Garo tribes. *Econ. Bot.* 35: 4-9.

Rao, R.R. & Jamir, N.S. 1982. Ethnobotanical studies in Nagaland – I: Medicinal plants. *J. Econ. Bot.* 36: 176 – 181.

Rao, R.R. & Jamir, N.S. 1982. Ethnobotanical studies in Nagaland – II: 54 Medicinal plants used by Nagas. *J. Econ. Taxon. Bot.*3: 11 – 17.

Rao, K.P. & Sreeramulu, S.H. 1985. Ethnobotany of selected medicinal plants of Srikakulam district, Andhra Pradesh. *Ancient Sci. life* 4: 238-244.

Rao, R.R. & Haridasan, K. 1991. An ethnobotanical survey of medicinal and other useful plants form North-East India. *J. Econ. Taxon. Bot.* 15: 423-436.

Rao, V.L., Busi, B.R., Dhrama Rao, B., Rao, C.S., Bharathi, K. & Venkaiah, M. 2006. Ethnomedicinal practices among Khonds of Visakhapatanam district. Andhra Pradesh. 5(2): 217-219.

Rastogi, R. P. & Mehrotra, B. N. 1997. *Compendium of Indian medicinal Plants*, Vol. I, II, III, IV & V. CDRI, Lucknow and PID, Delhi.

Ravikumar, R.K., Sudhakar Rao, B., Bose, S.C. & Sudhakar, K. 2004. Validity of ethno-veterinary practices adopted by farmers in Dindigul district of Tamilnadu. *Asian Agri-History* 8(1): 55-61.

Reddy, K.J. & Sudarsanam, G. 1987. Plants used as veterinary medicine in Chittoor district of Andhra Pradesh, India. *Int. J. Crude Drug Res.*25(3): 145-152.

Reddy, M.B., Reddy, K.R. & Reddy, M.N. 1991. Ethnobotany of Cuddapah district. Andhra Pradesh, India. *Int. J. Pharmacog.* 29: 273-280.

Reddy, R.V., Lakshmi, N.V.N. & Venkata Raju, R.R. 1998. Folk veterinary medicinal plants in Cuddapha Hills of Andhra Pradesh, India. *Fitoterapia* 69(4): 322-328.

Roy, N. 1955. Basketry and domestic utensils of the Adis. *Vanyajati* 5: 241-257.

Roy, J.K.,Rao, R.K. & Biswas, S.K. 1956. Diet surveys amongst the tribes of Madhya Pradesh (Baigas and Gonds). Part I. *Bull. Dept. Anthrop.* 5: 1-10.

Roy, J.K. & Rao, R.K. 1957. Investigation on the diet of the Muria of Bastar district. *Bull. Dept. Anthrop.* 6: 33-45.

Roy, J.K., Rao, R.K. & Roy, B.C. 1959. Mahua spirit and the chemical composition of the raw material (Mahua flowers) *J. Inst. Chem.* 31: 64.

Roy, J.K. 1978. Alcoholic beverages in tribal India and their nutritional role. *Man in India* 58: 298-326.

Roy, G.P. &. Chuturvedi, K.K. 1986. Plants used as tonic by the tribals of Abujh-Marh Reserve area, Bastar district in Madhya Pradesh. *Vanyajati* 34: 27 – 32.

Roy, G.P. & Chaturvedi, K.K. 1986. Ethnomedicinal trees of Abujh-Marh area, Madhya Pradesh. *Folklore* 27: 95 – 100.

Roy, G.P. & Chaturvedi, K.K. 1986. Ethnomedicinal plants of Abujmarh Reserve Area, Bastar district in Madhya Pradesh. *Folklore* 27(5): 95-98.

Roy, G.P. & Chaturvedi, K.K. 1987. Less known medicinal uses of rare and endangered plants of Abujh Marh reserve area, Bastar (Madhya Pradesh). *J. Econ. Taxon. Bot.*9: 325 – 328.

Roy, S. & Das, A.P. 2005. An effective remedy for jaundice with *Thespesia populnea* from Purulia district of West Bengal (India). *Ethnobotany* 16(1&2): 50-51.

Sadhale, A., Mujumdar, A.M. & Pendse, G.S. 1991. Ethnobotanical studies of sacred grove at Ajiwali, Pune district, *J. Econ. Taxon. Bot.* 15: 167-172.

Sahay, K.N. 1964. Tree-cult in tribal culture. *Folklore* 5(7): 241-257.

Sahoo, A.K. & Bahali, D.D. 2003. Plants used in less known ehtnomedicines and medico-religious beliefs in Phulbani district of Orissa. *J. Econ. Taxon. Bot.* 27(2): 500-604.

Sahu, P. 2000. Conservation based interaction between biological and cultural diversity amongst Maria tribals of Bastar. *JMAPS* (22/4A & 23/1A): 511-513.

Sahu, S.K. 1980. Health culture of the oraons in the context of different health institution. *Ehtnomedicine* 6: 167-182.

Sahu, T.R. 1982. An ethno-botanical study of Madhya Pradesh. I: Plants used against various disorders among tribal women. *Ancient Sci. Life* 1(3): 178-181.

Sahu, T.R. 1983. Further contributions towards the ethnobotany of Madhya Pradesh-II: Plants used against diarrhoea and dysentery. *Ancient Science Life* 2(3): 169-170.

Sahu, T.R. 1984. Less known uses of weeds as medicinal plants. *Ancient Science Life* 3(4): 245-249.

Sahu, T.R. 1996. Life support promising food plants among aboriginals of Bastar (Madhya Pradesh). In Jain, S.K. (ed.) *Ethnobiology for human welfare,* pp. 26-30. Deep Pub., New Delhi.

Sahu, T.R. 1999. Biodiversity conservation of Bastar (Madhya Pradesh) in traditional and religious context. In Tandon, R.K. and Singh, P. (Ed.). *Biodiversity, Taxonomy and Ecology* pp. 55-64. Scientific Publishers, Jodhpur.

Sahu, T.R., Sahu, B.K., Sahu, P. & Dubey, Gayatri 2003. Uses and conservation of the plants on folklore in Abujhmaria tribe dominating Abujhmarh area (Bastar-Chhatisgarah). *Vanyajati* XLXI (2): 46-52.

Sahu, T.R., Sahu, P. & Dubey, G. 2004. Popular beverage in Abujhmaria tribal dominating in Abujhmarh region, Bastar, Chhattisgarh. *Vanyataji* 52(1): 47-52.

Saini, V.K. 1996. Plants in the welfare of tribal women and children in certain areas of Central India. In Jain, S.K. (ed.). *Ethnobiology for human welfare,* pp 140-144. Deep Publication, New Delhi.

Samati, H. 2004. Kitchen garden plants of Pnar tribe in Jaintia Hills district, Meghalaya. *Ethnobotany* 16(1&2): 125-130.

Samvatsar, S. 1996. *The flora of western tribal Madhya Pradesh*, Scientific Publishers, Jodhpur.

Samvastar, S. & Diwanji, V.B. 1996. Plants used in snake and scorpion insect-bites/stings by Adibasis of Jhabua (Madhya Pradesh), India. *J. Econ. Taxon Bot. Addl Ser.* 12: 199-200.

Samvastar, S. & Diwanji, V.B. 2000a. Plant sources for the treatment of Jaundice in the tribals of western, Madhya Pradesh of India. *J. Ethnopharmac.* 73(1-2): 313-316.

Samvastar, S. & Diwanji, V.B. 2000b. Plants used for rheumatism by the tribals of western Madhya Pradesh. In Maheshwari, J.K. (ed.) *Ethnobotany and Medicinal Plants of Indian Sub-Continent,* pp. 305-314. Scientific Publishers, Jodhpur.

Samvatsar, S. & Diwanji, V.B. 2004. Plants used for the treatment of different types of fevers by Bhils & its sub tribes in India. *Indian J. Trad. Knowl.* 3: 101-104.

Sarangi, N. & Sahu, R.K. 2004. Ethnomedicinal plants used in veneral and gynaecological disorders in Kalahandi, Orissa. *Ethnobotany* 16(1&2): 16-20.

Sarma, S.K., Bevi,B. & Bhattacharjya, D.K. 2006. Ethnomedicinal uses of plants by the Sarania tribe of Nalbari district of Assam. *J. Econ. Taxon. Bot.* 30(Suppl.): 133-139.

Sasidharan, N. & Augustine, J. 2006. Ethnobotany of the tribes living in and around the Periyar Tiger Reserve, Southern Western Ghats, India. *J. Econ. Taxon. Bot.* 30(Suppl.): 45 - 58.

Santapau, H. 1955. *Botanical collectors Manual: A Handbook of instructions for the collections and preservation of plants for the use of Botanical Survey of India.* New Delhi.

Saxena, H.O. & Dutta, P.K. 1975. Studies on the ethnobotany of Orissa. *Bull. Bot. Surv. India* 17: 124 – 131.

Saxena, A.P. & Vyas, K.M. 1981. Ethnobotanical records on infections diseases from tribals of Banda district. *J. Econ. Taxon. Bot.*2: 191 – 194.

Saxena, H.O., Brahman, M. & Dutta, P.K. 1931. Ethnobotanical studies in Orissa. In Jain, S.K. (Ed.). *Glimpses of Indian Ethnobotany,* pp 232 – 244. Oxford & Indian IBH, New Delhi.

Saxena, A.P. & Vyas, K.M. 1983. Ethnobotany of Dhasan Valley. *J. Econ. Taxon. Bot.* 4: 121 – 128.

Saxena, A.P. & Vyas, K.M. 1986. Antimicrobial activity of seeds of some ethnomedicinal plants. *J. Econ. Taxon. Bot.*8: 291 – 299.

Saxena, H.O. 1986. Observations on the ethnobotany of Madhya Pradesh. *Bull. Bot. Surv. India* 28: 149 – 156.

Sebestian, M.K. & Bhandari. M.M. 1984a. Magico-religious beliefs about plants among the Bhils of Udaipur district of Rajasthan. *Folklore* April:77-80.

Sebestian, M.K. & Bhandari. M.M. 1984b. Some plants used as veterinary medicine by Bhils. *Inter. J. Trop. Agric.* 11(4): 307-310.

Sebastian, M.K. 1984c. Plants used as veterinary medicines-Galactagogues and fodder in the forest area of Rajasthan. *J. Econ. Taxon. Bot.* 5(4): 785-788.

Sebastian, M.K. & Bhandari, M.M. 1984. Medico-ethnobotany of Mount Abu, Rajasthan. India. *J. Ethnopharmacol.* 12: 223 – 230.

Sebastian, M.K. & Bhandari, M.M. 1988. Medicinal plant lore of Udaipur district, Rajasthan. *Bull. Med. Ethnobot. Res.* 5(3-4): 122-134.

Sen Gupta, A.K. 1965. Economic organization of an oraon village. *Vanyajati* 13(2): 41-45.

Sen, R., Pal, D. C. & Roy, B. 1984. Botany and ethnobotany of Parashi. *J. Econ. Taxon. Bot.* 5: 857 – 858.

Sen, R.S.,Bandopadhayay, E.,Biswas & A. Guha 1985. Ethnobotanical uses of Herbaria – IV. *J. Econ. Taxon. Bot.* 6: 331 – 335.

Sen, S.K. & Pradhan, N.B. 1999. Conservation of ethnomedicinal plants of Baragarh district in Orissa. *Adv. Plant Sci.* 12: 207-213.

Sen, S.K., Panda, H.S. & Behera, M. 2005. *Cordia macleodii* (Ethreticeae) – A wonderful wound – healer form Baragarh district, Orissa. *Ethnobotany* 17: 191-192.

Shah, N.C. & Joshi, M.C. 1971. An ethnobotanical study of Kumaon region of India. *Econ. Bot.* 25: 414 – 422.

Shah, N.C. & Kapoor, L.D. 1977. Ethnobotany of *Acorus calamus* L. *Indian Pharm.* 28-155.

Shah, G.L., Menon, A.R. & Gopal, G.V.1981. An account of the ethnobotany of Saurashtra in Gujarat State (India). *J. Econ. Taxon. Bot.*2: 173 – 182.

Shah, N.C. 1982. Herbal folk medicines in northern India. *J. Ethnopharmacol.*6: 293 – 301.

Shah, G.L. & Gopal, G.V. 1982. An ethnobotanical profile of the Dangies. *J. Econ. Taxon. Bot.* 3: 355-364.

Shah, G.L., Yadav, S.S. & Badri, N. 1983. Medicinal plants from Dahanu Forest Division in Maharashtra State. *J. Econ. Taxon. Bot.* 4: 141 – 151.

Shah, G.L. 1984. Some economically important plants of Salsette Island near Bombay. *J. Econ. Taxon. Bot.* 5: 753 – 765.

Shah, G.L. & Gopal, G.V. 1986. Folklore medicines of Vasavas, Gujarat, India. *Acta. Bot. Indica* 14: 48 – 53.

Shah, N.C. 1990. The role of ethnobotany in relation to medicinal plants in India. In Jain, S.K. (Ed.) *Contribution to Indian Ethnobotany,* pp 305-315. Scientific Publishers, Jodhpur.

Shah, N.C. & Singh, S.C. 1990a. Ethno-medicobotanical studies in Madhya Pradesh, India: An Appraisal. *Curr. Res. Med. & Arom. Plants (CROMAP)* 12(4): 201-217.

Shah, N.C. & Singh, S.C. 1990b. Hitherto unreported phytotherapeutical uses from tribal pockets of Madhya Pradesh, India. *Ethnobotany* 2: 91-95.

Shah, G.L. & Gopal, G.V. 1992. An ethnobotanical profile of the Dangies. *J. Econ. Taxon. Bot.*3: 355 – 364.

Shah, M. A. U. 2005. Ethnomedicinal plants of Bhakkar Tehsil in Punjab, Pakistan. Ethnobotany17: 171-175.

Shah, N.C. 2006. A historical and ethnobotanical study of *Nardostachys jatamansi* : An ancient incense & aromatic medicinal herb form Uttaranchal, India. *Ethnobotany* 18: 37-45.

Sharma, P.V. 1972. Bhava Mishra: A landmark in history of Indian medicine. *J. Res. Ind. Med.* 7(1): 63-75.

Sharma, P.V. 1973. Drugs as Landmark of the history of Indian medicine. *J. Res. Ind. Med.* 8(4): 86-93.

Sharma, R. K. & Bhagvan Dash 1976. *Charaka Samhita*, Vol. I & II, Chaukhamba Sanskrit Series, Varanasi, India.

Sharma, P.K., Dhyani, S.K. & Shanker, V. 1979. Some useful and medicinal plants of the district Dehradun and Siwalik. *J. Sci. Res. Pl. Med.*1: 17 – 43.

Sharma, B.D. & Malhotra, S.K. 1984. A contribution to the ethnobotany of tribal areas in Maharashtra. *J. Econ. Taxon. Bot.* 5: 533 – 537.

Sharma, P.K. & Singh, D. 1988. Ethnobotanical studies of some common plants in Kanker range Bastar. *Geobios, New Reports* 7(1): 69-71.

Sharma, D.C. & Chandra, U. 1998. Prophylactic uses of some medicinal plants in Bastar district of Madhya Pradesh. *Ancient Science Life* 17(4): 284-289.

Sharma, S. & Kumar, A. 2002. Ethnobotanical studies on medicinal plants Chitrak *(Plumbago zeylanica)*. *Int. J. Mendel.* 19(3): 129-130.

Sharma, P.V. 2003. *Dravyaguna-Vijnana*. Chaukhamba Bharti Academy Varanasi.

Sharma, B. & Maheshwari, S. 2005. Trditional medical practices of Gaddi tribes in Kangra district, Himachal Pradesh. *Indian J. Trad. Knowledge* 4(2): 169-172.

Sharma, H.M., Devi, A.R. & Sharma, B.M. 2005. Vegetable dyes used by the Meiti community of Manipur. *Indian J. Trad. Knowledge* 4(1): 39-45.

Sharma, S.C., Sharma, G.N. & Aggarwal, R.K. 2006. Traditional significant medicinal plants of Nagaur district (Rajasthan). *J. Econ. Taxon. Bot.* 30(Suppl.): 320-325.

Shekhawat, G.S. & Sunil Anand 1984. An ethnobotanical profile of Indian Desert. *J. Econ Taxon. Bot.* 5: 591 – 598.

Shrivastva, J.L., Jain, Seema & Dubey, Abhilasha 1999. Ethno-medicine for antifertility uses by the tribals in Bastar district (Madhya Pradesh). *J. Econ. Taxon. Bot.* 23(2): 197-198.

Shubhangi, Pawar & Patil, D.A. 2006. Ethnomedicinal aspects of naturalised exotic plants of Jalgaon district, Maharashtra, India. *J. Non-Timber Forest Prod.* 13(1):55-59

Siddiqui, M.B. & Husian, W. 1991. Traditional treatment of diarrhoea and dysentery through herbal drugs in rural India. *Fitoterapia* 62: 325 – 329.

Siddque, M.A.A., Jhon, A.Q. & Paul, T.M. 1995. Status of some important medicinal and aromatic plants of Kashmir Himalaya. *Adv Plant Sci.* 8: 134-139.

Sikarwar, R.L.S. 1992. Ethnomedicines of north Surguja forest division, Ambikapur, Madhya Pradesh. *Aryavaidyan* 6(2): 97-100.

Sikarwar, R.L.S. 1993. Ethnogynaecological use of plants by tribals of Madhya Pradesh. *Vanyajati* 41(2): 28-31.

Sikarwar, R.L.S. & Kaushik, J.P. 1993. Folk medicines of the Morena district, Madhya Pradesh. India. *Int. J. Pharmacog.* 31: 283-287.

Sikarwar, R.L.S. & Painuli, R.M. 1993. Studies on medicinal uses of plants by the Sahariya tribe of Madhya Pradesh, *Aryavaidyan* 7(1): 51-53.

Sikarwar, R.L.S., Bajpai, A.K. & Painuli, R.M. 1994. Plants used as veterinary medicines by aboriginals of Madhya Pradesh. India. *Int. J. Pharmacom.* 32(3): 251-252.

Sikarwar, R.L.S. 1995. Bastar Janpad ke Adivasiyon ka Kalpvriksh-Khajur (In Hindi). *Ethnobotany* 7: 137-138.

Sikarwar, R.L.S. 1996. Ethnoveterinary herbal medicnies in Morena district of Madhya Pradesh, India. In Jain, S.K. (Ed) *Ethnobiology in Human Welfare,* pp 194-196. Deep Publicaitons, New Delhi.

Sikarwar, R.L.S. 1997. Ethnobotany of Sahariya tribe Madhya Pradesh, India– A review. *Appl. Bot. Abstr.* 17(2): 129-140.

Sikarwar, R.L.S. 1998. Role of tribals in environmental conservation. *Ethnobotany* 10: 138-141.

Sikarwar, R.L.S. 1998a. Bastar ki Janjatiyon ke utthan hetu laghu vanopajon par adharit kuteer udhyog. In Solanki, S.S. (Ed.) *Adunik Prodhyogiki Tatha Gramin Vikas.* pp. 71-80. Bhagirath Sewa sansthan, Gaziabad.

Sikarwar, R.L.S. 1998b. Paryavaran Sanrakshan men Janjatiyon ka Yogdan (In Hindi). *Ethnobotnay* 10: 138-141.

Sikarwar, R.L.S. 2001. Ethnobotany of Madhya Pradesh, India. *Applied Bot. Abst.* 21(2): 133-147.

Sikarwar, R.L.S., Brijlal & Maheshwari, J.K. 2002. Traditional phytotherapy among the tribals in Raigarh district of Chattisgarh. *J. Non-Timber Forest Prod.* 9(1&2): 22-25.

Sikarwar, R.L.S. 2002. Mahua {*Madhuca longifolia* (Koen.) MacBride} – A paradise tree for the tribals of Madhya Pradesh. *Indian J. Trad. Knowledge* 1(1): 87-92.

Sikarwar, R.L.S., Kumar, V., Rawat, A.K.S., Mehrotra, S. & Pushpangadan, P. 2003. *Radermachera xylocarpa* (Roxb.) K. Schum. (Bignoniaceae)-An Endangered and Interesting tree of Central India. *Phytotaxonomy* 3: 32-34.

Sikarwar, R.L.S. & Kumar, V. 2005. Ethnoveterinary knowledge and practices prevalent among the tribals of Central India. *J. Natural Remedies* 5(2): 147-152.

Singh, Harbhajan 1945. Home useful wild plants of the Delhi Province. *India. J. Agr. Sci.*15: 297 – 307.

Singh, R. 1971. Famine & Folk songs of the Bhils of Rajasthan. *Folklore* March: 95-98.

Singh, K.K., Palvi, S.K. & Singh, H.B. 1980. Survey of some medicinal plants of Dharchula block in Pithoragarh district of U.P. *Bull. Medico-ethnobot. Res.* 1: 1– 7.

Singh, V. & Pandey, R.P. 1980. Medicinal plantlore of the tribals of eastern Rajasthan. *J. Econ. Taxon. Bot.* 1: 137 – 147.

Singh, V. & Pandey, R.P. 1982. Plant used in religion and magico-religious beliefs in Rajasthan. *J. Econ. Taxon. Bot.* 3: 273-298.

Singh, K.K. & Maheshwari, J.K. 1983. Traditional phytotherapy amongst the tribals of Varanasi district, Uttar Pradesh. *J. Econ. Taxon. Bot.* 4: 829 – 838.

Singh, K.K., Saha, S. & Maheshwari, J.K. 1985. Ethnobotany of *Helicteres isora* L. in Kheri district, Uttar Pradesh. *J. Econ. Taxon. Bot.* 7: 487 – 492.

Singh, K.K. & Maheshwari, J.K. 1985. Forest in the life and economy of the tribals of Varanasi district, Uttar Pradesh. *J. Econ. Taxon. Bot.* 6: 109-116.

Singh, A.K., Singh, R.N. & Singh, S.K. 1987. Some ethnobotanical plants of Terai region of Gorakhpur, district- I. *J. Econ. Taxon. Bot.* 9: 407 – 410.

Singh, K.K., Saha, S. & Maheshwari, J.K. 1987. Observation on the ethnobotany of Bhoxa tribe of Bajpur block of Nainital district., U.P. *Himalayan Res. Dev.* 6(1&4): 25-29.

Singh, H. 1988. Ethnobiological treatment of piles by Bhoxas of Uttar Pradesh. *Anc. Sci. Life* 8: 167 – 170.

Singh, H. & Maheshwari, J.K. 1992. Traditional remedies for snake bite & scorpion – sting among the Bhoxas of Nainital district. U.P. *Aryavaidyah.*6: 120 -123.

Singh, L.B. & Singh, C.L. 1992. An Ethno-medico-botanical study of Deoghar district (Bihar). *Biojournal* 4: 83 – 86.

Singh, P.B. & Aswal, B.S. 1992. Medicinal plants of Himachal Pradesh used in pharmaceutical industry. *Bull. Med. Ethno. Res.* 13: 172 – 208.

Singh, P.B. 1993. Medicinal plants of Ayurvedic importance from Mandi district of Himachal Pradesh. *Bull. Med. Ethnobot. Res.*14: 126 – 136.

Singh, K.K. & Prakash, A. 1994. Indigenous phytotherapy among the Gond tribes of Uttar Pradesh, India. *Ethnobotany* 6: 37 – 41.

Singh, K.K. 1997. Studies on native medicine of Jaunsari tribe of Dehradun district, U.P. India. *Int. J. Pharmacog.* 35: 105 – 110.

Singh, H.B.K. 1997. Studies on medico-botany of Meitel community in Manipur State, India. *Adv. Plant. Sci.* 10: 13 – 18.

Singh, V.K., Ali, Z.A. & Siddiqui, M.K. 1997. Medicinal plants used by the forest ethnics of Gorakhpur district (U.P.). *Int. J. Pharmacog.* 35: 194 – 206.

Singh, V.K., Ali, Z.A. & Siddique, M.K. 1997. Folk medicinal plants of the Garwal and Kumaon forest of U.P. India. *Hamdard Med.* 40(4): 35-47.

Singh, S.K., Das. P.K. & Das, M.N. 1998. Ethnobotanical studies on some medicinal plants of Rajgir, Bihar. *Int. J. Mendeol.* 15: 145 – 148.

Singh, V. & Pandey, R.P. 1998. *Ethnobotany of Rajasthan.* Scientific Publishers (India), Jodhpur.

Singh, K.K., Anand Prakash & Palvi, S.K. 1999. Observations on some energy plants among the tribals of Madhya Pradesh. *J. Econ. Taxon. Bot.* 23(2): 291-296.

Singh, K.K. & Kumar, K. 2000. Ethnotherapeuties of some medicinal plants as used as antipyretic agents among the tirbals of India. In Maheshwari, J.K. (ed.) *Ethnobotany and medicinal plants of Indian sub-contient,* pp. 135-141. Scientific Publishers, Jodhpur.

Singh, N.P., Khanna, K. K., Mudgal, V. & Dixit, R.D. 2001. *Flora of Madhya Pradesh,* Vol.-III. BSI, Kolkata.

Singh, B.K., Srivastava, S.K. & Dixit, R.D. 2002. Economic legumes of Chhattisgarh & Madhya Pradesh, India. *J. Econ. Taxon. Bot.* 26(3): 587-596.

Singh, K.K., Anand Prakash & Palvi, S.K. 2003. Observations on some energy plants among the tribals of Madhya Pradesh. In pp 291-296. J.K. Maheshwari (Ed.). *Ethnobotany and Medicinal Plants of Indian Subcontinent.* Sci. Pub. Jodhpur.

Singh, Neeta & Chauhan, S.V.S. 2004. Studies on some leaves (Patra) uses for worship in Brijmandal. *Ethnobotany* 16: 69-71.

Sinha, R.K. 1979. Pandos and the Forest: A study of nature and man interation. *Vanyajati* 27(2): 30-33.

Sinha, A. M. 1994. *Madhya Pradesh District Gazetteers: Ratlam District.* Directorate of Gazetteers, Department of Culture, Madhya Pradesh, Bhopal.

Sinha, B.K., Shukla, B.K. & Sharma, P. 2006. Ethnomedicinal plants of Chhattisgarh with sporadic distribution. *J. Econ. Taxon. Bot.* 30(Suppl.):70-109.

Sood, S.K. & Thakur, S. 2004. *Ethnobotany of Rewalsar Himalaya.* Deep Publications, New Delhi.

Srivastava, D.K. & Varma, S.K. 1981. An ethnobotanical study of Santhal Pargana, Bihar. *Indian Forester* 107: 30 – 41.

Srivastava, R. 1984. Tribals of Madhya Pradesh and forest bill of 1980. *Man in India* 64(3): 320-327.

Srivastava, R.K. 1985. *Herbal remedies used by the Bhils of Madhya Pradesh*, pp. 389-392. Oriential Med. Kyoto, Japan.

Srivastava, S.C. & Rout, N. 1994. Some plants of ethnopaediatric importance in district Koraput, Orissa. *Bull. Bot. Surv. India* 36: 166 – 168.

Srivastava, J. L., Jain, S. & Dubey, A. 1999. Ethnomedicine for anti-fertility used by the tribals of Bastar district of Madhya Pradesh. *J. Econ. Taxon. Bot.* 23(2): 297-300.

Srivastava, J. L., Jain, S. & Dubey, A. 2000. Ethnomedicine for anti-fertility used by the tribals of Bastar district of Madhya Pradesh. In Maheshwari, J.K. (ed.) *Ethnobotany and Medicinal plants of Indian subcontinent,* pp. 297-300. Scientific Publishers, Jodhpur.

Subramanian, K. & Jain, S.K. 1972. Conservation of Rare plants: A timely awakening. *Souv. Nat. On Fair,* pp. 73-79. Calcutta.

Subramanian, A., Mohan, V.R., Kumaresan, S. & Chelladurai, V. 2003. Medicinal plants used by the Valaiyans of Madurai district, Tamilnadu. *J. Econ. Taxon. Bot.,* 27(4): 785-787.

Sudhakar, S. & Rolla, R.S. 1985. Medicinal plants of Upper East Godavari district, Andhra Pradesh, and need for establishment of Medicinal Farm. *J. Econ. Taxon. Bot.*7: 399 – 406.

Sudhakar, A. & Chetty, K.M. 1998. Medicinal importance of some angiospermic weeds used by the rural people of Chittoor district of Andhra Pradesh, India. *Fitoterapia* 69: 390 400.

Tag, H. & Das, A.K. 2004. Ethnobotanical notes on the Hill Miri tribe of Arunachal Pradesh. *Indian J. Trad. Knowl.* 3(1): 80-85.

Tarafder, C. R.1978. Some interesting new uses of the well known plant *Vitex negundo* L. (NISINDA). *Bull. Bot. Surv. India* 20: 176-177.

Tarafder, C.R. & Rai Chaudhuri, H.N. 1981. Less known medicinal uses of plants among the tribals of Hazaribagh district of Bihar. In Jain, S.K. (Ed). *Glimpses of Indian Ethnobotany,* pp. 208-217. Oxford & Indian IBH, New Delhi.

Tarafder, C.R. 1983. Traditional medicinal plants used by the tribals of Ranchi & Hazaribagh district, Bihar- Plants used in stomach troubles. *J. Econ. Taxon. Bot.*4: 891 – 896.

Tarafder, C.R. 1983a. Ethnogynaecology in relation to plants used in antifert-ility and conception. *J. Econ. Taxon. Bot.* 4(2): 483-490.

Tarafder, C.R. 1983b. Ethnogynaecology in relation to plants used in abortion. *J. Econ. Taxon. Bot.,* 4(2): 507-516.

Tarafder, C.R. 1984. Ethnogynaecology in relation to plants used to accelerate delivery and in pre and post-natal complaints. *J. Econ. Taxon. Bot.* 5: 572 – 576.

Tarafder, C.R. 1984a. Less known nine medicinal plants used by the tribals for curing gonorrhoea in Ranchi and Hazaribagh district, Bihar. *Folklore* 25: 47-49.

Tarafder, C.R. 1984b. Less known twenty three medicinal plants used by the tribals for curing boils in Ranchi and Hazaribagh district. Bihar. *Vanyajati* 32: 14 – 19.

Tarafder, C.R. 1985. *Acacia catechu* (L.f.) Willd. (Mimosaceae) - Indigenous and less known uses of the plant. *Folklore* 26: 158 – 159.

Tarafder, C.R. 1986. Ethnobotany of Chhotanagpur (Bihar). *Folklore* 27: 95 – 100.

Tarafder, C.R. 1987. Some traditional knowledge about tribal health. *Folklore* 28: 37 – 42.

Tewari, L.C., Agrawal, R.G.,Pandey, M.J., Uniyal, M.R. & Pandey, G. 1990. Some traditional folk medicine from the Himalayas (U.P. region), *Aryavaidyan* 4: 49-57.

Thothathri, K.,Sen, R., Pal, D.C. & Molla, H.A. 1985. *Selected Poisonous Plants from tribal areas of India*. BSI. Calcutta.

Tirkey, A., Khan, F. & Khan, S.S. 2000a. Medicinal plants used in the treatment of hair in Raigarh district of Madhya Pradesh, (India) In Khan *et al.* (Ed.) *Vistas in Ethnobotany* 1: 104-110.

Tirkey, A., Khan, F. & Khan, S.S. 2000b. Medicinal plants used by tribals of Raigarh district of Madhya Pradesh (India) for jaundice. In Khan *et al.* (Ed.) *Vistas in Ethnobotany* 1: 163-171.

Tirkey, A., Khan, S.S. & Khan, F. 2002. Plants of Asteraceae an ehtnomedicinal study of Raigarh Distrct, (C.G.) *Indian J. App. Pure Biol.* 17(2): 166-169.

Tirkey, A. 2004. Some ethnomedicinal plant species of Chhattisgarh State. *Ethnobotany* 16: 118-124.

Tirkey, A., Khan, S.S. & Khan, F. 2004. Ethnomedicinally important plants used in the treatment of skin infections by the tribals of Chhattisgarh. In Khan, A. & Khanum, A. (ed.) *Ethnomedicine and Human Welfare,* pp. 68-90. Ukaaz publications, Hydrabad (Andhra Pradesh) India.

Tiwari, K.C., Majumdar, R. & Bhattacharjee, S. 1979. Folklore medicine from Assam and Arunachal Pradesh (district Tirap). *Quart. J. Crude Drug Res.* 17: 61 – 67.

Tiwari, K.C. & Majumdar, R. 1996. Medicinal plant from upper Assam borders having specific folk uses. *Sachitra Ayurved* 49: 207-214.

Tiwari, R. & Tiwari, R. 2003. Indigenous medicines for treatment of livestock in Tarai region of Uttar Pradesh. *Asian Agri. History* 7(2): 153-160.

Tiwari, L. & Pande, P.C. 2006. Ethnoveterinary plants of Uttarkashi district, Uttaranchal, India. *Ethnobotany* 18: 139-144.

Tondon, J.S. 1959. The Bhatra of Bastar. *Bull. Anthrop.* 8(2): 33-54.

Tribedi, G.N., Kayal, R.N. & Rai Chaudhuri, H.N. 1982. Some medicinal plants of Mayurbhanj (Orissa). *Bull. Bot. Surv. India* 24: 117 – 120.

Tribedi, G.N., Molla, H.A. & Pal, D.C. 1985. Some uses of plants from the tribal areas of Chotanagpur, Bihar. *Nagarjun.* 29: 15 – 18.

Tripathi, Y.C., Prabhu, V.V., Pal, R.S. & Mishra, R.N. 1996. Medicinal plants of Rajasthan in Indian system of medicine. *Ancient Sci. Life* 15: 190 – 212.

Udayan, P.S., George, S., Tushar, K.V. & Balachandran, I. 2005. Medicinal plants used by the Kaadar tribes of Sholayar forest. Thrissur district, Kerela. *Indian J. Trad. Knowledge* 4(2): 159-163.

Udayan, P.S., George, S., Tushar, K.V. & Balachandran, I. 2005. Some common medicinal plants used by the Kuruba community of Thamadihalli forest near Tiptur Tumkur district, Karnataka, India. *J. Non-Timber Forest Prod.* 11(3): 215-219.

Udayan, P.S., George, S., Tushar, K.V., Sudhakar Raja, S. & Balachandran, I. 2006a. A note on the ethnomedicinal plants used by the Kaadar Tribes, Kariyan Shola, Top Slip, Tamil Nadu. *J. Econ. Taxon. Bot.* 30(Suppl.): 5 – 12.

Udayan, P.S., George, S., Tushar, K.V., Sudhakar Raja, S. & Balachandran, I. 2006b. Ethnobotanical information from the Muduga tribe of Mukkali, near Silent Valley National Park, Palakkad district, Kerala. *J. Econ. Taxon. Bot.* 30(Suppl.):27 –30.

Uniyal, M.R. 1968. Medicinal plants of Bhagirathi valley lying in Uttarkashi forest division. *Indian Forester* 94: 407 – 420.

Uniyal, M.R. & Chauhan, M.S. 1973. Traditionally important medicinal plants of Kangra Valley in Dharamsala forest circle, Himachal Pradesh. *J. Res. Indian. Med.*8: 76 – 85.

Upadhaya, A.S., Kumbhojkar, M.S. & Kulkarni, D.K. 1997. Ethnomedicobotany of some sacred plants of Western Maharashtra. *Ethnobotany* 9: 65-66.

Upadhay, O.P., Kumar, K. & Tiwari, R.K. 1998. Ethnobotanical study of skin treatment uses of medicinal plants of Bihar, *Pharmaceut. Biol.* 36: 161 – 172.

Upadhaya, R. & Chauhan, S.V.S. 2000. Ethnobotanical observations on Koya tribe of Gundala mandal of Khamman district Andhra Pradesh. *Ethnobotany* 12: 93-99.

Upadhya, A.S. & Kumbhojkar, M.S. 2003. Ethnobotany of *Butea monosperma* (Lam.) Taub. *J. Econ. Taxon. Bot.* 27(1): 94-98.

Upadhyaya, R. & Chauhan, S.V.S. 2003. Ethnobotanical uses of plant gums by the tribals. *J. Econ. Taxon. Bot.* 27(3): 601-602.

Upadhyay, R. & Jaswant Singh. 2005. Ethnomedicinal uses of plants from Tikri forest of Gonda district. *Ethnobotany* 17: 167– 170.

Upadhye, A.,Kumbhojkar, M.S. & Vartak, V.D. 1986. Observations on wild plants used in folk medicine in the rural areas of the Kolhapur district. *Anc. Sci.Life* 6: 119 – 121.

Vaidya, R.R. & Dhumal, K.N. 2000. Studies on nutritive values in some vegetables used by tribals of Mahabaleshwar. *J. Med. Arom. Sci.* 22(4A), 23(1A):312-315.

Varghese, E., Jain, S.K. & Bose, N. 1993. A quantitative approach to establish the efficacy of herbal remedies: A case study on the Kharias: *Ethnobotany* 5: 149-154.

Varghese, E. 1996. *Applied ethnobotany: a case study among the Kharias of central India.* Deep. Publ. New Delhi.

Varghese, E., Hembrom, P.P. & Mallokunnel, S. 1998. An ethnomedicinal alternative in treatment of Kala-Azar. *Ethnobotany* 10: 50-55.

Varghese, E. & Hembrom, P.P. 2000. *Ethnomedicinal lore of the Paharias.* Central Human Ecol, Mumbai.

Vartak, V.D. 1981. Observations on wild edible plants from hilly regions of Maharashtra and Goa: Resume and future prospects. In Jain, S.K. (Ed.) *Glimpses of Indian ethnobotany,* 261-271. Oxford & Indian IBH, New Delhi.

Ved Prakash & Jain, S.K. 1984. Poaceae: Tribe - Isachneae. *Fasc. Fl. India.*14: 41 – 42. Bot. Surv. India, Calcutta.

Ved Prakash & Mehrotra, B.N. 1987. Ethnobotanical studies on the flora of Khandala, Maharashtra State. *J. Econ. Taxon. Bot.* 9: 205 – 208.

Ved Prakash & Mehrotra, B.N. 1988. Unrecorded traditional medicines-I. *Anc. Sci. Life* 3: 110-112.

Ved, D.K., Kinhal, G.A., Ravikumar, K., Karnat, M., Vijaya Sankar, R. & Indresha, J.H. 2003. *Threat assessment and management prioritization for the medicinal plants of Chhattisgarh and Madhya Pradesh.* FRLHT, Banglore.

Venugopal, K. & Raghunathan, K. 1982 (Ed.). Medicobotanical-cum-ethnomedicinal survey of district Jhabua (Madhya Pradesh). *Centr. Counc. Res. Ayurv. Siddha Newsletter* 4 - 5.

Verma, D.M. & Hanif, M.I. 1988. Consepectus of the Family – Malvaceae in Madhya Pradesh. *J. Econ. Taxon. Bot.* 12(1): 211-223.

Verma, D.M., Balakrishan, N.P. & Dixit, R.D. 1993. *Flora of Madhya Pradesh,* Vol. I. BSI, Calcatta.

Vijaya Kumar, R. & Pullaiah, T. 1998. Medicinal plants used by the tribals of Prakasham district, Andhra Pradesh. *Ethnobotany* 10: 97-102.

Vir Jee, Dar, G.H., Kachroo, P. & Bhat, G.M. 1984. Taxo – ethnobotanical studies of the rural areas in district Rajouri (Jammu). *J. Econ. Taxon. Bot.* 5: 831 – 838.

Viswanathan, K. 1995. Survey on medicinal spicies of the Nilgiris. *Ancient Sci. Life* 14: 258-267.

Waring, E.J. 1868. *The Pharmacopoeia of India.* W.H. Allen and Co., London.

Warrier, P.K., Nambiar, V.P.K. & Ramankutty, C. 1994. *Indian Medicinal Plants,* Vol. I- V. Oriental Longum, Chennai.

Yesodharan, K. & Sujana, K.A. 2006. Ethnomedicinal plants used by the trbals of Parambikulam Wildlife Sanctuary, Kerala to cure cuts and wounds. *J. Econ. Taxon. Bot.* 30(Suppl.): 365-369.

APPENDIX - I

INDEX TO VALID BOTANICAL NAMES (Attached local names also given)

APPENDIX - II
PLANTS USED FOR DIFFERENT DISEASES

I. Plants used for different human diseases

- **Appendicitis:** *Tamarindus indica.*
- **Asthma:** *Achyranthes aspera; Calotropis gigantea; Cannabis sativa; Cayratia trifolia; Celosia argentea; Cocos nucifera; Datura metel; Datura stramonium; Gossypium herbaceum; Opuntia elatior; Zingiber officinale.*
- **Backache:** *Butea monosperma; Calotropis gigantea.*
- **Blood sugar:** *Coccinia grandis; Momordica charantia*
- **Bodyache:** *Azadirachta indica; Boerhavia repens* var. *diffusa; Butea monosperma; Curuma longa; Cuscuta reflexa; Pongamia pinnata; Sonchus brachyotus.*
- **Boils:** *Argemone mexicana; Barleria prionitis; Bombax ceiba; Butea monosperma; Calotropis gigantea; Calotropis procera; Datura stramonium; Erythrina suberosa; Euphorbia neriifolia; Ipomoea carnea* ssp. *fistulosa; Jasminum sambac; Jatropha curcas; Kalanchoe pinnata; Phyllanthus fraternus; Quiscualis indica; Ricinus communis; Senna tora.*
- **Bone fracture:** *Capparis zeylanica; Cissus quadrangularis; Curcuma amada; Gloriosa superba; Madhuca indica; Triumfetta rhomboidea.*
- **Bronchitis:** *Justicia adhatoda.*
- **Burn/Burning sensation/Sunstroke:** *Aloe vera; Azadirachta indica; Lepidagathis cristata.*
- **Cervical pain:** *Brassica rapa* ssp. *campestris; Pongamia pinnata.*
- **Cholera:** *Drimia indica.*
- **Constipation:** *Ricinus communis.*

- **Corn:** *Achyranthes aspera; Allium cepa; Barleria prionitis; Calotropis procera; Tridax procumbens.*
- **Cough & Cold:** *Justicia adhatoda; Aegle marmelos; Aloe vera; Calotropis gigantea; Plectranthus amboinicus; Curcuma longa; Echinops echinatus; Euphorbia neriifolia; Ocimum tenuiflorum; Piper betle; Plumbago zeylanica; Psidium guajava; Zingiber officinale.*
- **Cuts & Wounds/injury:** *Aloe vera; Capsicum annuum; Convolvulus arvensis; Eclipta prostrata; Holoptelea integrifolia; Ipomoea carnea* ssp. *fistulosa; Jatropha curcas; Madhuca indica; Nicotiana tabacum; Portulaca oleracea; Senna tora; Tectona grandis; Tridax procumbens; Typha elephantina.*
- **Deafness:** *Aegle marmelos.*
- **Diarrhoea/Dysentery:** *Acacia leucophloea; Acacia nilotica* ssp. *indica; Aegle marmelos; Aerva lanata; Annona squamosa; Anogeissus latifolia; Asparagus racemosus; Azadirachta indica; Bauhinia variegata; Bombax ceiba; Butea monosperma; Capparis zeylanica; Citrus limon; Colocasia esculenta; Corchorus olitorius; Dalbergia sissoo; Echinops echinatus; Ficus racemosa; Holarrhena pubescens; Lantana camara; Lawsonia inermis; Madhuca indica; Mangifera indica; Moringa oleifera; Neptunia oleracea; Phyllanthus emblica; Pithecellobium dulce; Solanum virginianum; Syzygium cumini; Tectona grandis; Terminalia cuneata; Tinospora cordifolia; Zea mays; Ziziphus mauritiana; Z. nummularia.*
- **Dogbite:** *Balanites aegyptiaca; Calotropis gigantea; Calotropis procera; Cayratia trifolia; Chrozophora rottleri; Datura metel; Gloriosa superba; Mucuna pruriens; Opuntia elatior; Tinospora cordifolia; Triticum aestivum.*
- **Epilepsy:** *Balanites aegyptiaca; Cuscuta reflexa; Withania somnifera.*
- **Eye & Ear diseases:** *Acacia nilotica* ssp. *indica; Allium cepa; Argemone mexicana; Bauhinia racemosa; Calotropis gigantea; Cleome viscosa; Amischophacelus axillaris; Enicostema axillare; Ficus benghalensis; Gossypium herbaceum; Jasminum officinale* var. *grandiflorum; Jatropha curcas; Ocimum basilicum; Ocimum tenuiflorum; Papaver somniferum; Quiscualis indica; Ziziphus nummularia.*
- **Gastric trouble:** *Vernonia anthelmintica; Coriandrum sativum; Ferula asa-foetida; Mentha spicata.*
- **Gonorrhoea:** *Butea monosperma.*

- **Hair tonic/Shampoo:** *Ficus religiosa; Sida rhombifolia.*
- **Headache:** *Acacia leucophloea; Acacia nilotica* ssp. *indica; Azadirachta indica; Brassica nigra; Coriandrum sativum; Enicostema axillare; Erythrina suberosa; Merremia gangetica; Moringa oleifera; Plumbago zeylanica; Ricinus communis; Tridax procumbens; Vitex negundo; Xanthium strumarium.*
- **Heel cracking:** *Diospyros melenoxylon; Eclipta prostrata; Erythrina suberosa; Ficus benghalensis; Lawsonia inermis.*
- **Hoarseness:** *Piper nigrum.*
- **Hysteria:** *Balanites aegyptiaca; Echinops echinatus.*
- **Indigestion:** *Cuminum cyminum.*
- **Intermittent/Fever:** *Acacia leucophloea; Achyranthes aspera; Anisomeles indica; Azadirachta indica; Barleria prionitis, Boerhavia repens* var. *diffusa; Cleome viscosa; Clitoria ternatea; Echinops echinatus; Eclipta prostrata; Enicostema axillare; Indigofera trita; Nyctanthes arbor-tristis; Phyllanthus fraternus; Physalis minima; Piper longum; Plumbago zeylanica; Tinospora cordifolia; Tribulus terrestris; Ziziphus mauritiana.*
- **Intsetinal worms/Anthelmintic:** *Acacia leucophloea; Annona squamosa; Balanites aegyptiaca; Bombax ceiba; Butea monosperma; Ficus benghalensis; Senna tora; Syzygium cumini.*
- **Itch/Eczema:** *Azadirachta indica; Lepidagathis cristata; Nyctanthes arbor-tristis; Ocimum basilicum; Orthosiphon pallidus; Plumbago zeylanica; Pongamia pinnata; Senna tora; Tinospora cordifolia.*
- **Jaundice:** *Ailanthus excelsa; Ampelocissus latifolia; Asparagus racemosus; Calotropis gigantea; Carica papaya; Cicer arietinum; Cuscuta reflexa; Enicostema axillare; Foeniculum vulgare; Lantana camara; Musa paradisiaca; Oroxylum indicum; Phyllanthus fraternus; Raphanus sativus; Saccharum officinarum; Senna tora; Syzygium aromaticum.*
- **Joint diseases:** *Acacia nilotica* ssp. *indica; Argyreia nervosa; Asparagus racemosus; Butea monosperma; Calotropis gigantea; Curcuma amada; Curuma longa; Cuscuta reflexa; Cyperus rotundus; Datura metel; Pongamia pinnata; Senna tora; Sesbania bispinosa; Tinospora cordifolia; Trigonella foenum-graecum; Vitex negundo; Withania somnifera.*
- **Kidney stone:** *Albizia procera; Carica papaya; Linum usitatissimum; Tridax procumbens.*

- **Leprosy:** *Nyctanthes arbor-tristis; Tinospora cordifolia.*
- **Liver disorders:***Phyllanthus fraternus.*
- **Loose motion:** *Anisomeles indica; Bauhinia racemosa; Erythrina suberosa; Mangifera indica; Musa paradisiaca; Triumfetta rhomboidea.*
- **Men diseases:** *Abelmoschus esculentus; Aegle marmelos; Bombax ceiba; Carica papaya; Dioscorea bulbifera; Ficus benghalensis; Hibiscus rosa–sinensis; Hygrophila schulli; Lawsonia inermis; Sida cordifolia; Sida rhombifolia; Triumfetta rhomboidea.*
- **Mental weakness:** *Dendrocalamus strictus; Opuntia elatior.*
- **Muscular pain:** *Brassica rapa* ssp. *campestris; Holoptelea integrifolia; Paspalum scrobiculatum; Santalum album.*
- **Night blindness:** *Medicago sativa.*
- **Paralysis:** *Datura metel.*
- **Piles:** *Acacia leucophloea; Acacia nilotica* ssp. *indica; Aloe vera; Azadirachta indica; Bombax ceiba; Cannabis sativa; Cocculus hirsutus; Convolvulus arvensis; Crinum defixum; Datura metel; Lawsonia inermis; Nyctanthes arbor-tristis; Vigna mungo.*
- **Pimples:** *Butea monosperma.*
- **Pneumonia:** *Barlaria prionitis; Cassia fistula; Plectranthus amboinicus; Datura metel; Datura stramonium; Mucuna pruriens; Piper longum; Plumbago zeylanica.*
- **Psoriasis:** *Tinospora cordifolia.*
- **Rickets:** *Anisomeles indica; Piper betle.*
- **Ringworm:** *Allium sativum; Argemone mexicana; Momordica charantia; Pongamia pinnata; Tinospora cordifolia.*
- **Sciatica:** *Datura metel; Nyctanthes arbor-tristis.*
- **Scorpion sting:** *Achyranthes aspera; Balanites aegyptiaca; Boerhavia repens* var. *diffusa; Calotropis procera; Cassia fistula; Eclipta prostrata; Ipomoea carnea* ssp. *fistulosa; Madhuca indica; Mangifera indica; Martynia annua; Mucuna pruriens; Solanum virginianum; Tamarindus indica; Tridax procumbens; Ziziphus mauritiana.*
- **Small pox/Chiken pox:** *Dioscorea bulbifera; Euphorbia thymifolia; Crotalaria hebecarpa; Oroxylum indicum; Tectona grandis.*
- **Snake bite:** *Acacia nilotica* ssp. *indica; Albizia procera; Butea monosperma; Cayratia trifolia; Clitoria ternatea; Cocculus*

hirsutus; Cuscuta reflexa; Limonia acidissima; Madhuca indica; Nicotiana tabacum; Senna occidentalis; Senna tora; Tinospora cordifolia; Triumfetta rhomboidea.

- **Stomach trouble:** *Azadirachta indica; Caesalpinia bonduc; Capparis zeylanica; Plectranthus amboinicus; Cuminum cyminum; Curcuma longa; Dendrocalamus strictus; Diospyros melenoxylon; Ferula asafoetida; Helicteres isora; Indigofera cordifolia; Mentha spicata; Ocimum tenuiflorum; Piper betle; Tridax procumbens.*
- **Swelling:** *Ailanthus excelsa; Butea monosperma; Calotropis gigantea; Calotropis procera; Citrus limon; Curuma longa; Physalis minima.*
- **Tetanus:** *Balanites aegyptiaca.*
- **Throat infection:** *Echinops echinatus; Euphorbia neriifolia; Quiscualis indica.*
- **Toothache:** *Acacia nilotica* ssp. *indica; Achyranthes aspera; Albizia procera; Allium sativum; Azadirachta indica; Calotropis procera; Carica papaya; Cordia dichotoma; Euphorbia neriifolia; Ferula asafoetida; Ficus benghalensis; Ficus racemosa; Jatropha curcas; Mangifera indica; Psidium guajava; Ricinus communis; Solanum virginianum; Tridax procumbens; Triumfetta rhomboidea.*
- **Tuberculosis:** *Calotropis gigantia; Calotropis procera; Cayratia trifolia.*
- **Tumour:** *Carica papaya; Cayratia trifolia; Cocculus hirsutus; Convolvulus arvensis; Limonia acidissima.*
- **Typhoid:** *Abrus precatorius; Ailanthus excelsa; Asparagus racemosus; Azadirachta indica; Balanites aegyptiaca; Calotropis gigantea; Diplocyclos palmatus; Echinops echinatus; Enicostema axillare; Oroxylum indicum; Orthosiphon pallidus; Pandanus odoratissimus; Senna tora; Terminalia cuneata; Tinospora cordifolia; Tinospora cordifolia; Tribulus terrestris.*
- **Ulcer:** *Abrus precatorius; Acacia catechu; Commiphora wightii; Dalbergia sissoo; Ficus racemosa; Jasminum officinale* var. *grandiflorum; Psidium guajava; Solanum virginianum; Syzygium cumini.*
- **Urinary diseases:** *Achyranthes aspera; Butea monosperma; Citrus limon; Foeniculum vulgare; Phoenix sylvestris; Tinospora cordifolia; Triticum aestivum.*
- **Urticaria:** *Pongamia pinnata.*

- **Vomiting:** *Anisomeles indica; Azadirachta indica; Lantana camara; Lawsonia inermis; Tectona grandis; Trachyspermum ammi.*

- **Women diseases:** *Abelmoschus esculentus; Abrus precatorius; Abutilon indicum; Albizia procera; Anethum graveolens; Annona squamosa; Anogeissus latifolia; Asparagus racemosus; Azadirachta indica; Balanites aegyptiaca; Boerhavia repens* var. *diffusa; Bombax ceiba; Butea monosperma; Carica papaya; Chlorophytum tuberosum; Cleome gynandra; Clerodendrum multiflorum; Cocculus hirsutus; Cocos nucifera; Corchorus olitorius; Coriandrum sativum; Dalbergia sissoo; Diplocyclos palmatus; Dioscorea bulbifera; Echinops echinatus; Euphorbia hirta; Foeniculum vulgare; Gossypium herbaceum; Haldina cordifolia; Hibiscus rosa–sinensis; Limonia acidissima; Mangifera indica; Musa paradisiaca; Nerium indicum; Orthosiphon pallidus; Oryza sativa; Pandanus odoratissimus; Phoenix sylvestris; Piper longum; Plumbago zeylanica; Syzygium aromaticum; Tectona grandis; Terminalia cuneata; Tinospora cordifolia; Trachyspermum ammi; Trapa natans* var. *bispinosa; Tribulus terrestris; Triticum aestivum; Triumfetta rhomboidea; Zingiber officinale.*

II. Plants used for Animal diseases:

Allium cepa; Ammannia baccifera; Brassica rapa ssp. *campestris; Calotropis gigantea; Cassia fistula; Cayratia trifolia; Vernonia anthelmintica; Cissus quadrangularis; Convolvulus arvensis; Datura stramonium; Erythrina suberosa; Ficus benghalensis; Hordeum vulgare; Launaea procumbens; Luffa acutangula; Mangifera indica; Moringa oleifera; Musa paradisiaca; Mucuna pruriens; Pithecellobium dulce; Sorghum bicolor; Tinospora cordifolia.*

APPENDIX - III
PLANTS USED FOR OTHER PURPOSES

1. Plants used for Agriculture implements, Handle, Tools, Boats

Acacia nilotica ssp. *indica; Azadirachta indica; Bombax ceiba; Diospyros melenoxylon; Tectona grandis.*

2. Plants used as edible plants

i. **Rhizome/Tuber/Corm/bulb:** *Allium cepa; Colocasia esculenta; Dioscorea bulbifera; Ipomoea batatas; Raphanus sativus.*

ii. **Leaves and tender shoots:** *Amaranthus spinosus; Anethum graveolens; Brassica nigra; Celosia argentea; Chenopodium album; Chlorophytum tuberosum; Cicer arietinum; Coccinia grandis; Commelina benghalensis; Corchorus olitorius; Digera muricata; Leucas cephalotes; Medicago sativa; Portulaca oleracea; Raphanus sativus; Senna tora; Trianthema portulacastrum; Trigonella foenum-graecum.*

iii. **Inflorescence:** *Morus alba.*

iv. **Flowers:** *Cassia fistula; Cordia dichotoma; Cordia gharaf.*

v. **Fruits:** *Abelmoschus esculentus; Adansonia digitata; Aegle marmelos; Annona squamosa; Artocarpus heterophyllus; Azadirachta indica; Bauhinia racemosa; Carica papaya; Carissa carandas; Cordia dichotoma; Cordia gharaf; Diospyros melenoxylon; Ficus racemosa; Grewia tiliifolia; Lantana camara; Limonia acidissima; Mangifera indica; Melothria madraspatana; Mimusops elengi; Manilkara hexandra; Momordica charantia; Momordica dioica; Moringa oleifera; Musa paradisiaca; Phoenix sylvestris; Phyllanthus emblica; Physalis minima; Pithecellobium dulce; Psidium guajava; Raphanus sativus; Solanum nigrum; Syzygium cumini; Tamarindus indica; Terminalia bellirica; Trapa natans* var. *bispinosa; Ziziphus mauritiana; Z. nummularia.*

vi. **Seeds:** *Bombax ceiba, Holoptelea integrifolia, Martynia annua.*

3. **Plants used for beverage and drinks:** *Madhuca indica; Phoenix sylvestris; Senna tora.*

4. **Plants used as fodder/Forage:** *Acacia nilotica* ssp. *Indica; Azadirachta indica; Cassia fistula; Cicer arietinum; Cordia dichotoma; Ficus benghalensis; Ficus racemosa; Holarrhena pubescens; Madhuca indica; Mangifera indica; Nyctanthes arbor-tristis; Oryza sativa; Vigna mungo; Saccharum officinarum; Tectona grandis; Terminalia bellirica; Trigonella foenum-graecum; Triticum aestivum.*

5. **Plants used for gum and resin:** *Acacia nilotica* ssp. *indica; Anogeissus latifolia; Butea monosperma.*

6. **Plants used for tannin:** *Acacia nilotica* ssp. *Indica; Acacia pennata; Acacia leucophloea; Anogeissus latifolia; Bauhinia racemosa; Cassia fistula; Haldina cordifolia; Nyctanthes arbortristis; Tectona grandis; Terminelia cuneata; T. bellirica; Zizyphus mauritiana.*

7. **Plants used for dye:** *Butea monosperma; Curcuma longa; Lawsonia inermis.*

8. **Plants used for oil:** *Brassica rapa* ssp. *campestris; Butea monosperma; Jatropha curcas; Madhuca indica; Pongamia pinnta; Ricinus communis.*

9. **Plants used for House, Building construction, Thatching and Furniture:** *Acacia leucophloea; Acacia nilotica* ssp. *indica; Anogeissus latifolia; Azadirachta indica; Dendrocalamus strictus; Diospyros melenoxylon; Eucalyptus globulus; Nyctanthes arbor-tristis; Phoenix sylvestris; Syzygium cumini; Tectona grandis.*

10. **Plants used for fencing purposes:** *Acacia nilotica* ssp. *indica; Acacia leucophloea; Acacia pennata; Annona squamosa; Butea monosperma; Euphorbia neriifolia; Jatropha curcas; Lantana camara; Lawsonia inermis; Ricinus communis*

11. **Plants used for ropes, mats and basketry**

    - **Ropes:** *Abelmoschus esculentus; Butea monosperma; Cocos nucifera; Hibiscus cannabinus.*
    - **Mats:** *Dendrocalamus strictus; Phoenix sylvestris.*
    - **Basketry:** *Dendrocalamus strictus; Lantana camara; Nyctanthes arbor-tristis; Phoenix sylvestris; Vitex negundo.*

12. **Plants used for taboos and totems:** *Acacia catechu; Artocarpus heterophyllus; Bauhinia racemosa; Benincasa hispida; Madhuca indica; Mangifera indica.*

13. **Plants used for musical instruments:** *Acacia nilotica; Ailanthus excelsa; Bombax ceiba; Butea monosperma; Cocos nucifera; Dendrocalamus strictus; Ficus racemosa; Haldina cordifolia; Mangifera indica; Moringa oleifera; Tectona grandis.*

14. **Plants used as fuel:** *Acacia leucophloea; Acacia nilotica* ssp. *indica; Azadirachta indica; Butea monosperma; Jatropha curcas; Phyllanthus emblica.*

15. **Plants used as detergent:** *Albizia procera; Balanites aegyptiaca; Momordica dioica.*

16. **Plants used in Festival and Magico-religious beliefs:** *Abrus precatorius; Acacia leucophloea; Acacia nilotica* ssp. *indica; Aegle marmelos; Ailanthus excelsa; Allium cepa; Allium sativum; Ampelocissus latifolia; Asparagus racemosus; Azadirachta indica; Benincasa hispida; Boerhavia repens* var. *diffusa; Brassica nigra; Calotropis gigantea; Capsicum annuum; Cassia fistula; Clitoria ternatea; Cocos nucifera; Datura metel; Dendrocalamus strictus; Diplocyclos palmatus; Echinops echinatus; Ficus religiosa; Lepidagathis cristata; Senna occidentalis; Xanthium strumarium.*

17. **Other Economic plants**

    - **For wrapping tobbaco and as narcotics:** *Cannabis sativa; Diospyros melenoxylon; Nicotiana tabacum.*
    - **For making plates and bowls:** *Butea monosperma.*
    - **Plants used for fibre:** *Gossypium herbaceum; Hibiscus cannabinus.*
    - **For comb:** *Haldina cordifolia.*
    - **Plants are used for fish poisoning:** *Acacia pennata; Albizia procera; Balanites aegyptiaca; Butea monosperma; Diospyros melenoxylon; Holarrhena pubescens; Jatropha curcas; Pongamia pinnata; Syzygium cumini; Terminalia cuneata.*